VOLUME FOUR HUNDRED AND THIRTY-ONE

Methods in
ENZYMOLOGY

Translation Initiation:
Cell Biology,
High-Throughput Methods,
and Chemical-Based
Approaches

METHODS IN ENZYMOLOGY

Editors-in-Chief

JOHN N. ABELSON AND MELVIN I. SIMON

Division of Biology
California Institute of Technology
Pasadena, California

Founding Editors

SIDNEY P. COLOWICK AND NATHAN O. KAPLAN

VOLUME FOUR HUNDRED AND THIRTY-ONE

METHODS IN
ENZYMOLOGY

Translation Initiation: Cell Biology, High-Throughput Methods, and Chemical-Based Approaches

EDITED BY

JON LORSCH
*Department of Biophysics and Biophysical Chemistry
Johns Hopkins University School of Medicine
Baltimore, Maryland*

AMSTERDAM • BOSTON • HEIDELBERG • LONDON
NEW YORK • OXFORD • PARIS • SAN DIEGO
SAN FRANCISCO • SINGAPORE • SYDNEY • TOKYO
Academic Press is an imprint of Elsevier

ELSEVIER

Academic Press is an imprint of Elsevier
525 B Street, Suite 1900, San Diego, California 92101–4495, USA
84 Theobald's Road, London WC1X 8RR, UK

This book is printed on acid-free paper.

For information on all Elsevier Academic Press publications
visit our Web site at www.books.elsevier.com

ISBN: 978-0-12-373964-3

PRINTED IN THE UNITED STATES OF AMERICA
07 08 09 10 9 8 7 6 5 4 3 2 1

Contents

14. Isolation and Identification of Eukaryotic Initiation Factor 4A as a Molecular Target for the Marine Natural Product Pateamine A   303

Woon-Kai Low, Yongjun Dang, Tilman Schneider-Poetsch, Zonggao Shi,
Nam Song Choi, Robert M. Rzasa, Helene A. Shea, Shukun Li, Kaapjoo Park,
Gil Ma, Daniel Romo, and Jun O. Liu

CONTRIBUTORS

Paul Anderson
Division of Rheumatology, Immunology, and Allergy, Brigham and Women's Hospital, Boston, Massachusetts

Yoav Arava
Department of Biology, Technion—Israel Institute of Technology, Haifa, Israel

Mark P. Ashe
Faculty of Life Sciences, The University of Manchester, Manchester, United Kingdom

Traude H. Beilharz
Molecular Genetics Program, Victor Chang Cardiac Research Institute (VCCRI), Sydney, Australia; and
St. Vincent's Clinical School, University of New South Wales, Sydney, Australia

Letizia Brandi
Biotechnomics, Insubrias BioPark, Gerenzano, Italy

Susan G. Campbell
Faculty of Life Sciences, The University of Manchester, Manchester, United Kingdom

Marcello Carotti
Department of Biology MCA, University of Camerino, Camerino, Italy

Regina Cencic
Department of Biochemistry, McGill University, Montreal, Quebec, Canada

Nam Song Choi
Department of Chemistry, Texas A&M University, College Station, Texas

Jennifer L. Clancy
Molecular Genetics Program, Victor Chang Cardiac Research Institute (VCCRI), Sydney, Australia

Yongjun Dang
Department of Pharmacology, Johns Hopkins School of Medicine, Baltimore, Maryland

Edward Darzynkiewicz
Division of Biophysics, Institute of Experimental Physics, Faculty of Physics, Warsaw University, Warsaw, Poland

Naama Eldad
Department of Biology, Technion—Israel Institute of Technology, Haifa, Israel

Attilio Fabbretti
Department of Biology MCA, University of Camerino, Camerino, Italy

Edith Gomez
Faculty of Life Sciences, The University of Manchester, Manchester, United Kingdom

Ewa Grudzien-Nogalska
Division of Biophysics, Institute of Experimental Physics, Faculty of Physics, Warsaw University, Warsaw, Poland; and
Department of Biochemistry and Molecular Biology, Louisiana State University Health Sciences Center, Shreveport, Louisiana

Claudio O. Gualerzi
Department of Biology MCA, University of Camerino, Camerino, Italy

Raphaël Haddad
Faculty of Life Sciences, The University of Manchester, Manchester, United Kingdom

Yi-Shuian Huang
Division of Neuroscience, Institute of Biomedical Sciences, Academia Sinica, Taipei, Taiwan

David T. Humphreys
Molecular Genetics Program, Victor Chang Cardiac Research Institute (VCCRI), Sydney, Australia

Jacek Jemielity
Division of Biophysics, Institute of Experimental Physics, Faculty of Physics, Warsaw University, Warsaw, Poland

Nancy Kedersha
Division of Rheumatology, Immunology, and Allergy, Brigham and Women's Hospital, Boston, Massachusetts

Shukun Li
Department of Chemistry, Texas A&M University, College Station, Texas

Jun O. Liu
Department of Pharmacology; Solomon H. Synder Department of Neuroscience; and Department of Oncology, Johns Hopkins School of Medicine, Baltimore, Maryland

Woon-Kai Low
Department of Pharmacology, Johns Hopkins School of Medicine, Baltimore, Maryland

Gil Ma
Department of Chemistry, Texas A&M University, College Station, Texas

Daniel Melamed
Department of Biology, Technion—Israel Institute of Technology, Haifa, Israel

Pohl Milon
Department of Biology MCA, University of Camerino, Camerino, Italy

Sarah S. Mohammad-Qureshi
Faculty of Life Sciences, The University of Manchester, Manchester, United Kingdom

Christopher V. Nicchitta
Department of Cell Biology, Duke University Medical Center, Durham, North Carolina

Klaus H. Nielsen
Department of Molecular Biology, University of Aarhus, Aarhus, Denmark

Marco Nousch
Molecular Genetics Program, Victor Chang Cardiac Research Institute (VCCRI), Sydney, Australia

Karren S. Palmer
Faculty of Life Sciences, The University of Manchester, Manchester, United Kingdom

Kaapjoo Park
Department of Chemistry, Texas A&M University, College Station, Texas

Graham D. Pavitt
Faculty of Life Sciences, The University of Manchester, Manchester, United Kingdom

Jerry Pelletier
Department of Biochemistry and McGill Cancer Center, McGill University, Montreal, Quebec, Canada

Cynthia L. Pon
Department of Biology MCA, University of Camerino, Camerino, Italy

Thomas Preiss
Molecular Genetics Program, Victor Chang Cardiac Research Institute (VCCRI), Sydney, Australia;
St. Vincent's Clinical School; and School of Biotechnology and Biomolecular Sciences, University of New South Wales, Sydney, Australia

Christopher G. Proud
Department of Biochemistry and Molecular Biology, University of British Columbia, Vancouver, Canada

Robert E. Rhoads
Department of Biochemistry and Molecular Biology, Louisiana State University Health Sciences Center, Shreveport, Louisiana

Jonathan P. Richardson
Faculty of Life Sciences, The University of Manchester, Manchester, United Kingdom

Joel D. Richter
Program of Molecular Medicine, University of Massachusetts Medical School, Worcester, Massachusetts

Francis Robert
Department of Biochemistry, McGill University, Montreal, Quebec, Canada

Daniel Romo
Department of Chemistry, Texas A&M University, College Station, Texas

Robert M. Rzasa
Department of Chemistry, Texas A&M University, College Station, Texas

Tilman Schneider-Poetsch
Department of Pharmacology, Johns Hopkins School of Medicine, Baltimore, Maryland

Helene A. Shea
Department of Chemistry, Texas A&M University, College Station, Texas

Zonggao Shi
Department of Pharmacology, Johns Hopkins School of Medicine, Baltimore, Maryland

Samuel B. Stephens
Department of Cell Biology, Duke University Medical Center, Durham, North Carolina

Janusz Stepinski
Division of Biophysics, Institute of Experimental Physics, Faculty of Physics, Warsaw University, Warsaw, Poland

Ryszard Stolarski
Division of Biophysics, Institute of Experimental Physics, Faculty of Physics, Warsaw University, Warsaw, Poland

Leoš Valášek
Laboratory of Regulation of Gene Expression, Institute of Microbiology, Prague, the Czech Republic

Xuemin Wang
Department of Biochemistry and Molecular Biology, University of British Columbia, Vancouver, Canada

Belinda J. Westman
Division of Gene Regulation & Expression, School of Life Sciences, Wellcome Trust Biocentre, University of Dundee, Dundee, United Kingdom

Joanna Zuberek
Division of Biophysics, Institute of Experimental Physics, Faculty of Physics, Warsaw University, Warsaw, Poland

PREFACE

Over the past 15 years, it has become clear that translation initiation is a key regulatory point in the control of gene expression. Loss-of-control of protein synthesis has been implicated in a variety of diseases ranging from cancer to viral infection, and there is increasing interest in the development of new drugs that target translation initiation. Despite the profound biological and medical importance of this key step in gene expression, we are only beginning to understand the molecular mechanics that underlie translation initiation and its control, and much work remains to be done.

These MIE volumes (429, 430, and 431) are a compilation of current approaches used to dissect the basic mechanisms by which bacterial, archaeal, and eukaryotic cells assemble, and control the assembly of, ribosomal complexes at the initiation codon. A wide range of methods is presented from cell biology to biophysics to chemical biology. It is clear that no one approach can answer all of the important questions about translation initiation, and that major advances will require collaborative efforts that bring together various disciplines. I hope that these volumes will facilitate cross-disciplinary thinking and enable researchers from a wide variety of fields to explore aspects of translation initiation throughout biology.

Initially, we had planned to publish a single volume on this subject. However, the remarkable response to my requests for chapters allowed us to scale up to three volumes. I would like to express my sincerest appreciation and admiration for the contributors to this endeavor. I am impressed with the outstanding quality of the work produced by the authors, all of whom are leaders in the field. I am especially grateful to John Abelson for giving me the opportunity to edit this publication and for his support and advice throughout the project. Finally, I am indebted to Cindy Minor and the staff at Elsevier for their help and wisdom along the way.

JON LORSCH

METHODS IN ENZYMOLOGY

Purification of FLAG-Tagged Eukaryotic Initiation Factor 2B Complexes, Subcomplexes, and Fragments from Saccharomyces cerevisiae

Sarah S. Mohammad-Qureshi, Raphaël Haddad, Karren S. Palmer, Jonathan P. Richardson, Edith Gomez, *and* Graham D. Pavitt

Contents

Abstract

The eukaryotic initiation factor 2B (eIF2B) is a five-subunit guanine nucleotide exchange factor, that functions during translation initiation to catalyze the otherwise slow exchange of GDP for GTP on its substrate eIF2. Assays to measure substrate interaction and guanine nucleotide release ability of eIF2B require the complex to be purified free of interacting proteins. We have also found that a subcomplex of two subunits, γ and ε or the largest one, ε alone, promotes this activity. Within eIF2Bε, the catalytic center requires the C-terminal 200 residues only. Here, we describe our protocols for purifying

Faculty of Life Sciences, The University of Manchester, Manchester, United Kingdom
[1] Sarah S. Mohammad–Qureshi and Raphaël Haddad contributed equally to this work

Methods in Enzymology, Volume 431
ISSN 0076-6879, DOI: 10.1016/S0076-6879(07)31001-X

the *Saccharomyces cerevisiae* eIF2B complexes and the catalytic subunit using FLAG-tagged proteins overexpressed in yeast cells. Using commercially available FLAG-affinity resin and high salt buffer, we are able to purify active eIF2B virtually free of contaminants.

1. INTRODUCTION

eIF2B is the guanine-nucleotide exchange factor (GEF) that converts eIF2 from its inactive GDP-bound state to a GTP-containing complex that has a higher affinity for initiator methionyl-tRNA. It is an important protein that catalyzes one of the rate-limiting and regulated steps in protein synthesis. One of the earliest described forms of translational control was phosphorylation of the alpha subunit eIF2 (Farrell *et al.*, 1977) by the heme-regulated inhibitor (HRI) in rabbit reticulocyte lysates. This is now known to be one of a family of protein kinases (Gebauer and Hentze, 2004) that respond to different cellular stresses and each phosphorylate serine 51 on eIF2α, to convert eIF2 from a substrate to an inhibitor of guanine nucleotide exchange (Pavitt *et al.*, 1998; Rowlands *et al.*, 1988).

eIF2B was originally purified and characterized from a variety of mammalian cell sources in the early 1980s by several research groups, which each used different names for the factor, including GEF, reversing factor-RF, and anti-HRI (Goss *et al.*, 1984; Konieczny and Safer, 1983; Matts *et al.*, 1983; Panniers and Henshaw, 1983) and, more recently, from yeast (Cigan *et al.*, 1993). eIF2B is a heteropentamer with subunits designated α–ε (from smallest to largest). Interest in eIF2B function and regulation has spread beyond those studying translation and its control, as mutations in each of the five structural genes have been found responsible for a group of leukodystrophies, now collectively called eIF2B-related disorders (Fogli and Boespflug-Tanguy, 2006; Fogli *et al.*, 2004a; van der Knaap *et al.*, 2003). The eIF2B mutations are widespread throughout the subunits, and they lower eIF2B activity and alter stress responses (Fogli *et al.*, 2004b; Kantor *et al.*, 2005; Li *et al.*, 2004; Richardson *et al.*, 2004; van Kollenburg *et al.*, 2006).

Current understanding of eIF2B structure and function has derived mainly from a combination of yeast genetic experiments and biochemistry studies from yeast and mammalian cell systems (Pavitt, 2005). These studies have determined that the largest subunit (eIF2Bε) is the catalytic subunit that enhances the rate of nucleotide exchange for eIF2-bound nucleotide (Fabian *et al.*, 1997; Gomez and Pavitt, 2000; Pavitt *et al.*, 1998). The C-terminal 200 residues contain the catalytic domain (Boesen *et al.*, 2004; Gomez *et al.*, 2002) that can interact with eIF2β (Asano *et al.*, 1999) and eIF2γ (Alone and Dever, 2006; Mohammad-Qureshi *et al.*, 2007) and the N-terminal portion shares extensive sequence homology with eIF2Bγ-to which it binds (Li *et al.*, 2004; Mikami *et al.*, 2006; Pavitt *et al.*, 1998). Collectively, eIF2Bγε are called the "catalytic" subcomplex. The remaining three subunits (eIF2Bαβδ all share sequence and structural similarity both with each other and a number of phosphate- or sulfate-binding enzymes found in archaea, bacteria, and some eukaryotes (Bumann et al., 2004; Kakuta et al., 2004). The reason that eIF2B subunits share structural similarity with these enzymes is not clear but has caused some researchers to suggest that the archebacterial enzymes could be eIF2B orthologues (Kyrpides and Woese, 1998). Yeast eIF2Bαβδ interact with each other independently of the other subunits and recognize the phosphorylation status of serine 51 of eIF2α; they have therefore been referred to as the "regulatory" subcomplex (Krishnamoorthy et al., 2001; Pavitt et al., 1997; Yang and Hinnebusch, 1996). The ability to bind phosphate or structurally similar sulfate may therefore be a factor important in the structural similarity between these eIF2B subunits and the other enzymes because together these three subunits bind with higher affinity to phosphorylated eIF2α (Krishnamoorthy et al., 2001; Pavitt et al., 1998).

The development of recombinant systems to express and purify eIF2B, its constituent subcomplexes, subunits, and domains has facilitated improved understanding of the structure, function, and kinetic mechanism of this multisubunit protein. Because eIF2B is a multisubunit phosphoprotein (Wang et al., 2001), eukaryotic expression systems have been more commonly used, including yeast and insect cell cultures. Baculovirus transfer vectors and Sf 9 cells were developed by the Jefferson group to express and purify FLAG-tagged rat eIF2B holoprotein and subunit combinations. This facilitated identification of eIF2Bε as the catalytic subunit of eIF2B (Fabian et al., 1997) and provided some evidence of interprotein interactions between eIF2Bε and other subunits (Anthony et al., 2000). In addition, it was possible to purify eIF2B complexes containing a mutated form of the delta subunit analogous to alleles described in yeast that have reduced sensitivity to eIF2 phosphorylation (Kimball et al., 1998; Pavitt et al., 1997, 1998). In 2006, a similar approach using human cDNAs helped establish an efficient mammalian cell-free translation system (Mikami et al., 2006).

We and other researchers have purified yeast eIF2B from yeast cells engineered to overexpress the desired subunits. The purified materials have been used for enzyme kinetic studies (Nika et al., 2000), analysis of catalytic activity of mutated proteins to define residues critical for nucleotide exchange function (Gomez and Pavitt, 2000) and the effects of the mutations of the human eIF2B–related disorders (Richardson et al., 2004). We have also determined the minimal catalytic domain (Gomez et al., 2002) and examined protein–protein interactions with eIF2 (Gomez and Pavitt, 2000; Mohammad-Qureshi et al., 2007).

In this chapter, we describe our methods for purification of eIF2B holocomplexes as well as the catalytic subcomplex and the catalytic eIF2Bε subunit (encoded by GCD6 in yeast) alone. We typically epitope tag only one eIF2B subunit, over-express the desired subunit combination, and use a single affinity chromatography step to purify the desired complex. In early work, a

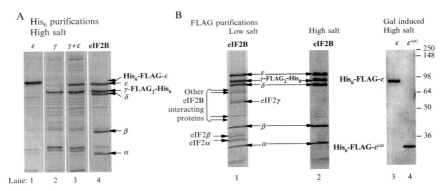

Figure 1.1   Examples of eIF2B protein preparations. Purified eIF2B complexes separated by SDS-PAGE and stained with Coomassie blue. (A) Examples of nickel affinity chromatography. After a single step, each target factor is highly enriched but is not purified free from contaminants. Loading: lanes 1 and 4 (2 μg) lanes 2 and 3 (4 μg). (B) Examples of FLAG affinity chromatography. Lane 1: Use of low-salt lysis buffer results in isolation of a complex containing eIF2 and eIF2B and other proteins (left). Lanes 2 to 4500 mM salt buffers generate pure proteins (middle and right panels). Loading: lanes 1 and 2 (2 μg), lanes 3 and 4 (1 μg).

hemagglutinin (HA) or hexa–histidine (His$_6$) tag was added to the C-terminus of eIF2Bγ (*GCD1*) and found not to affect its function *in vivo* (Cigan *et al.*, 1993; Pavitt *et al.*, 1998). Genes were overexpressed from multicopy yeast shuttle vectors and proteins were purified using nickel affinity chromatography (Fig. 1.1A). Because yeast contains other natural polyhistidine tagged proteins, further chromatographic steps were required to obtain highly purified protein (Nika *et al.*, 2000). The FLAG tag system (Sigma–Aldrich) was found to have greater specificity and to enable purification from a cell lysate in a single affinity step and there is now a wide variety of commercial reagents to assist purification of FLAG-tagged proteins. The methods described here therefore focus on our use of the FLAG system.

2. PLASMID VECTORS AND YEAST STRAINS USED

To ensure a high yield of purified protein product, the genes encoding the five subunits of eIF2B were cloned into 2μ yeast high–copy vectors. With the exception of pAV1427 and 1689, each gene is expressed from its authentic yeast promoter and two or three genes were cloned into a single vector so that just two plasmids are required to overexpress the eIF2B holoenzyme complex. Also, different versions of each plasmid are available to maximize compatible pairs (see Table 1.1). pAV1427 and its derivative, pAV1689, drive expression of the cloned insert from a hybrid *CYC-GAL* promoter and therefore require specialized growth conditions (see later). To

Table 1.1 Plasmids used to express and purify eIF2B subunits, sub- and holo-complexes

Plasmid name	eIF2B[a] genes	subunits	Tagged gene[b]	Selectable marker	Expression[c]	Reference
pAV1136	GCN3	α	None	URA3	Con	(Richardson et al., 2004)
	GCD7	β				
	GCD2	δ				
pAV1492	GCN3	α	None	TRP1	Con	(Gomez et al., 2002)
	GCD7	β				
	GCD2	δ				
pAV1494	GCN3	α	None	LEU2	Con	(Gomez and Pavitt, 2000)
	GCD7	β				
	GCD2	δ				
pAV1427	GCD6	ε	Fl-His$_6$–GCD6	URA3	Gal	(Gomez and Pavitt, 2000)
pAV1533	GCD6	ε	GCD1-Fl$_2$-His$_6$	URA3	Con	(Gomez et al., 2002)
	GCD1	γ				
pAV1689	GCD6^{cat}	ε^{cat}	Fl-His$_6$–GCD6^{cat}	URA3	Gal	(Gomez et al., 2002)
pTK11.1	GCD6	ε	GCD1-Fl$_2$-His$_6$	LEU2	Con	(Krishnamoorthy et al., 2001)
	GCD1	γ				

[a] GCN3, GCD7, GCD1, GCD2, and GCD6 encode eIF2Bα,β,γ,δ, and ε, respectively. GCD6^{cat} (or ε^{cat}) is the catalytic domain only (residues 518–712).
[b] Fl-His$_6$– indicates N-terminal tandem FLAG and hexahistidine tags, while -Fl$_2$-His$_6$ indicates C-terminal tandem FLAG (twice) and hexahistidine tags.
[c] Con = constitutive (gene from natural promoter)/ Gal = galactose-inducible GAL1–10 promoter; must use medium containing galactose for expression.

affinity purify the eIF2B holoenzyme complex or the catalytic eIF2Bγε subcomplex with the FLAG affinity gel, one gene must bear a FLAG (Fl) tag. We generated two tagged versions: an amino-terminally tandem (Fl, then His$_6$) Fl-His$_6$-*GCD6* and a carboxyl-terminal tagged *GCD1*-Fl-Fl-His$_6$ (Table 1.1). Plasmids are transformed into yeast strains using a standard lithium acetate procedure (Gietz and Woods, 2002).

We have successfully purified eIF2B from several yeast strains, but have found those where one or more vacuolar protease genes have been deactivated to be beneficial; for example, BJ1995 (*MATα prb1-1122 pep4-3 leu2 trp1 ura3-52 gal2*) (Jones, 1991) or *GAL2* permease positive strain GP3889 (*MATα trp1Δ-63 ura3-52 leu2-3 leu2-112 GAL2$^+$ gcn2Δ pep4::LEU2*) or its Leu$^-$ derivative GP4597 (*MATα trp1Δ-63 ura3-52 leu2-3 leu2-112 GAL2$^+$ gcn2Δ pep4::hisG*).

3. Expression and Purification of eIF2B

A step-by-step protocol is described here and is summarized graphically in Fig. 1.2.

3.1. Buffers for protein purification

The purification procedures use the buffers listed later. The main features are the use of a high-salt lysis buffer because this disrupts the interaction between eIF2B and other proteins, notably, eIF2 (see Fig. 1.1B for comparison of low- and high-salt preparations) and to include protease inhibitors and a reducing agent compatible with the purification resin. During washing, the salt concentration is lowered to 100 mM and the competitive elution uses the 3×FLAG peptide (sequence: MDYKDHDGDYKDHDIDYKDDDDK, Sigma) because this is more effective than the shorter FLAG peptide.

Lysis buffer: 100 mM Tris–HCl (pH 8.0), 500 mM KCl; 5 mM MgCl$_2$, 5 mM NaF, 2.5 mM PMSF, 7 mM β-mercaptoethanol[2], 10% ($^v/_v$) glycerol, 0.1% ($^v/_v$) Triton X-100, with 1 μg/ml pepstatin A, 1 μg/ml leupeptin, 5 μg/ml aprotinin, and 1 complete EDTA-free protease inhibitor tablet (Roche) added per 50 ml

Wash buffer: identical to lysis buffer except KCl concentration reduced to 100 mM

Elution buffer: identical to wash buffer, but additionally containing 3×FLAG peptide (Sigma) to a final concentration of 0.1 mg/ml

---

[2] Although reducing agents are not recommended by the ANTI-FLAG-resin manufacturer, we have found that their addition is critical for maintaining protein–protein interactions between FLAG-eIF2Bγ and the other eIF2B subunits during purification.

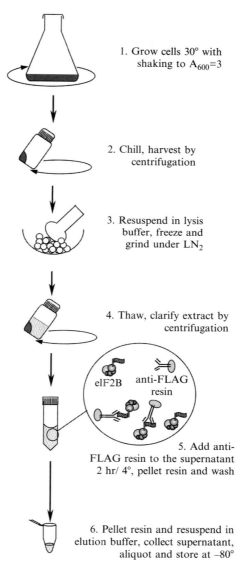

1. Grow cells 30° with shaking to $A_{600}=3$

2. Chill, harvest by centrifugation

3. Resuspend in lysis buffer, freeze and grind under LN_2

4. Thaw, clarify extract by centrifugation

eIF2B   anti-FLAG resin

5. Add anti-FLAG resin to the supernatant 2 hr/ 4°, pellet resin and wash

6. Pellet resin and resuspend in elution buffer, collect supernatant, aliquot and store at –80°

Figure 1.2 Schematic outlining the purification protocol. The principal steps of purifying FLAG-tagged eIF2B proteins from yeast whole cell extracts are outlined in the diagram; for details, refer to the main text.

Dialysis buffer: 20 mM Tris–HCl (pH 7.5), 100 mM KCl; 0.1 mM MgCl$_2$, 10% ($\frac{v}{v}$) glycerol

3.2. Cell growth and harvest

Cell cultures are grown in synthetic complete media with plasmid selection nutrients dropped out. This maintains selection for each plasmid while allowing a relatively rapid growth rate. Each strain is grown in a two-stage culture; the second larger culture media is varied depending on whether constitutive or *CYC-GAL*-promoter vectors are used.

1. Inoculate 50 ml of synthetic complete dropout (SCD⁻) medium (1.7 g yeast nitrogen base, 5 g ammonium acetate, 20 g glucose, and 2 g amino acid dropout-mix[3] per liter (Adams *et al.*, 1998)) with the strain over-expressing eIF2B, and incubate overnight at 30° with shaking at ∼200 rpm.
2. For constitutive expressing plasmids: Use the starter culture to inoculate four larger (2-liter) flasks, each containing 800 ml of SCD⁻, to a starting A_{600} of 0.02. Incubate overnight (16–18 h) at 30° with shaking (∼200 rpm) until the culture reaches an A_{600} of 1.5 to 3.

   For galactose-regulated expression plasmids: Use this starter culture to inoculate a 2-liter flask containing 800 ml of SCDGal⁻ (synthetic complete dropout media containing 0.4% glucose, 2% galactose[4]), to a starting OD_{600} of 0.2. Incubate at 30° with shaking until the culture reaches an OD_{600} of 3 (approximately 24 h).
3. Harvest cells by centrifugation at $3500 \times g$ (4000 rpm) in a Sorvall Evolution RC centrifuge, rotor SLC-6000 for 10 min at 4°, resuspend pelleted cells in ice-cold water, and centrifuge at $3500 \times g$ as previously. Cell pellet wet weight is ∼16 g/800 ml medium.

3.3. Cell lysis

We have lysed cells successfully utilizing glass-bead disruption in the past, but now favor grinding cells under liquid nitrogen (LN₂) using a mortar and pestle. This can be semi-automated, using an electric grinder model RM100 from Retsch (Schultz, 1999).

1. Suspend cell pellet with minimal volume of lysis buffer and pack into the back of a plastic syringe. Dispense yeast through syringe into LN_2[5] to make frozen yeast beads or noodles (Schultz, 1999). Frozen yeast can be stored at −80° or processed immediately.

---

[3] Dropout mix is a dry mix of 10 g L-leucine, 2 g each of every other L-amino acid, 2 g uracil, 0.5 g adenine, and 0.2 g para-aminobenzoic acid, where any component not required is omitted from the mix (dropped out). All are ground together with a porcelain mortar and pestle to a fine powder and stored dry in a foil-wrapped bottle.

[4] Addition of the low percentage of glucose helps cells transition quickly to the new medium and is used quickly before the cells switch to fermenting galactose.

[5] LN₂ is a potential asphyxiant and appropriate local safety procedures should be followed to ensure that it is used in a well-ventilated environment.

2. Lyse by mechanical grinding using a porcelain mortar and pestle. Add further LN_2 as necessary during the grinding, until a fine, whitish powder is formed. The LN_2 must not be allowed to evaporate completely or the cells will start to thaw and form an unmanageable paste.

3. Transfer powder into a 50-ml Falcon-style tube, measure the volume of powder, and add an equal volume of lysis buffer to suspend cell proteins in buffer. Centrifuge at $3500 \times g$ (4000 rpm in a Sigma 4K15 centrifuge, rotor 11150) for 5 min at $4°$ to pellet the cell debris.

4. Centrifuge the supernatant in a 50-ml polycarbonate tube (Beckman) at $18,000 \times g$ (12,250 rpm in a Heraeus Biofuge Stratos centrifuge, rotor 3335) for 15 min at $4°$, and collect the clarified lysate.

3.4. Purification using anti-FLAG M2 affinity resin

1. Binding. Pre-equilibrate 250 μl of 50% EZview Red ANTI-FLAG® M2 affinity gel (Sigma) with $2 \times$ 1-ml washes in lysis buffer to remove the storage buffer. Collect the resin by centrifugation at $400 \times g$ for 2 min at $4°$ in a microcentrifuge (2000 rpm in an Eppendorf 5415R) between each equilibration step. Add the prepared affinity matrix to the clarified lysate from the previous step in a 50-ml Falcon-style tube and incubate for 2 h at $4°$ under constant rotation (Stuart Roller Mixer).

2. Washing. Collect the FLAG resin by centrifugation at $3500 \times g$ (4000 rpm in a Sigma 4K15 centrifuge, rotor 11150) for 5 min at $4°$. Wash the resin once in 5 ml of lysis buffer and twice in 5 ml of wash buffer for 5 min at $4°$. Transfer the resin to a microcentrifuge tube.

3. Elution. Incubate the resin for 30 min with rolling in 500 μl Elution Buffer to competitively elute retained proteins. Collect the supernatant after centrifugation at $9300 \times g$ for 5 min at $4°$ in a microcentrifuge (10,000 rpm in an Eppendorf 5415R).

3.5. Dialysis to remove 3XFLAG peptide

For some downstream applications of the purified protein, the contaminating 3XFLAG peptide (mol. wt. 2862) must be removed. This is achieved by dialyzing with an appropriate buffer.

1. Dialyze eluted proteins twice in 1 liter Dialysis Buffer using Slide-A-Lyzer dialysis cassettes (Pierce) with a 30,000 molecular weight cutoff (MWTco). Because the eIF2Bε^{cat} domain is 25.9 KDa, dialysis of small domains requires a lower MWTco membrane. Therefore, 15,000 MWTco Float-A-Lyzer dialysis membranes (Spectrum) are used.

2. Freeze in 50 μl aliquots and store at $-80°$.

4. FUNCTIONAL ANALYSIS OF PURIFIED PROTEINS

As indicated in the introduction, eIF2B proteins purified have been used in a variety of assays. It is most critical to determine functionality as a GEF. The methods for this have been described extensively elsewhere (Asano *et al.*, 2002). Briefly, purified eIF2 is mixed with radiolabeled GDP, such as [³H]GDP, and incubated in buffer with a physiological magnesium concentration. In the presence of excess unlabeled GDP, labeled nucleotide is slowly released from the eIF2•[³H]GDP complex. Addition of purified eIF2B accelerates the rate of GDP release. The excess unlabeled GDP prevents rebinding of the removed labeled nucleotide. Aliquots removed from each reaction at various time intervals can be passed through protein-binding filters (e.g., nitrocellulose) under vacuum pressure. Filters are rinsed in high magnesium buffer to fix the protein-bound nucleotide and remove unbound nucleotide. Filters are dried and counted by liquid scintillation to determine the protein (eIF2)-bound nucleotide amount in each sample.

5. CONCLUSIONS

The development of methods to purify yeast eIF2B complexes from yeast cells and subunits has enhanced our understanding of eIF2B biology and provided some insight into how mutations in eIF2B affect its catalytic function (Gomez and Pavitt, 2000; Gomez *et al.*, 2002) and its ability to interact with phosphorylated eIF2 (Krishnamoorthy *et al.*, 2001). Identification of the small minimal catalytic domain was critical for initiating X-ray structural determination (Boesen *et al.*, 2004) and finding that eIF2Bε^cat has an almost identical structure to the C-terminal domain of eIF5 (Bieniossek *et al.*, 2006; Wei *et al.*, 2006), which has implications for the functions of these eIF2 regulatory proteins (Singh *et al.*, 2006). In addition, purifying mutant complexes provided clues to better understanding the molecular defects responsible for fatal brain diseases caused by eIF2B mutations (eIF2B-related disorders) (Richardson *et al.*, 2004). These methods lay the groundwork required for future studies to provide a greater insight into eIF2B structure and function.

ACKNOWLEDGMENTS

The work described was supported by grants from The Wellcome Trust (UK) to G. D. P. We thank members of the Pavitt laboratory for critical comments on the manuscript.

REFERENCES

Adams, A., Gottschling, D. E., Kaiser, C. A., and Stearns, T. (1998). "Methods in Yeast Genetics: A Cold Spring Harbor Laboratory Course Manual." Cold Spring Harbor Laboratory Press, Cold Spring Harbor, NY.

Alone, P. V., and Dever, T. E. (2006). Direct binding of translation initiation factor eIF2gamma-G domain to its GTPase-activating and GDP–GTP exchange factors eIF5 and eIF2B epsilon. *J. Biol. Chem.* **281,** 12636–12644.

Anthony, T. G., Fabian, J. R., Kimball, S. R., and Jefferson, L. S. (2000). Identification of domains within the epsilon-subunit of the translation initiation factor eIF2B that are necessary for guanine nucleotide exchange activity and eIF2B holoprotein formation. *Biochim. Biophys. Acta* **1492,** 56–62.

Asano, K., Krishnamoorthy, T., Phan, L., Pavitt, G. D., and Hinnebusch, A. G. (1999). Conserved bipartite motifs in yeast eIF5 and eIF2Bepsilon, GTPase-activating, and GDP–GTP exchange factors in translation initiation, mediate binding to their common substrate eIF2. *EMBO J.* **18,** 1673–1688.

Asano, K., Phan, L., Krishnamoorthy, T., Pavitt, G. D., Gomez, E., Hannig, E. M., Nika, J., Donahue, T. F., Huang, H. K., and Hinnebusch, A. G. (2002). Analysis and reconstitution of translation initiation *in vitro*. *Methods Enzymol.* **351,** 221–247.

Bieniossek, C., Schutz, P., Bumann, M., Limacher, A., Uson, I., and Baumann, U. (2006). The crystal structure of the carboxy-terminal domain of human translation initiation factor eIF5. *J. Mol. Biol.* **360,** 457–465.

Boesen, T., Mohammad, S. S., Pavitt, G. D., and Andersen, G. R. (2004). Structure of the catalytic fragment of translation initiation factor 2B and identification of a critically important catalytic residue. *J. Biol. Chem.* **279,** 10584–10592.

Bumann, M., Djafarzadeh, S., Oberholzer, A. E., Bigler, P., Altmann, M., Trachsel, H., and Baumann, U. (2004). Crystal structure of yeast Ypr118w, a methylthioribose-1-phosphate isomerase related to regulatory eIF2B subunits. *J. Biol. Chem.* **279,** 37087–37094.

Cigan, A. M., Bushman, J. L., Boal, T. R., and Hinnebusch, A. G. (1993). A protein complex of translational regulators of *GCN4* is the guanine nucleotide exchange factor for eIF-2 in yeast. *Proc. Natl. Acad. Sci. USA* **90,** 5350–5354.

Fabian, J. R., Kimball, S. R., Heinzinger, N. K., and Jefferson, L. S. (1997). Subunit assembly and guanine nucleotide exchange activity of eukaryotic initiation factor-2B expressed in Sf9 cells. *J. Biol. Chem.* **272,** 12359–12365.

Farrell, P. J., Balkow, K., Hunt, T., Jackson, R. J., and Trachsel, H. (1977). Phosphorylation of initiation factor eIF-2 and the control of reticulocyte protein synthesis. *Cell* **11,** 187–200.

Fogli, A., and Boespflug-Tanguy, O. (2006). The large spectrum of eIF2B-related diseases. *Biochem. Soc. Trans.* **34,** 22–29.

Fogli, A., Schiffmann, R., Bertini, E., Ughetto, S., Combes, P., Eymard-Pierre, E., Kaneski, C. R., Pineda, M., Troncoso, M., Uziel, G., Surtees, R., Pugin, D., *et al.* (2004a). The effect of genotype on the natural history of eIF2B-related leukodystrophies. *Neurology* **62,** 1509–1517.

Fogli, A., Schiffmann, R., Hugendubler, L., Combes, P., Bertini, E., Rodriguez, D., Kimball, S. R., and Boespflug-Tanguy, O. (2004b). Decreased guanine nucleotide exchange factor activity in eIF2B-mutated patients. *Eur. J. Hum. Genet.* **12,** 561–566.

Gebauer, F., and Hentze, M. W. (2004). Molecular mechanisms of translational control. *Nat. Rev. Mol. Cell Biol.* **5,** 827–835.

Gietz, R. D., and Woods, R. A. (2002). Transformation of yeast by lithium acetate/single-stranded carrier DNA/polyethylene glycol method. *Methods Enzymol.* **350,** 87–96.

Gomez, E., Mohammad, S. S., and Pavitt, G. D. (2002). Characterization of the minimal catalytic domain within eIF2B: The guanine-nucleotide exchange factor for translation initiation. *EMBO J.* **21,** 5292–5301.

Gomez, E., and Pavitt, G. D. (2000). Identification of domains and residues within the epsilon subunit of eukaryotic translation initiation factor 2B (eIF2Bepsilon) required for guanine nucleotide exchange reveals a novel activation function promoted by eIF2B complex formation. *Mol. Cell Biol.* **20**, 3965–3976.

Goss, D. J., Parkhurst, L. J., Mehta, H. B., Woodley, C. L., and Wahba, A. J. (1984). Studies on the role of eukaryotic nucleotide exchange factor in polypeptide chain initiation. *J. Biol. Chem.* **259**, 7374–7377.

Jones, E. W. (1991). Tackling the protease problem in *Saccharomyces cerevisiae*. *Methods Enzymol.* **194**, 428–453.

Kakuta, Y., Tahara, M., Maetani, S., Yao, M., Tanaka, I., and Kimura, M. (2004). Crystal structure of the regulatory subunit of archaeal initiation factor 2B (aIF2B) from hyperthermophilic archaeon *Pyrococcus horikoshii* OT3: A proposed structure of the regulatory subcomplex of eukaryotic IF2B. *Biochem. Biophys. Res. Commun.* **319**, 725–732.

Kantor, L., Harding, H. P., Ron, D., Schiffmann, R., Kaneski, C. R., Kimball, S. R., and Elroy-Stein, O. (2005). Heightened stress response in primary fibroblasts expressing mutant eIF2B genes from CACH/VWM leukodystrophy patients. *Hum. Genet.* **118**, 99–106.

Kimball, S. R., Fabian, J. R., Pavitt, G. D., Hinnebusch, A. G., and Jefferson, L. S. (1998). Regulation of guanine nucleotide exchange through phosphorylation of eukaryotic initiation factor eIF2alpha. Role of the alpha- and delta-subunits of eIF2B. *J. Biol. Chem.* **273**, 12841–12845.

Konieczny, A., and Safer, B. (1983). Purification of the eukaryotic initiation factor 2-eukaryotic initiation factor 2B complex and characterization of its guanine nucleotide exchange activity during protein synthesis initiation. *J. Biol. Chem.* **258**, 3402–3408.

Krishnamoorthy, T., Pavitt, G. D., Zhang, F., Dever, T. E., and Hinnebusch, A. G. (2001). Tight binding of the phosphorylated alpha subunit of initiation factor 2 (eIF2alpha) to the regulatory subunits of guanine nucleotide exchange factor eIF2B is required for inhibition of translation initiation. *Mol. Cell Biol.* **21**, 5018–5030.

Kyrpides, N. C., and Woese, C. R. (1998). Archaeal translation initiation revisited: The initiation factor 2 and eukaryotic initiation factor 2B alpha-beta-delta subunit families. *Proc. Natl. Acad. Sci. USA* **95**, 3726–3730.

Li, W., Wang, X., Van Der Knaap, M. S., and Proud, C. G. (2004). Mutations linked to leukoencephalopathy with vanishing white matter impair the function of the eukaryotic initiation factor 2B complex in diverse ways. *Mol. Cell Biol.* **24**, 3295–3306.

Matts, R. L., Levin, D. H., and London, I. M. (1983). Effect of phosphorylation of the alpha-subunit of eukaryotic initiation factor 2 on the function of reversing factor in the initiation of protein synthesis. *Proc. Natl. Acad. Sci. USA* **80**, 2559–2563.

Mikami, S., Masutani, M., Sonenberg, N., Yokoyama, S., and Imataka, H. (2006). An efficient mammalian cell-free translation system supplemented with translation factors. *Protein Expr. Purif.* **46**, 348–357.

Mohammad-Qureshi, S. S., Haddad, R., Hemingway, E. J., Richardson, J. P., and Pavitt, G. D. (2007). Critical contacts between the eukaryotic initiation factor 2B (eIF2B) catalytic domain and both eIF2β and 2γ mediate guanine nucleotide exchange. *Mol. Cell. Biol.* **27**, in press.

Nika, J., Yang, W., Pavitt, G. D., Hinnebusch, A. G., and Hannig, E. M. (2000). Purification and kinetic analysis of eIF2B from *Saccharomyces cerevisiae*. *J. Biol. Chem.* **275**, 26011–26017.

Panniers, R., and Henshaw, E. C. (1983). A GDP/GTP exchange factor essential for eukaryotic initiation factor 2 cycling in Ehrlich ascites tumor cells and its regulation by eukaryotic initiation factor 2 phosphorylation. *J. Biol. Chem.* **258**, 7928–7934.

Pavitt, G. D. (2005). eIF2B, a mediator of general and gene-specific translational control. *Biochem. Soc. Trans.* **33**, 1487–1492.

Pavitt, G. D., Ramaiah, K. V., Kimball, S. R., and Hinnebusch, A. G. (1998). eIF2 independently binds two distinct eIF2B subcomplexes that catalyze and regulate guanine-nucleotide exchange. *Genes Dev.* **12,** 514–526.

Pavitt, G. D., Yang, W., and Hinnebusch, A. G. (1997). Homologous segments in three subunits of the guanine nucleotide exchange factor eIF2B mediate translational regulation by phosphorylation of eIF2. *Mol. Cell Biol.* **17,** 1298–1313.

Richardson, J. P., Mohammad, S. S., and Pavitt, G. D. (2004). Mutations causing childhood ataxia with central nervous system hypomyelination reduce eukaryotic initiation factor 2B complex formation and activity. *Mol. Cell Biol.* **24,** 2352–2363.

Rowlands, A. G., Panniers, R., and Henshaw, E. C. (1988). The catalytic mechanism of guanine nucleotide exchange factor action and competitive inhibition by phosphorylated eukaryotic initiation factor 2. *J. Biol. Chem.* **263,** 5526–5533.

Schultz, M. C. (1999). Chromatin assembly in yeast cell-free extracts. *Methods* **17,** 161–172.

Singh, C. R., Lee, B., Udagawa, T., Mohammad-Qureshi, S. S., Yamamoto, Y., Pavitt, G. D., and Asano, K. (2006). An eIF5/eIF2 complex antagonizes guanine nucleotide exchange by eIF2B during translation initiation. *EMBO J.* **25,** 4537–4546.

van der Knaap, M. S., van Berkel, C. G., Herms, J., van Coster, R., Baethmann, M., Naidu, S., Boltshauser, E., Willemsen, M. A., Plecko, B., Hoffmann, G. F., Proud, C. G., Scheper, G. C., et al. (2003). eIF2B-related disorders: Antenatal onset and involvement of multiple organs. *Am. J. Hum. Genet.* **73,** 1199–1207.

van Kollenburg, B., van Dijk, J., Garbern, J., Thomas, A. A., Scheper, G. C., Powers, J. M., and van der Knaap, M. S. (2006). Glia-specific activation of all pathways of the unfolded protein response in vanishing white matter disease. *J. Neuropathol. Exp. Neurol.* **65,** 707–715.

Wang, X., Paulin, F. E., Campbell, L. E., Gomez, E., O'Brien, K., Morrice, N., and Proud, C. G. (2001). Eukaryotic initiation factor 2B: Identification of multiple phosphorylation sites in the epsilon-subunit and their functions *in vivo*. *EMBO J.* **20,** 4349–4359.

Wei, Z., Xue, Y., Xu, H., and Gong, W. (2006). Crystal structure of the C-terminal domain of *S. cerevisiae* eIF5. *J. Mol. Biol.* **359,** 1–9.

Yang, W., and Hinnebusch, A. G. (1996). Identification of a regulatory subcomplex in the guanine nucleotide exchange factor eIF2B that mediates inhibition by phosphorylated eIF2. *Mol. Cell Biol.* **16,** 6603–6616.

CHAPTER TWO

In Vivo Deletion Analysis of the Architecture of a Multiprotein Complex of Translation Initiation Factors

Klaus H. Nielsen* *and* Leoš Valášek†

Contents

Abstract

Protein complexes play a critical role in virtually all cellular processes that have been studied to date. Comprehensive knowledge of the architecture of a protein complex of interest is, therefore, an important prerequisite for understanding its role in the context of a particular pathway in which it participates. One of the possible approaches that has proven very useful in characterizing a protein

*Department of Molecular Biology, University of Aarhus, Denmark
†Laboratory of Regulation of Gene Expression, Institute of Microbiology, Prague, Czech Republic

Methods in Enzymology, Volume 431
ISSN 0076-6879, DOI: 10.1016/S0076-6879(07)31002-1

complex is outlined in this chapter using the example of the eukaryotic initiation factor 3 (eIF3) and some of its binding partners. eIF3 is one of the major players in the translation initiation pathway because it orchestrates several crucial steps that ultimately conclude with formation of the 80S ribosome where the anticodon of methionyl-tRNA$_i^{Met}$ base-pairs with the AUG start codon of the mRNA in the ribosomal P-site. We previously demonstrated that, in the budding yeast *Saccharomyces cerevisiae*, eIF3 closely cooperates with several other eIFs to stimulate recruitment of methionyl-tRNA$_i^{Met}$ and mRNA to the 40S ribosome and that it forms, together with eIFs 1, 2, and 5, an important intermediate in translation initiation called the multifactor complex (MFC). Here, we summarize the fundamental procedure that allowed in-depth characterization of the MFC composition and identification of protein–protein interactions among its constituents. Primarily, we describe in detail *in vivo* purification techniques that, in combination with systematic deletion analysis, produced a 3D subunit interaction model for the MFC. Site-directed clustered-10-alanine-mutagenesis (CAM) employed to investigate the physiological significance of individual interactions is also presented. The general character of the entire procedure makes it usable for first-order structural characterization of virtually any soluble protein complex in yeast.

1. INTRODUCTION

Numerous polypeptides orchestrate assembly of the 80S initiation complex in which the anticodon of the methionyl initiator tRNA (Met-tRNA$_i^{Met}$) is base-paired with the AUG start codon of the mRNA in the ribosomal P-site. This is a prerequisite for the onset of the actual translation of information encoded in a particular gene into its protein product. Of the 12 eukaryotic translation initiation factors (eIFs), three occur in multiprotein complexes: (I) eIF2 is a GTP-binding trimeric complex that, when bound to GTP, forms a ternary complex (TC) with Met-tRNA$_i^{Met}$ and ensures its delivery to the 40S ribosome. (II) Heteropentameric eIF2B serves as a nucleotide exchange factor for eIF2. (III) eIF3 with 6 to 13 subunits (depending on the species) is the most complex initiation factor and, thus, it is not surprising that eIF3 is thought to participate in virtually every step of the initiation phase of translation and maybe even beyond that (for a review, see Dong and Zhang, 2006; Hershey and Merrick, 2000; Hinnebusch, 2006). Ever since its identification in the early 1970s, much effort has been spent in elucidating the precise composition of eIF3 as well as the contribution of its individual subunits to particular initiation reactions at the molecular level. Since that kind of information is only slowly beginning to emerge for eIF3 from higher eukaryotic organisms, much of our knowledge about how eIF3 works comes from a series of studies conducted with budding yeast *Saccharomyces cerevisiae* over the last decade. Certainly, at least two properties make yeast an ideal model organism for studies of the translation initiation mechanism. *S. cerevisiae* appears to contain homologues of the majority of eIFs that occur in higher eukaryotes, and it provides researchers with the powerful combination of genetic tools and various *in vivo* and *in vitro* biochemical assays for partial reactions of the initiation pathway.

Yeast eIF3 contains five essential core subunits—eIF3a/TIF32, eIF3b/PRT1, eIF3c/NIP1, eIF3g/TIF35, and eIF3i/TIF34—and the nonessential subunit eIF3j/HCR1 (for a summary, see Table 2.1). Although it is less complex than mammalian eIF3, thought to be composed of ~13 polypeptides, all five essential subunits of yeast eIF3 are believed to represent the core of mammalian eIF3 and, in addition, yeast eIF3 was shown to possess the hallmark functions of its mammalian counterpart (Phan *et al.*, 1998). One of the major findings has been that yeast eIF3 can be isolated together with eIF1, eIF5, and TC in a multifactor complex (MFC) free of 40S subunits (Asano *et al.*, 2000). This raised the possibility that the assembly of these factors in the MFC, either in solution or on the surface of the 40S subunit, facilitates their cooperative binding to the ribosome to form the 43S pre-initiation complex (PIC). Detailed *in vitro* and *in vivo* mapping of mutual protein–protein interactions among all MFC components followed by genetic analysis of random or site-directed mutagenesis of critical contact points, combined with biochemical measurements, have provided compelling evidence that the contacts among factors in the MFC enhance 43S PIC assembly, mRNA association with the 43S PIC, as well as the post-assembly functions of scanning, AUG recognition, and GTP hydrolysis on the TC upon AUG recognition (Asano *et al.*, 1998, 2001; Nielsen *et al.*, 2004, 2006; Singh *et al.*, 2004, 2005; Valášek *et al.*, 2001, 2002, 2003, 2004). The techniques outlined herein have greatly contributed to these achievements by identifying the web of protein–protein interactions that stabilize the MFC and by elucidate the importance of these interactions for MFC functions.

It should be emphasized that the nature of all presented protocols is very general and, thus, their application for a comprehensive characterization of your favorite multiprotein complex (YFMPC) in yeast might require only minor modifications. The logical sequence of all required steps is schematically shown in Fig. 2.1. The initial large-scale Ni affinity isolation of eIF3 followed by mass spectrometry (MS) of its subunit composition has already been described (Asano *et al.*, 2002), and methods for identification of protein–protein interactions such as yeast two-hybrid (Y2H) and *in vitro* glutathione-S-transferase (GST) pull-down analysis are presented in volume 429. This chapter focuses on a description of the small-scale one-step *in vivo* affinity purification techniques that were used to determine the effects of deletions and

Table 2.1 Overview of homologues eIF3 subunits across the species with potential functions assigned

Subunit	S. cerevisiae-MW	H. sapiens	S. pombe	A. thaliana	Wheat	Domain(s)	Function(s) of S. c.
eIF3a	TIF32–110 kD	p170	p107	p114	p116	PCI/HLD	MFC assembly, 40S binding, TC and mRNA recruitment
eIF3b	PRT1–90 kD	p116	p84	p82	P83	RRM/WD repeat	MFC assembly, 40S binding, TC and mRNA recruitment, scanning
eIF3c	NIP1–93 kD	p110	p104	p105	p107	PCI	MFC assembly, 40S binding, TC and mRNA recruitment, AUG recognition
eIF3d	–	p66	Moe1	p66	P87	binds RNA	–
eIF3e	?PCI8?–51 kD	INT–6/p48	Int6	p52	P45	PCI	–
eIF3f	–	p47	Csn6	p32	P34	MPN	–
eIF3g	TIF35–33 kD	p44	Tif35	p33	P36	RRM/Zn fingers	Unknown
eIF3h	–	p40	p40	p38	p41a	MPN	–
eIF3i	TIF34–39 kD	TRIP-1/p36	Sum1	p36	p41b	WD repeat	Unknown
eIF3j	HCR1–35 kD	p35	–	–	–	–	MFC assembly, 40S binding, 40S biogenesis
eIF3k	–	p28	–	p25	P28	PCI / MPN	–
eIF3l	–	p67	–	p60	P56	PCI	–
eIF3m	–	GA17	Csn7B	?	?	PCI	–

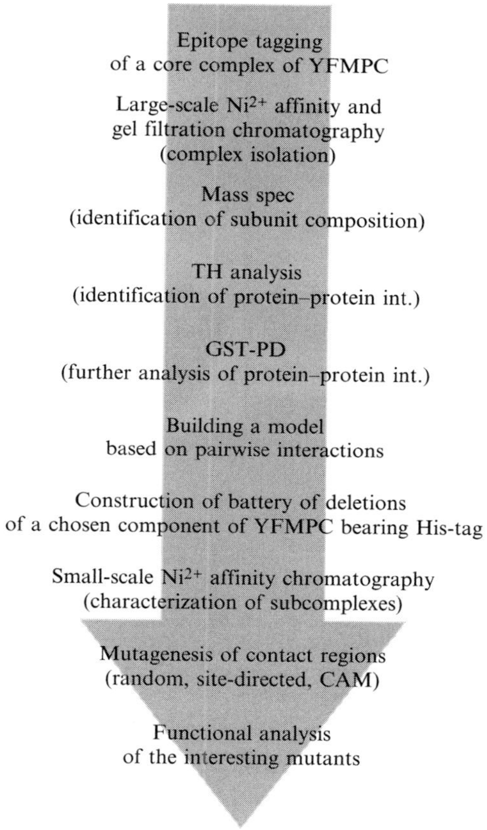

Epitope tagging
of a core complex of YFMPC

Large-scale Ni^{2+} affinity and
gel filtration chromatography
(complex isolation)

Mass spec
(identification of subunit composition)

TH analysis
(identification of protein–protein int.)

GST-PD
(further analysis of protein–protein int.)

Building a model
based on pairwise interactions

Construction of battery of deletions
of a chosen component of YFMPC bearing His-tag

Small-scale Ni^{2+} affinity chromatography
(characterization of subcomplexes)

Mutagenesis of contact regions
(random, site-directed, CAM)

Functional analysis
of the interesting mutants

Figure 2.1   Schematic illustrating the ideal sequence of all steps to be taken toward a comprehensive characterization of a multiprotein complex of interest.

CAMs on the subunit composition of the mutant MFC. This approach allowed us to significantly refine our knowledge regarding the architecture of the yeast MFC as we extended the *in vitro* deletion mapping of regions required for pairwise interactions of isolated subunits (Asano *et al.*, 1998; Phan *et al.*, 1998; Valášek *et al.*, 2001) to the mapping of regions in each subunit required for its interactions with all other components of the MFC within the living cells (Valášek *et al.*, 2002) (Fig. 2.2). The last part of this chapter is devoted to a description of our modification of alanine scanning mutagenesis (Wertman *et al.*, 1992), which we designated clustered–10–alanine–mutagenesis (CAM), that has proven to be a valuable tool for rapid identification of functionally important residues in a given subunit of the MFC.

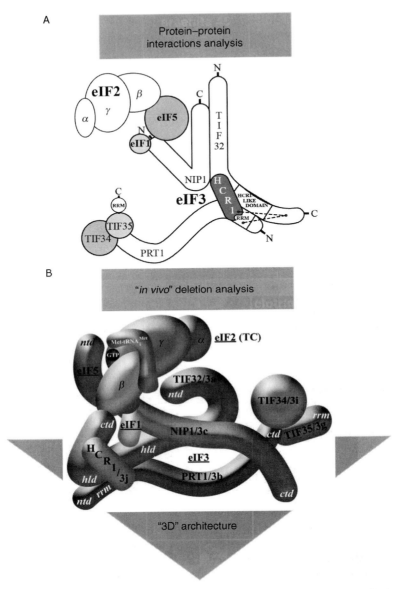

Figure 2.2 Schematic illustrating the gain of knowledge after employing the *in vivo* deletion analysis approach. (A) Summary of protein–protein interactions within the yeast eIF3 complex (adopted with permission from Valášek *et al.*, 2001). The eIF3 subunits, as well as eIF5, eIF2, and eIF1 factors, are shown as various shapes with sizes roughly proportional to their molecular weights. Points of overlap between the various shapes indicate sites of known protein–protein interaction. (B) A 3D model of the MFC based on a comprehensive deletion analysis of subunit interactions (adopted with permission from Valášek *et al.*, 2003). The labeled protein subunits are shown roughly in proportion to their molecular weights. The degree of overlap between two different subunits depicts the extent of their interacting surfaces. (See color insert.)

2. Ni²⁺ Affinity Purification of eIF3 Using a Polyhistidine Tag

Nickel affinity chromatography was chosen as the primary purification technique because it is a fast and reliable one-step assay and purified complexes can often be used in downstream applications without the necessity of removing the polyhistidine tag. In addition, the polyhistidine tag is smaller than many other affinity tags targeted by commercially available affinity resins and, in most cases, does not seem to interfere with the structure and function of the recombinant protein.

2.1. Affinity-tagging

The placement of the tag is worth careful consideration. Although it may be difficult to predict at the beginning of the study, a true subunit of YFMPC should be chosen for tagging. Although there might be a weak bias for using a C-terminal tag, because this position ensures that only full-length proteins are purified, it is advisable to create two alleles with one having the epitope-tag placed at its N-terminus and the other at its C-terminus. Either of the termini can mediate vital interactions in YFMPC, and the presence of an artificial sequence may interfere with the interaction. For example, in the initial attempt to isolate eIF3 on a large scale, the PRT1 subunit was chosen and the His_8-tag was inserted at its C-terminus immediately in front of the stop codon. Despite the ambiguity regarding the composition of yeast eIF3 in the mid-1990s, PRT1, with the RNA-recognition motif (RRM) present at its extreme N-terminus, was uniformly identified as a true component of eIF3 in all purification schemes (Danaie *et al.*, 1995; Naranda *et al.*, 1994). For subsequent His_8-tagging of other core eIF3 subunits that followed identification of their protein–protein contacts, we always tried to select the terminus that was less likely to contain a binding domain. It is worth noting that, contrary to the routinely used His_6-tag, we found that using eight histidines instead of only six greatly increased specificity and yield without any detectable impairment of the activity of the tagged proteins. Presumably, the slightly longer stretch of histidine residues increases the chance that more of them will be exposed on the surface, allowing stronger binding to the resin.

The coding sequences for His_8-tag can be easily fused with your gene (YFG) using a variety of methods including *in vitro* mutagenesis, PCR, or *in vivo* homologous recombination. When a unique restriction site X exists or is inserted immediately prior to the stop codon by site-directed mutagenesis, two complementary oligonucleotides His-A (5′ xxxxxCACCACCACCACCACCACCACTAA 3′) and His-B

(5′ xxxxxTT-AGTGGTGGTGGTGGTGGTGGTGGTG 3′) can be used to form a DNA duplex containing eight consecutive histidine codons and the nucleotides (x) at each end capable of annealing with digested X-site ends. The ends of the duplex are then phosphorylated with T4 kinase and inserted into the X-digested vector containing YFG, dephosphorylated with calf intestine phosphatase beforehand. The insert is confirmed by PCR where one of the oligonucleotide primers is complementary to the His_8–tag sequence and the other is complementary to the coding sequence of YFG. Alternatively, the tag can be engineered into the sequence of YFG itself by employing PCR. The upstream PCR primer contains a unique endogenous restriction site which is located within YFG, the downstream primer includes sequences corresponding to the very 3′ end of YFG, followed by eight CAC triplets, the stop codon, and sequences from the 3′ UTR, including a unique restriction site, which may have to be introduced by site-directed mutagenesis. The PCR product and the original construct are then both cut with the appropriate restriction enzymes and ligated together. It is also possible to introduce affinity tags at the 3′ end of YFG by homologous recombination in yeast cells, as described thoroughly in a previous report (Longtine *et al.*, 1998).

2.2. Whole-cell extract (WCE) preparation

A derivative of yeast strain W303 (*MATa ade2-1 trp1-1 can1-100 leu2-3, 112 his3-11, 15 ura3*) is employed carrying a single-copy plasmid encoding a His_8–tagged eIF3a/TIF32 in the presence of chromosomal wild-type *TIF32*.

1. The strain expressing TIF32-His_8 from a single-copy plasmid is grown in 100 ml of SD medium supplemented with the required amino acids in a 0.5-L Erlenmeyer flask overnight to an OD_{600} of ~1.0 at 30° (200 rpm).
2. Cells are harvested by centrifugation for 5 min at $3000 \times g$ (e.g., 2500 rpm in a Beckman JS-4.2 rotor) and washed with 10 ml of ice-cold deionized water in 15 ml Falcon tubes. The remaining steps are carried out at 4°.
3. Cells are collected by centrifugation as has been described and resuspended in 1 ml per gram (wet weight) of cells in breaking/binding buffer BB (20 mM Tris-HCl [pH 7.5], 100 mM KCl, 5 mM MgCl$_2$, 0.5 mM β-mercaptoethanol, 1 mM phenylmethylsulfonyl fluoride, 20 mM imidazole, 10% [v/v] glycerol, 1 × EDTA-free complete Protease Inhibitor Mix tablets [Roche]).
4. Cells are broken by adding 4 mm diameter acid-washed glass beads (Thomas Scientific) equal to one-half of the total volume of resuspended cells and rigorously vortexed for 30 s, followed by 1 min on ice, conducted 5 times. Less than 5 cycles were not sufficient for complete cell lysis, whereas more than 5 cycles of vortexing resulted in a significant loss of co-purifying components.

The resulting WCE is shortly spun down (1 min at 3000×*g*) and the beads free lysate is transfered to pre-cooled Eppendort tubes.
5. The WCE is further clarified by two successive centrifugations at 16,100×*g* for 2 and 10 min, respectively, using an Eppendorf F 45-24-11 rotor in a table centrifuge. After each centrifugation, the supernatant is carefully transferred to a new pre-cooled Eppendorf tube avoiding the lipid layer, and the total protein concentration (mg/ml) is estimated using the Bradford method (Biorad).

2.3. Batch Ni²⁺ affinity purification of eIF3 (Ni pull-down [Ni PD])

1. One gram of Ni²⁺-NTA-silica resin (Qiagen) is suspended in 5 ml of sterile deionized water resulting in a 50% slurry. The resin can be kept in this form for several months at 4°, providing that evaporation is prevented.
2. The required volume of 50% Ni-silica slurry (7.5 μl/100 ml of starting yeast culture) is equilibrated just before use in buffer BB by two washes each of 500 μl. The resin is collected by centrifugation at 1000 rpm for 2 min in an Eppendorf F 45-24-11 rotor to complete each wash.
3. 0.5 to 1 mg of the total protein is brought to a total volume of 200 μl with buffer BB and mixed with 7.5 μl of 50% slurry of Ni-silica resin in an Eppendorf tube and incubated overnight with slow rotation at 4°, making sure the resin is kept in suspension and does not sediment.
4. An aliquot of 3% of the total protein used in each reaction is stored at −20° to provide the "input" reference sample in the subsequent analysis.
5. Centrifuge the sample for 2 min at 500×*g* to collect the resin, ensuring that it is pelleted evenly at the bottom of the tub. Remove the "flow-through" supernatant by pipetting (reserving a 3% aliquot) and wash the resin 3 times with 1 ml of buffer BB each.
6. Resuspend the resin in 50 μl of elution buffer E (20 mM Tris–HCl [pH 7.5], 100 mM KCl, 5 mM MgCl$_2$, 0.5 mM β-mercaptoethanol, 1 mM phenylmethylsulfonyl fluoride, 250 mM imidazole, 10% (v/v) glycerol, 1× Complete Protease Inhibitor Mix tablets [Roche]), briefly vortex, and elute the bound proteins by mixing on a rotator at 4° for 20 min. Please note that a concentration of imidazole and the kind of Roche tablets differs from BB buffer.
7. Collect the resin by centrifuging, as described previously, and carefully remove and save the "eluate" supernatant.

2.4. Analysis of the purified proteins

1. The aliquot containing 3% "input," the entire eluted fraction, and the aliquot of 3% flow-through supernatant is mixed with 4× sample loading buffer (1 M Tris–HCl [pH 6.8], 40% glycerol, 8% SDS, 0.06%

bromophenol blue [sodium salt], 1.47 % β-mercaptoethanol) and boiled for 3 min at 95°.

2. The boiled samples are resolved by sodium dodecyl sulfate polyacrylamide gel electrophoresis (SDS-PAGE) using 4 to 20% polyacrylamide gels, loaded in the following order: 3% input −10 μl of eluate (15% of the total volume of the boiled eluate; E1); −20 μl of eluate (30%; E2); −3% supernatant, respectively, and transferred to nitrocellulose membranes (Novex or Bio-Rad).

3. Immunoblot analysis is conducted with antibodies to the His_8-tag epitope (to detect the TIF32-His_8 subunit; Santa Cruz) or with antibodies to the other factors.

4. Immunodetection is performed by chemiluminescence (ECL^{TM}, Amersham Pharmacia Biotech) using horseradish peroxidase-conjugated secondary antibodies (Amersham Pharmacia Biotech).

2.5. Troubleshooting

The major concern is the binding of endogenous proteins with stretches of histidine residues that interact with the Ni–NTA groups and co-purify nonspecifically with eIF3. This can be reduced by using a slightly higher 20–30 mM imidazole concentration in the buffer BB. (The imidazole ring is part of the structure of histidine that can bind to the nickel ions, although with lower affinity, and disrupt the binding of dispersed histidine residues in nontagged background proteins.) Alternatively, detergents such as Triton X-100 and Tween 20 (up to 2%), or high salt concentrations (up to 0.5 M KCl) may reduce nonspecific binding to the resin by disrupting hydrophobic or ionic interactions without impairing binding of the TIF32-His_8 and associated proteins to the resin. Similarly, a mild decrease in the pH (to 6.5) may increase the stringency of the purification scheme. And finally, an increased concentration of β-mercaptoethanol (to 7 mM) in buffer BB can be used to prevent the co-purification of proteins that have formed disulfide bonds with the proteins of interest during cell lysis. It should be noted that using Dithiotreitol (DTT) in place of β-mercaptoethanol is not recommended because DTT reduces Ni^{2+} ions.

3. DELETION/MUTATIONAL ANALYSIS OF EIF3 SUBUNITS AND NI^{2+} AFFINITY PURIFICATION OF THEIR SUBCOMPLEXES

Large-scale purification of eIF3 using PRT1-His_8 followed by MS analysis provided a clear answer regarding the composition of yeast eIF3 (Phan *et al.*, 1998). Following extensive studies of the interactions among the yeast eIF3 subunits and associated factors by yeast two-hybrid analysis and *in vitro* binding assays, a subunit interaction model for the MFC was constructed (Asano *et al.*, 1998; Valášek *et al.*, 2001) (Fig. 2, upper panel). In order to test the *in vivo* relevance of this model and to reveal other potential contacts among the MFC components, we made deletions of predicted binding domains for various MFC components in affinity-tagged forms of the three largest eIF3 subunits, and determined the compositions of the Ni-affinity-purified complexes. The results obtained not only provided *in vivo* confirmation of the subunit interactions depicted in Fig. 2.2 (upper panel) but also revealed previously undetected contacts between NIP1 and PRT1 and between TIF32 and eIF1 (Fig. 2.2, lower panel). Most importantly, we also uncovered the first direct contact between eIF2 and eIF3 (Valášek *et al.*, 2002), the existence of which was expected based on the role of eIF3 in stimulation of the TC recruitment to the 40S ribosome. In addition, this technique enabled direct comparison of binding affinities between the wild-type and mutant forms of tested interacting domains, as will be demonstrated (Nielsen *et al.*, 2006; Valášek *et al.*, 2004).

In general, each truncated version of an eIF3 subunit bearing a His$_8$-tag at the same end as the original full-length protein was expressed in the presence of the wild-type untagged protein in the W303 strain and subjected to Ni affinity chromatography, as previously described, with minor modifications. The presence of the wild-type untagged eIF3 subunit was a necessity, because most deletions or truncations rendered the resulting protein incapable of supporting growth. In addition, some truncations reduced protein stability, as might be expected and, in these cases, high copy number vectors were used to compensate for the lower expression level of the mutant form. All strains expressing tagged mutant proteins were always analyzed in parallel with those containing untagged and tagged versions of the wild-type protein as controls. If required, the Western signals of individual co-purifying proteins in the Eluate 1 and Eluate 2 fractions were quantified using the NIH image program (version beta 3b), as described in Fig. 2.3.

3.1. Typical example

Using Ni affinity chromatography, we were able to show that the extreme N-terminal 205 amino acid residues of His$_8$-tagged eIF3c/NIP1 co-purified *in vivo* with eIF5, eIF1, and eIF2, but not with any of the eIF3 subunits in agreement with the structural model for the MFC shown in Figs. 2.2 and 2.3A (Valášek *et al.*, 2002). Substitution of the 10 mostly acidic aa residues (11–20) in the NIP1-NTD with a stretch of 10 alanines (referred to as CAM herein; see Section 5) resulted in a nearly complete impairment of its interaction with eIF5 *in vitro* (Valášek *et al.*, 2004). By assaying formation of the NIP1-NTD-eIF1-eIF2-eIF5 subcomplex *in vivo*, we confirmed that this so-called Box2 mutation has a much stronger effect on the interaction of NIP1-NTD with eIF5 than eIF1. As shown in Fig. 2.3C, a fraction of

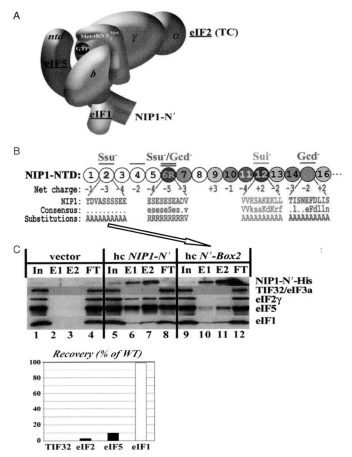

Figure 2.3 An example of the Ni-affinity chromatography showing that the *NIP1-Box2* mutation diminishes binding of eIF5 and eIF2 to the NIP1-NTD *in vivo* (adopted with permission from Valášek *et al.*, 2004). (A) A 3D model of the NIP1-NTD subcomplex with TC, and eIF1 and eIF5. (B) Schematics illustrating the application of clustered 10-Ala mutagenesis (CAM) to the example of the N-terminal domain of NIP1. The sequence of the first 160 amino acids of NIP1 is shown as numbered circles (Boxes 1–16), each of them composed of 10 residues substituted with a stretch of 10 alanines. Different shades of gray indicate the degree of identities between the NIP1-NTD and the N-termini of its various homologues. Color-coded bars above the circles indicate the phenotypes associated with amino acid substitutions in the corresponding boxes: Ssu⁻ (*suppressor of Sui⁻*), Gcd⁻ (*general control derepressed*), and Sui⁻ (*suppressor of initiation codon*). Blown-up segments in blue, green, and yellow indicate the amino acid sequences, a consensus sequence derived from sequence alignments, and the substitutions made in the corresponding boxes of the NIP1-NTD. Net charge of individual boxes is indicated below each of them. (C) WCEs prepared from the cells overexpressing either wild-type or mutant form of the NIP1-NTD were incubated with Ni²⁺-NTA-silica resin, and the bound proteins were eluted and subjected to Western blot analysis using

eIF2, eIF5, and eIF1 co-purified specifically with WT His_8-tagged NIP1-NTD in Ni^{2+} chelation chromatography of whole cell extracts (WCEs) (cf. lanes 6–7 and 2–3). The Box2 mutation greatly reduced the proportions of eIF5 and eIF2 that co-purified with His_8-NIP1-NTD without affecting the association with eIF1 (Fig. 2.3C, cf. lanes 10–11 and 6–7). Because eIF5 bridges the interaction between the NIP1-NTD and eIF2β, it is not surprising that Box2 reduces the binding of NIP1-NTD to both eIF2 and eIF5.

4. eIF3 Purification Using Other Epitope Tags

To provide the reader with a more complete picture, this section describes the use of affinity tags other than a His_8-tag which were also used successfully for purification of eIF3 and its binding partners (Asano *et al.*, 2000; Valášek *et al.*, 2003).

4.1. Immunoaffinity purification of eIF3 containing hemagglutinin-tagged eIF3i/TIF34 (HA pull-down)

1. Yeast strain carrying a single HA-tagged subunit of eIF3i/TIF34 as the only copy of this subunit is grown in 100 ml of SD medium to early log phase (OD_{600} of ~1.0–2.0) at 30° (200 rpm), harvested, and suspended in 0.2 ml of ice-cold lysis buffer HA (20 mM Tris-HCl, pH 7.4, 1 mM magnesium acetate, 100 mM KCl, 0.1% Triton X-100, Complete™ protease inhibitor (Roche), 1 mM phenylmethylsulfonyl fluoride, 40 μg/ml Aprotinin, 20 μg/ml Leupeptin, 10 μg/ml Pepstatin) after washing with 10 ml of the same buffer.
2. WCE is prepared by homogenizing the washed cells by rigorously vortexing with 4 mm diameter acid-washed glass beads (Thomas Scientific) corresponding to 1/2 of the total volume of resuspended cells for 30 s, followed by 1 min on ice, conducted 5 times. Cells were placed on ice for 1 min between each cycle of vortexing.

---

antibodies against the His_8 epitope (to detect the NIP1-NTD polypeptides) or with antibodies against the other factors listed to the right of the blots. Lanes 1, 5, and 9 contained 3% of the input WCEs (In); lanes 2, 6, and 10 contained 15% of the first fractions eluted from the resin (E1); lanes 3, 7, and 11 contained 30% of the same fractions as lanes 2, 6, and 10 (E2); and lanes 4, 8, and 12 contained 3% of the flow-through fractions (FT). The Western signals for eIF2, eIF1, and eIF5 in the E1 and E2 fractions for the Box2 mutant (lanes 10–11) were quantified, combined, normalized for the amounts of the NIP1-NTD-Box2 fragment in these fractions, and the averaged values from 3–5 independent experiments were plotted in the histogram on the right as percentages of the corresponding values calculated for the WT NIP1-NTD (fractions 6–7). (See color insert.)

3. The WCE is clarified by two successive centrifugations at $16,100 \times g$ for 2 and 10 min at $4°$, respectively, using an Eppendorf F 45-24-11 rotor in a table centrifuge. After each centrifugation, the supernatant is carefully transferred to a new precooled Eppendorf tube, avoiding the lipid layer, and the total protein concentration (mg/ml) is estimated using the Bradford method (Biorad).

4. Approximately 1 mg of total protein is pre-incubated with 25 μl of A-Sepharose beads CL-4B (Amersham Pharmacia Biotech) in 300 μl of binding buffer HA for 1 h at $4°$ with gentle rocking. The sample is then spun down at 1000 rpm for 3 min in an Eppendorf F 45-24-11 rotor, preserving the supernatant for further use. An aliquot of 3% of the total protein from each reaction is stored at $-20°$ to provide the input reference sample in the subsequent analysis.

5. In the meantime, 2.5 μl of mouse monoclonal 12CA5 anti-HA antibodies (Roche) are incubated with gentle rocking with 40 μl of protein A-Sepharose beads CL-4B (Amersham Pharmacia Biotech) in 300 μl of binding buffer HA for 1 h at $4°$. The sample is then spun down at 1000 rpm for 3 min in an Eppendorf F 45-24-11 rotor, the pellet ("activated" beads) is washed once with 500 μl of binding buffer and saved for further use. Including both of the latter steps is optional but recommended because it should greatly reduce nonspecificity in co-immunoprecipitation reactions.

6. Activated beads from step 5 are mixed with the supernatant from step 4 and incubated with gentle rocking for 2 hrs at $4°$.

7. Immune complexes attached to the beads are then spun down at 1000 rpm for 3 min in an Eppendorf F 45-24-11 rotor (preserving a 3% aliquot), and the beads are washed 3 times with 1 ml of HA buffer each.

8. Bound proteins are released by boiling for 2 min at $95°$ in the $4\times$ sample loading buffer and separated by SDS-PAGE for immunoblot analyses as described for Ni affinity chromatography in Section 2.4.

4.2. Immunoaffinity purification of eIF3 containing FLAG-tagged eIF3g/TIF35 (FLAG pull-down)

1. Yeast strain carrying a FLAG-tagged subunit of eIF3g/TIF35 as the only copy of this subunit is grown in 100 ml of SD medium to early log phase (OD_{600} of $\sim 1.0 - 2.0$) at $30°$ (200 rpm), harvested, and suspended in 0.2 ml of ice-cold lysis buffer F (20 mM Tris-HCl, pH 7.5, 5 mM magnesium chloride, 100 mM KCl, 0.1 mM EDTA, 5 mM NaF, Complete^{TM} protease inhibitor (Roche), 1 mM phenylmethylsulfonyl fluoride, 40 μg/ml Aprotinin, 20 μg/ml Leupeptin, 10 μg/ml Pepstatin, 7 mM β-mercaptoethanol) after washing with 10 ml of ice-cold water.

2. WCE is prepared by homogenizing the washed cells by rigorously vortexing with 4 mm diameter acid-washed glass beads (Thomas Scientific) corresponding to 1/2 of the total volume of resuspended cells for 30 s, followed by 1 min on ice, conducted 5 times. Cells were placed on ice for 1 min between each cycle of vortexing.

3. The WCE is clarified by two successive centrifugations at $16,100 \times g$ for 2 and 10 min at $4°$, respectively, using an Eppendorf F 45-24-11 rotor in a table centrifuge. After each centrifugation, the supernatant is carefully transferred to a new precooled Eppendorf tube, avoiding the lipid layer, and the total protein concentration (mg/ml) is estimated using the Bradford method (Biorad).

4. Approximately 1 mg of WCE is mixed with 100 μl of M2 anti-FLAG-affinity resin (Sigma) and incubated at $4°$ for 2 hrs with gentle rocking. An aliquot of 3% of the total protein used in each reaction is stored at $-20°$ to provide the input reference sample in the subsequent analysis.

5. The bound proteins attached to the beads are then spun down at 1000 rpm for 3 min in an Eppendorf F 45-24-11 rotor (preserving a 3% aliquot), and the beads are washed 3 times with 1 ml of the F buffer each.

6. Co-purifying proteins are eluted in 200 μl of buffer F containing 400 ng/μl FLAG peptide (Sigma) at room temperature with gentle rocking.

7. The entire eluted fraction and the aliquots containing 3% of input and 3% of flow-through are then mixed with 4× sample loading buffer, boiled for 3 min at $95°$, and subjected to Western blot analysis, as has been described in Section 2.4.

5. CAM (CLUSTERED-10-ALANINE MUTAGENESIS)

Building a structural model of a multiprotein complex of interest with various domains organizing its overall architecture does not reveal anything about the physiological importance of its individual interactions. Several mutagenic approaches can be utilized to demonstrate that these interactions have functional significance *in vivo*. Site-directed mutagenesis is employed when a binding domain of interest (BDI) contains some well-characterized motif that is thought to play a vital role in mediating the interaction. When there is no such information available, random mutagenesis is the commonly used technique.

We have employed a more systematic "random" mutagenesis approach by dividing the BDI into consecutive clusters of 10 amino acid residues (Boxes) that are individually replaced with a string of 10 alanine residues in the full-length His_8-tagged allele to facilitate affinity purification of the mutant protein (Valášek *et al.*, 2004). Alternatively, clusters rich in negatively or positively charged residues, capable of mediating ionic interactions, can be substituted with amino acid residues of the opposite charge. These two approaches are alternatives to the previously published alanine-scanning mutagenesis approach that was designed to target clusters of charged residues (Longtine *et al.*, 1998). The resulting mutant alleles are then introduced into a yeast strain bearing a deletion of a particular wild-type gene by plasmid shuffling and tested for phenotypes such as slow growth or temperature sensitivity, which may indicate a defect caused by impairment of the interaction under study. Selected mutants displaying the strongest phenotypes can subsequently be analyzed for the integrity of the multiprotein complex using Ni-affinity chromatography. This technique is more elaborate than classical random mutagenesis but enables the researcher to systematically scan through a relatively large BDI for the segments of highest importance that can then be subjected to more detailed examination. We have successfully used this technique to demonstrate that interactions of particular segments of the NIP1-NTD with eIFs 1, 2, and 5 (all of which regulate selection of the AUG start codon [Donahue, 2000]) promote pre-initiation complex assembly and stringent AUG selection (Valášek *et al.*, 2004).

Replacing a 10 amino acid-segment in the BDI with a stretch of 10 alanines can be achieved by fusion PCR technique. Both end primers (5'End and 3'End) that delineate the BDI must contain a unique restriction site. The mutagenic primer 1 that is used in the first round of PCR in combination with 5'End is designed to contain a reverse complement sequence of 27 nucleotides that immediately precede the particular Box to be substituted. The mutagenic primer 2 that is used in combination with 3'End in the first round is designed to contain 27 nucleotides immediately preceding the Box, followed by GCA GCT GCT GCA GCA GCA GCA GCA GCA GCA and ending with 27 nucleotides that immediately follow the Box. The latter sequence contains *Pst*I and *Alw*NI restriction sites that can be used for insert confirmation. The PCR products obtained in the first round are purified and used in a 1:1 ratio as templates for the second round of PCR amplification using primers 5'End and 3'End. The resulting PCR product is digested with the unique restriction sites at the ends and subcloned into an appropriate construct to replace the wild-type sequence. When substituting with a stretch of arginines, using the CGT CGA CGA CGA sequence will insert a novel *Sal*I site.

ACKNOWLEDGMENTS

We thank Jon Lorsch for inviting us to contribute to this volume of *Methods in Enzymology*. We are also greatly indebted to Alan G. Hinnebusch, under whose supervision the previously presented techniques were developed, for his helpful comments on the manuscript. LV was supported by The Wellcome Trust's Grant 076456/Z/05/Z, Howard Hughes Medical Institute, NIH Research Grant R01 TW007271 funded by Fogarty International Center, Fellowship of Jan E. Purkyne from Academy of Sciences of the Czech Republic and Inst. Research Concept AV0Z50200510. KHN was supported by the Danish National Research Foundation.

REFERENCES

Asano, K., Clayton, J., Shalev, A., and Hinnebusch, A. G. (2000). A multifactor complex of eukaryotic initiation factors eIF1, eIF2, eIF3, eIF5, and initiator tRNA^{Met} is an important translation initiation intermediate *in vivo*. *Genes Dev.* **14**, 2534–2546.

Asano, K., Phan, L., Anderson, J., and Hinnebusch, A. G. (1998). Complex formation by all five homologues of mammalian translation initiation factor 3 subunits from yeast *Saccharomyces cerevisiae*. *J. Biol. Chem.* **273**, 18573–18585.

Asano, K., Phan, L., Krishnamoorthy, T., Pavitt, G. D., Gomez, E., Hannig, E. M., Nika, J., Donahue, T. F., Huang, H. K., and Hinnebusch, A. G. (2002). Analysis and reconstitution of translation initiation *in vitro*. *Methods Enzymol.* **351**, 221–247.

Asano, K., Shalev, A., Phan, L., Nielsen, K., Clayton, J., Valášek, L., Donahue, T. F., and Hinnebusch, A. G. (2001). Multiple roles for the carboxyl terminal domain of eIF5 in translation initiation complex assembly and GTPase activation. *EMBO J.* **20**, 2326–2337.

Danaie, P., Wittmer, B., Altmann, M., and Trachsel, H. (1995). Isolation of a protein complex containing translation initiation factor Prt1 from *Saccharomyces cerevisiae*. *J. Biol. Chem.* **270**, 4288–4292.

Donahue, T. (2000). Genetic approaches to translation initiation in *Saccharomyces cerevisiae*. *In* "Translational Control of Gene Expression" (N. Sonenberg, J. W. B. Hershey, and M. B. Mathews, eds.), pp. 487–502. Cold Spring Harbor Laboratory Press, Cold Spring Harbor, New York.

Dong, Z., and Zhang, J. T. (2006). Initiation factor eIF3 and regulation of mRNA translation, cell growth, and cancer. *Crit. Rev. Oncol. Hematol.* **59**, 169–180.

Hershey, J. W. B., and Merrick, W. C. (2000). Pathway and mechanism of initiation of protein synthesis. *In* "Translational Control of Gene Expression" (N. Sonenberg, J. W. B. Hershey, and M. B. Mathews, eds.), pp. 33–88. Cold Spring Harbor Laboratory Press, Cold Spring Harbor, New York.

Hinnebusch, A. G. (2006). eIF3: A versatile scaffold for translation initiation complexes. *Trends Biochem. Sci.* **31**, 553–562.

Longtine, M. S., McKenzie, A., III, Demarini, D. J., Shah, N. G., Wach, A., Brachat, A., Philippsen, P., and Pringle, J. R. (1998). Additional modules for versatile and economical PCR–based gene deletion and modification in *Saccharomyces cerevisiae*. *Yeast* **14**, 953–961.

Naranda, T., MacMillan, S. E., and Hershey, J. W. B. (1994). Purified yeast translational initiation factor eIF-3 is an RNA-binding protein complex that contains the PRT1 protein. *J. Biol. Chem.* **269**, 32286–32292.

Nielsen, K. H., Szamecz, B., Valášek, L., Jivotovskaya, A., Shin, B. S., and Hinnebusch, A. G. (2004). Functions of eIF3 downstream of 48S assembly impact AUG recognition and GCN4 translational control. *EMBO J.* **23**, 1166–1177.

Nielsen, K. H., Valášek, L., Sykes, C., Jivotovskaya, A., and Hinnebusch, A. G. (2006). Interaction of the RNP1 motif in PRT1 with HCR1 promotes 40S binding of eukaryotic initiation factor 3 in yeast. *Mol. Cell. Biol.* **26**, 2984–2998.

Phan, L., Zhang, X., Asano, K., Anderson, J., Vornlocher, H. P., Greenberg, J. R., Qin, J., and Hinnebusch, A. G. (1998). Identification of a translation initiation factor 3 (eIF3) core complex, conserved in yeast and mammals, that interacts with eIF5. *Mol. Cell. Biol.* **18**, 4935–4946.

Singh, C. R., Curtis, C., Yamamoto, Y., Hall, N. S., Kruse, D. S., He, H., Hannig, E. M., and Asano, K. (2005). Eukaryotic translation initiation factor 5 is critical for integrity of the scanning preinitiation complex and accurate control of GCN4 translation. *Mol. Cell. Biol.* **25,** 5480–5491.

Singh, C. R., He, H., Ii, M., Yamamoto, Y., and Asano, K. (2004). Efficient incorporation of eukaryotic initiation factor 1 into the multifactor complex is critical for formation of functional ribosomal preinitiation complexes *in vivo. J. Biol. Chem.* **279,** 31910–31920.

Valášek, L., Mathew, A., Shin, B. S., Nielsen, K. H., Szamecz, B., and Hinnebusch, A. G. (2003). The yeast eIF3 subunits TIF32/a and NIP1/c and eIF5 make critical connections with the 40S ribosome *in vivo. Genes & Dev.* **17,** 786–799.

Valášek, L., Nielsen, K. H., and Hinnebusch, A. G. (2002). Direct eIF2–eIF3 contact in the multifactor complex is important for translation initiation *in vivo. EMBO J.* **21,** 5886–5898.

Valášek, L., Nielsen, K. H., Zhang, F., Fekete, C. A., and Hinnebusch, A. G. (2004). Interactions of Eukaryotic Translation Initiation Factor 3 (eIF3) Subunit NIP1/c with eIF1 and eIF5 promote preinitiation complex assembly and regulate start codon selection. *Mol. Cell. Biol.* **24,** 9437–9455.

Valášek, L., Phan, L., Schoenfeld, L. W., Valášková, V., and Hinnebusch, A. G. (2001). Related eIF3 subunits TIF32 and HCR1 interact with an RNA recognition motif in PRT1 required for eIF3 integrity and ribosome binding. *EMBO J.* **20,** 891–904.

Wertman, K. F., Drubin, D. G., and Botstein, D. (1992). Systematic mutational analysis of the yeast ACT1 gene. *Genetics* **132,** 337–350.

CHAPTER THREE

An Approach to Studying the Localization and Dynamics of Eukaryotic Translation Factors in Live Yeast Cells

Susan G. Campbell *and* Mark P. Ashe

Contents

Abstract

The discovery of Green Fluorescent Protein (GFP) and the development of technology that allows specific proteins to be tagged with GFP has fundamentally altered the types of question that can be asked using cell biological methods. It is now possible not only to study where a protein is within a cell, but also feasible to study the precise dynamics of protein movement within living cells. We have exploited these technical developments and applied them to the study of translation initiation factors in yeast, focusing particularly on the

Faculty of Life Sciences, The University of Manchester, Manchester, United Kingdom

Methods in Enzymology, Volume 431
ISSN 0076-6879, DOI: 10.1016/S0076-6879(07)31003-3

key regulated guanine nucleotide exchange step involving eIF2B and eIF2. This chapter summarizes current methodologies for the tagging and visualization of GFP-tagged proteins involved in translation initiation in live yeast cells.

1. INTRODUCTION

The discovery of fluorescent tags, such as the green fluorescent protein, GFP, has revolutionized the study of proteins and RNA molecules in living cells. In the field of post-transcriptional control, the ability to locate individual proteins has resulted in many discoveries with regard to the cellular organization and dynamics of the various processes involved. For instance, observations showing that factors involved in mRNA decay localize to specific cytoplasmic processing bodies (P-bodies) are completely reliant on the ability to tag proteins with fluorescent markers (Sheth and Parker, 2003). The adaption of this technique to the study of mRNA localization has also had a dramatic impact upon the field of RNA localization. Several studies have revealed that certain mRNAs are transported to particular locations within cells prior to their translation (St. Johnson, 2005). Furthermore, these techniques have allowed the fate of mRNAs during stressful conditions to be determined because individual mRNAs have been identified associated with stalled initiation complexes in stress granules and also in P-bodies (Kedersha et al., 2005; Sheth and Parker, 2003).

We have used this technology to C terminally tag eukaryotic translation initiation factors to determine their localization within living yeast (*Saccharomyces cerevisiae*) cells (Campbell et al., 2005). The mechanism of protein synthesis in budding yeast is very similar to that found in higher eukaryotes (Hinnebusch, 2000), making *S. cerevisiae* a very appealing model organism to study the process of translation initiation. In addition, the underlying principles of translational regulation are conserved in yeast. For example, both eIF2α kinases and eIF4E binding proteins are utilized in yeast as mechanisms to control the rate of translation initiation (Dever, 2002). Finally, studies in yeast have historically served as a platform for mechanistic studies in higher eukaryotes, especially with regard to the regulation of translation initiation by eIF2α kinases (Hinnebusch, 2000).

1.1. Epitope tagging

Epitope tagging is a powerful tool for detection, purification, and functional studies on proteins. The "tagging" of the genomic copy of a yeast gene via directed genetic manipulation is a relatively straightforward yet elegant way to follow the fate of the resulting protein. In particular, the genomic environment, including the control regions of the gene of interest, can be preserved such that differences in expression between strains bearing tagged versus untagged genes will be minimized. In addition, if the gene of interest is essential, then the impact of the tag on the growth of the strain gives a clear indication of the functionality of the resulting tagged gene. For instance, many of the eukaryotic translation initiation factors are essential; therefore, the ability of these tagged strains to grow without any obvious phenotypes indicates that the tagged gene is functional as the sole source of this translation initiation factor.

Since 1998, there has been a plethora of publications demonstrating new strategies for epitope-tagging genes of interest (Janke *et al.*, 2004; Knop *et al.*, 1999; Longtine *et al.*, 1998; Sung *et al.*, 2005). These tags include the *Aequorea victoria* green fluorescent protein GFP, the several variants, such as cyan (CFP), yellow (YFP), blue (BFP), and the red fluorescent proteins DsRed from the coral *Dicosoma striata*. In addition, epitope tags such as HA or c–Myc may be added to a protein, thus allowing a more detailed biochemical analysis of the protein.

Genes of interest can be tagged at either the N or C terminal end. The decision to tag a protein at either the N or the C terminal depends upon the properties of the protein of interest. In our case, all the eukaryotic translation initiation factors were tagged C terminally to allow the endogenous promoter to influence the expression of the tagged protein.

1.2. Amplification of epitope tagging cassette

The tagging method referred to in this chapter is based on an original protocol (Knop *et al.*, 1999) and an updated version (Janke *et al.*, 2004). This two–step procedure involves, first, the PCR amplification of the tagging cassette along with a selection marker. The protocol described here is for C terminally tagged proteins. The full collection of tagging plasmids is described in Janke *et al.* and available through EUROSCARF (http://www.uni-frankfurt.de/fb15/mikro/euroscarf/index.html). Figure 3.1A shows a cartoon of the C and N terminal cassettes. For C terminal tags, PCR primers consist of 55 nucleotides immediately upstream and downstream of the stop codon and 21 nucleotides at the 3′ end complementary to vector sequences (see Fig. 3.1A and B). The advantage of this system is that the 21-nucleotide sequence that amplifies the tagging cassette is present on a family of tagging plasmids, allowing amplification of various tagging cassettes using the same primer pair. As previously noted (Janke *et al.*, 2004), the 21 nucleotide vector sequences have the ability to form primer dimers; therefore, to overcome this, the primers can be added during an initial 95° denaturing step. For a GFP-tagging cassette, the final PCR product is approximately 2.5 kb in length; hence, the DNA polymerase Expand HiFi DNA polymerase (Roche) can be used to both maximize production and reduce the likelihood of mutation. Reaction mixes are prepared with 100 ng plasmid template, $1 \times$ Expand HiFi reaction buffer and 200 μM for each dNTP. During an initial 95° step, 10 pmoles of

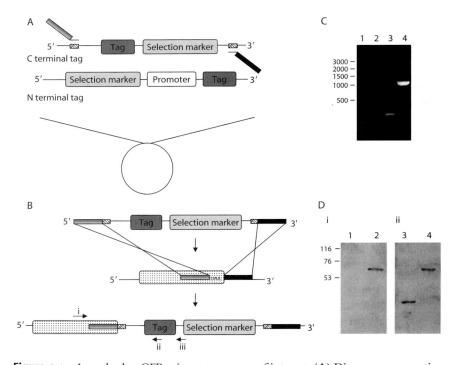

Figure 3.1 A method to GFP epitope tag a gene of interest. (A) Diagram representing the C and N terminal cassettes. The upstream and downstream primers representing 55 nucleotides upstream and downstream (but not including the stop codon) are depicted by ▬▬▬▬▬▬ and ▬▬▬▬▬ respectively, and represent the amplification sequence common to this family of tagging plasmids. (B) Cartoon representing the homologous recombination event that occurs between the PCR product and the genomic DNA. Example verification primers are indicated with arrows and are labeled i, ii, and iii. (C) PCR analysis to determine the tagging. Lanes 1 and 2 represent genomic DNA from the untagged parental strain: 1, using primers i and ii; 2, using primers i and iii. Lanes 3 and 4 represent genomic DNA from the *SUI2-GFP::KanMX6* tagged strain: 3, using primers i and ii; 4, using primers i and iii. (D) Western showing the tagging of *SUI2-GFP::KanMX6*; (i) anti GFP antibody lane 1 untagged strain, lane 2 tagged strain; (ii) anti Sui2 antibody lane 3 untagged strain, lane 4 tagged strain.

each primer is added to bring the final reaction volume to 50 μl. The PCR amplification program consists of 10 cycles (95°, 15 s; 55°, 30 s; and 68°, 2 min), then 20 cycles (95°, 45 s; 55°, 30 s; and 68°, 2 min, where the extension time is increased by 5 s each cycle), and finally a 7 min, 68° extension.

As a positive control, an Spc42p (a major component of the spindle pole body) can be GFP-tagged. To accomplish this, the *Spc42-GFP::KanMX6* PCR product needs to be generated (Janke *et al.*, 2004; Knop *et al.*, 1999).

2. YEAST STRAINS AND GROWTH CONDITIONS

Yeast strains are grown on either standard yeast extract, peptone, glucose media (YPD) (1% (w/v) yeast extract, 2% (w/v) bactopeptone, and 2% (w/v) glucose) and supplemented with the appropriate antibiotic, or in synthetic complete media (SCD media) (0.17% (w/v) yeast nitrogen base. 0.5% (w/v) ammonium sulphate, 2% (w/v) glucose, and supplemented with 20 mg/l arginine, 100 mg/l aspartic acid, 100 mg/l glutamine, 30 mg/l isoleucine, 30 mg/l lysine, 20 mg/l methionine, 50 mg/l phenylalanine, 400 mg/l serine, 200 mg/l threonine, 30 mg/l tyrosine, and 150 mg/l valine. When needed, the media was also supplemented with 20 mg/l adenine, 10 mg/l leucine, 60 mg/l histidine, 60 mg/l tryptophan, and 20 mg/l uracil).

For efficient live cell imaging of tagged cells, the yeast strains should ideally bear a wildtype *ADE2* gene because the accumulation of the adenine biosynthesis intermediate phosphoribosylaminoimidazole in the vacuoles of *ade2* mutants generates background signal for the fluorescence localization studies (Ishiguro, 1989; Stotz and Linder, 1990). This is a particular problem if the tagged protein of interest is of low abundance. To overcome this issue, most strains bearing commonly used alleles of *ade2* can be transformed with an *ADE2* wild type DNA fragment and selected on SCD minus adenine plates.

2.1. Yeast transformation

The PCR-amplified tagging cassette is transformed into yeast and inserted at either the 5′ or 3′ end of the gene of interest via homologous recombination. The yeast transformation protocol we use is based on the Lithium Acetate method (Gietz and Schiestl, 1995); however, a number of modifications have been applied. The modifications described here are for the efficient transformation of a *SUI2-GFP::KanMX6* cassette into an *ADE2* W303-1A strain. A 3-ml overnight culture of yeast is diluted into 50 ml of fresh YPD media to an OD_{600} of 0.25 and the cells are grown aerobically until they reach an OD_{600} of 0.7 to 1.0. The cells are harvested by centrifugation at 5500g for 5 min washed in 25 ml sterile distilled water, recentrifuged, and resuspended in 1 ml 100 mM LiOAc. The cells are pelleted at 16,000g and the LiOAc supernatant removed. The cells are resuspended to a final volume of 500 μl with 100 mM LiOAc and divided into 50-μl aliquots. A transformation mix consisting of 38% (w/v) PEG3350, 0.1 M LiOAc, 140 μg denatured salmon sperm ssDNA (single-stranded salmon sperm DNA), and 25 μl of the PCR product is added to the cells. We routinely transform 25 μl of the 50 μl PCR reaction into each strain. This PCR product can be added without any further processing; indeed, further purification of the PCR product seems to decrease the efficiency of the resulting transformation. After a 25-min heat shock at 42°, the cells are resuspended in 3 ml YPD and left at room temperature overnight to recover. The cells are then gently pelleted and resuspended in 200 μl YPD and plated onto YPD plates containing 300 μg/ml G418 (Melford labs). After 2–3 days at 30°, the transformed plates are replica-plated onto fresh G418 plates to remove any false positives. Resistant single colonies are isolated on a fresh G418 plate before verification of tag incorporation.

2.2. Confirmation of epitope-tagged proteins

The tagged gene can be confirmed in a number of ways. First, PCR-based assays or Southern blot analysis can be used to determine whether the tag has been incorporated into the correct location on the genome. In addition, direct DNA sequencing of these amplified PCR products from the genomic region confirms the absence of mutations. We routinely use a PCR-based approach assay that first involves preparation of genomic DNA from possible tagged clones. A 3-ml overnight culture of the presumptive tagged strain is harvested by centrifugation at 5500g for 5 min. The pellet is washed in 1 ml of Extraction Buffer (1 M sorbitol, 1 mM EDTA, 30 mM Dithiothreitol) and the cells are pelleted by centrifugation at 5500g for 5 min. The pellet is resuspended in 500 μl of Extraction Buffer containing 235 μg/ml lyticase (Sigma-Aldrich). Spheroplasts are generated by incubating the cells at 37° for a minimum of 20 min. The reaction is stopped by the addition of 55 μl of Stop Solution (3 M NaCl, 100 mM Tris pH 7.5, 20 mM EDTA). 30 μl of 20% (w/v) SDS is added and the tubes mixed. Following extraction with an equal volume of phenol:chloroform (1:1 (v/v) Tris·HCl pH 8.0 buffered phenol/chloroform), the samples are mixed and centrifuged at 16,000g for 2 min. The aqueous layer is then removed and extracted with an equal volume of chloroform. The tubes are mixed and centrifuged at 16,000g for 2 min. The aqueous layer is removed and nucleic acids are precipitated with 1 ml ethanol. The tubes are inverted a number of times and the samples are then centrifuged at 16,000g for 15 min. The pellets are air dried and resuspended in the desired volume of sterile distilled water.

2.3. PCR analysis to verify tagged genes

A number of PCR reactions can be carried out to verify the *SUI2-GFP::KanMX6* tagging. Some of the primers we use are highlighted Fig. 3.1B. A forward verification primer within the *SUI2* coding region and reverse primers within the *GFP* tag and the *KanMX6* selection marker generate products of 254 bp and 1087 bp, respectively, which are not observed in the negative control lanes containing genomic DNA from the original strain (Fig. 3.1C). A similar analysis is commonly performed using primers downstream of the gene of interest, thus ensuring that the precise site of both

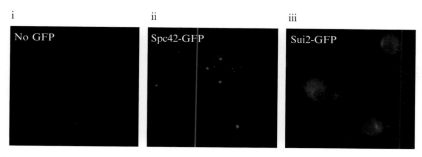

Figure 3.2 Live cell epifluorescent images of tagged strains: I, untagged cells (negative control); ii, Spc42–GFP tagged cells (positive control); and iii, Sui2–GFP tagged cells.

homologous recombination events is as predicted. These PCR products can also be sequenced to directly show the accuracy of both the recombination events and the PCR amplification process.

In addition, Western analysis can be used to verify that the tag has been efficiently added to the protein and that the resultant tagged protein's abundance is not altered by the presence of the tag. Using either an antibody specific to your protein of interested or an anti–GFP antibody (Clontech), a correctly tagged protein should be approximately 29 kDa heavier than the untagged protein (Fig. 3.1D).

2.4. Live cell imaging

Live cell imaging allows the visualization of GFP-tagged proteins in their native state. Yeast cells containing the tagged protein of interest are routinely grown in SCD media at 30° to an OD_{600} of 0.5–0.7. To adhere cells to the glass slide, slides are coated with 0.5% poly–L-lysine (Sigma) and left at room temperature for 3–5 min. After such time the excess poly–L-lysine is washed off and the slides are air dried; 2 μl of cells are applied onto the precoated glass slide and a cover slip is placed over the cells. The cells are visualized immediately. Live cell imaging of *SUI2-GFP::KanMX6* cells, the positive control *SPC42-GFP::KanMX6*, and untagged cells are shown in Fig. 3.2.

 ## 3. Application of Live Cell Imaging

3.1. FRAP analysis

FRAP (fluorescence recovery after photo bleaching) analysis was originally developed in the mid–1970s to study the diffusion of biomolecules in living cells (Edidin *et al.*, 1976). However, due to the increased availability of GFP tags and advances in the bleaching capabilities of confocal microscopes, this

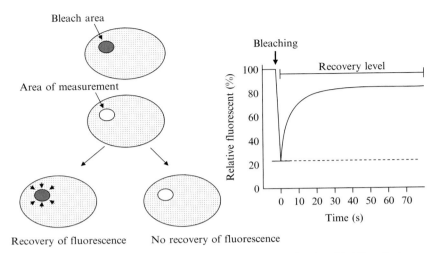

Figure 3.3   Cartoon and example recovery curve explaining FRAP analysis.

technique has become an invaluable method for determining the dynamics of GFP-tagged proteins in living cells.

FRAP analysis involves the bleaching of a small region of fluorescence within the cell with a high-intensity laser beam (Fig. 3.3). This photo-bleaching is irreversible, yet does not affect the function of the photo-bleached proteins (White and Stelzer, 1999). The bleached region is then monitored and any recovery of fluorescence into the photo-bleach region can be measured, thereby providing qualitative information about the behavior of molecules in the bleached region, such as whether they are mobile or immobile. Also, this method allows comparison of recovery times of different molecules or of identical molecules under different experimental conditions.

3.2. Microscopy and FRAP analysis

In studying the localization of eukaryotic translation initiation factors, we made use of the FRAP technique to determine whether the localized regions of eIF2/eIF2B represented dynamic centers of these proteins (Campbell et al., 2005).

3.2.1. Preparation of cells and slides for FRAP analysis

FRAP analysis was performed using live cells grown in SCD. Cells were routinely grown at 30° to an OD_{600} of 0.5–0.7. Cells were prepared as described previously and visualized immediately on a Zeiss LSM 510 confocal microscope (the specific settings mentioned here are for the Zeiss LSM 510 confocal; however, similar settings can be used with other confocal microscopes). The cells were visualized through a 100× Plan Apochromat oil objective (NA 1·4) and the photo-bleaching was carried out using an argon laser (488 nm). The region to be bleached was defined and a series of pre-bleach images (n = 3) was taken at 55% laser power. The photo-bleaching is performed by repetitively targeting the desired region with 100% laser power. The number of laser iterations required to give 80% photo-bleaching is entirely dependent on the intensity of the region to be bleached. In our case, for eIF2α–GFP, 12 iterations were required, whereas 20 iterations were necessary for eIF2Bγ–GFP. Recovery is then followed by recording images at 5-s intervals post bleaching at 4% laser transmission. Clearly, specific bleaching conditions should be optimized for each new experiment. As a critical control, identical FRAP experiments are carried out on cells fixed in 3.7% formaldehyde for 1 h before FRAP analysis (Rabut and Ellenberg, 2005).

3.3. Analysis of FRAP experiments

The image analysis used here is derived from that described by Rabut and Ellenberg in their chapter on photo-bleaching techniques in "Live Cell Imaging: A Laboratory Manual" (Rabut and Ellenberg, 2005). After the generation of pre-bleached, bleached, and recovery images (Fig. 3.4), the images generated from each FRAP experiment can be imported into the National Institutes of Health ImageJ software for analysis. Here, three measurements for each image should be calculated; these include the total fluorescence in the cell (Tot), the fluorescence across the bleached region (ROI), and the background fluorescence (BG). The acquisition of this background value is critical because the region of interest always contains an amount of background fluorescence derived from the medium, glass cover slip, and objectives. Once obtained, the ImageJ-generated values are exported into a Microsoft Excel worksheet. The background values are first subtracted from the total fluorescence and bleach region values:

$$\text{Tot} - \text{BG} = \text{Tot}[b]$$
$$\text{ROI} - \text{BG} = \text{ROI}[b] \tag{1}$$

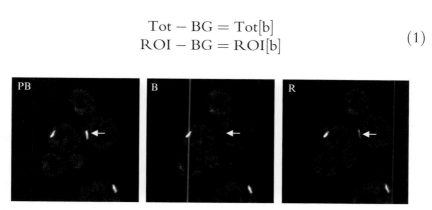

Figure 3.4 FRAP analysis of strain bearing Sui2–GFP: PB, pre-bleached image; B, bleached image; R, recovered image. The bleached focus is marked with an arrow.

These background subtracted values [b] should now reflect the fluorophore concentration in the region of interest and the intensity change within this region will represent the movement of unbleached proteins into the photo-bleached region. However, to correct for any photo-bleaching caused during the acquisition of the images or laser-intensity fluctuations, the background corrected fluorescence present in the region of interest is calculated as a fraction of the background corrected total cell value:

$$ROI[b]/Tot[b] \tag{2}$$

These values can be normalized to allow comparison with other FRAP experiments. This involves the generation of a relative fluorescence value expressed as a percentage of the total pre-bleached value.

Finally, the ½ value for each FRAP experiment can be calculated as the time taken for half of the unbleached proteins to recover in the bleached region using PRISM® Version 4 software (GraphPad Software). Comparing the ½ values for the same molecules under different experimental conditions can be used to determine the steps limiting the exchange rate.

3.4. RNA localization

An obvious extension of studies aimed at investigating the localization of translation initiation factors is an analysis of the localization of specific RNA species. For instance, with respect to the localization of eIF2, the localization of the initiator methionyl tRNA is an important consideration.

3.4.1. Fixed cell imaging of RNA

The imaging of RNA in fixed cells uses a technique known as FISH, or fluorescent *in situ* hybridization. We have used this technique to determine the location of initiator Met tRNAi in yeast. An oligonucleotide probe was designed based on the 3′end of *IMT1* (5′ GCCGCTCGGTTTCG-ATCCGA3′). As a control, a probe was generated specific to the 3′ end of *EMT1* (5′ CCAGGGGAGGTTCGAACTCT3′). The protocol used is derived from that used by the Hopper lab (Sarkar and Hopper, 1998).

Oligonucleotide probes were labeled at their 3′ end using terminal transferase and digoxigenin-11-UTP, according to the manufacturer's recommendations (Roche Pharmaceuticals). Ten ml of cells are pre-fixed in 3.7% formaldehyde for 15 min at room temperature. Five ml of cells are harvested and resuspended in 6 ml of 4% paraformaldehyde, 0.1 M KPO$_4$ (pH 6.5), and 5 mM MgCl$_2$. After 3 h, the cells are washed twice in solution B (1.2 M sorbitol, 0.1 M KPO$_4$ (pH 6.5)). The cells are then resuspended in 2.8 ml solution B containing 0.05% β-mercaptoethanol and 50 μl lyticase (1 mg/ml) and incubated at 30° for 20 min. The resultant spheroplasts are then washed three times in solution B and resuspended in 300 μl of solution B. The cells are then adhered to multi-well slides (Sigma) previously treated with 0.5% poly-L-lysine. After 2 to 3 min, the non-adhered cells are removed by aspiration. The cells are then treated with 70, 90, and 100% ethanol in succession for 5 min and then incubated for 2 h at 37° in pre-hybridization solution (50% formamide, 10% dextran sulphate, 0.2% bovine serum albumin, 2× SSC (0.3 M NaCl, 30 mM Sodium citrate pH 7.0)), 125 μg/ml $E.$ $coli$ tRNA, 500 μg/ml ssDNA, and 10 ml RNasin); 450 pg/ml of probe is then added and the cells are incubated at 37° overnight. After the hybridization, the cells are washed three times in 2× SSC at 45°, then briefly washed with 4× SSC containing 1% Triton X-100. The cells are then blocked for 2 h in 1% BSA containing 4× SSC. A secondary anti-DIG Rhodamine conjugated antibody (Roche) is diluted in 1% BSA containing 4× SSC and incubated for a further 2 h. Finally, the cells are washed twice with 4× SSC and mounted in 50% (v/v) glycerol, 0.5 μg/ml DAPI (4'-6-diamidino-2-phenylindole dihydrochloride), 1 mg/ml p-phenylenediamine, and 1× PBS, and the images are observed on a fluorescent microscope.

3.4.2. Live cell imaging of RNA

Since 1998, strategies have been described that allow the insertion of defined cis-acting sequences that direct the recruitment of highly specific RNA binding proteins to an mRNA of interest. For example, the bacteriophage MS2 coat protein has been fused to GFP and expressed as a tracer molecule for specific mRNAs within yeast and higher cells (Bertrand et $al.$, 1998). This technology offers the significant advantage over FISH analysis in that the dynamics of an mRNA species can be followed in living cells. To date, this technique has been documented only for mRNA localization studies and not for tRNA localization or the localization of rRNAs. This is most likely due to the implications for expression or function for these RNAs if six MS2-binding sites (each consisting of a 19-nucleotide RNA stem loop) are added to the 3' end of the tRNA or rRNA sequence.

4. ADDITIONAL *IN VIVO* TECHNIQUES

The potential of live cell imaging to address mechanisms of cellular biology is ever expanding. Directed protein-tagging techniques have been used to visualize nascent versus mature protein in $vivo$ (Rodriguez et $al.$, 2006). This technique involves the use of arsenic-based dyes, such as FiAsH or ReAsH, which bind to tetracysteine (TC) tags (Zhang et $al.$, 2002). In addition, photo-activatable variants of GFP have been shown to determine the kinetics of protein movement in live cells (Patterson and Lippincott-Schwarz, 2002). Furthermore, techniques such as FRET and the use of split GFP allow for an *in vivo* study of protein–protein interactions. Such techniques have been used in the study of both translational control and splicing to determine the interactions between eIF4E and eIF4E-T (Andrei *et al.*, 2005) and interactions between U2AF subunits (Chusainow *et al.*, 2005) respectively.

5. CONCLUSIONS

Recent progress the technology underlying cell biology means that it is now not only feasible to understand where protein and RNA are at a particular instant within a fixed cell, but it is possible to determine the kinetics and dynamics of factor localization and interaction within the highly complex structure of living cells. When applied to the study of protein synthesis and mRNA biogenesis, it is clear that the factors involved are not randomly distributed as a freely diffusible pool within the cytoplasm, but are, in fact, highly organized. Further studies on the organization of the translational machinery will almost certainly rely on future technological advances in the area of cell biology and will ultimately forge closer links between the fields of cell and molecular biology.

ACKNOWLEDGMENTS

This work and SGC were supported by a Wellcome Trust project grant 080349/Z/06/Z to MPA. We thank G. Pereira and G. Pavitt for reagents and D. Jackson and C. Tang for their assistance with the confocal microscopy.

REFERENCES

Andrei, M. A., Ingelfinger, R., Heintzmann, R., Aschel, T., Rivera-Pomar, R., and Luhrmann, R. (2005). A role for eIF4E and eIF4E-transporter in targeting mRNPs to mammalian processing bodies. *RNA* **11,** 717–727.

Bertrand, E., Chartrand, P., Schaefer, M., Shenoy, S. M., Singer, R. H., and Long, R. M. (1998). Localization of ASH1 mRNA particles in living yeast. *Mol. Cell* **2,** 437–445.

Campbell, S. G., Hoyle, N. P., and Ashe, M. P. (2005). Dynamic cycling of eIF2 through a large eIF2B-containing cytoplasmic body: Implications for translation control. *J. Cell Biol.* **170,** 925–934.

Chusainow, J., Ajuh, P. M., Trinkle-Mulcahy, L., Sleeman, J. E., Ellenberg, J., and Lamond, A. I. (2005). FRET analyses of the U2AF complex localize the U2AF35/U2AF65 interaction *in vivo* and reveal a novel self-interaction of U2AF35. *RNA* **11,** 1201–1214.

Dever, T. E. (2002). Gene-specific regulation by general translation factors. *Cell* **108,** 545–556.

Edidin, M., Zagyansky, Y., and Lardner, T. (1976). Measurement of membrane protein lateral diffusion in single cells. *Science* **191,** 466–468.

Gietz, R., and Schiestl, R. (1995). Transforming yeast with DNA. *Mol. Cell. Biol.* **5,** 255–269.

Hinnebusch, A. (2000). "Mechanism and Regulation of Initiator Methionyl-tRNA Binding to Ribosomes." Cold Spring Harbor Laboratory Press, Cold Spring Harbor, NY.

Ishiguro, J. (1989). An abnormal cell division cycle in an AIR carboxylase-deficient mutant of the fission yeast *Schizosaccharomyces pombe. Curr. Genet.* **15,** 71–74.

Janke, C., Magiera, M. M., Rathfelder, N., Taxis, C., Reber, S., Maekawa, H., Moreno-Borchart, A., Doenges, G., Schwob, E., Schiebel, E., and Knop, M. (2004). A versatile toolbox for PCR-based tagging of yeast genes: New fluorescent proteins, more markers, and promoter substitution cassettes. *Yeast* **21,** 947–962.

Kedersha, N., Stoecklin, G., Ayodele, M., Yacono, P., Lykke-Andersen, J., Fritzler, M. J., Scheuner, D., Kaufman, R. J., Golan, D. E., and Anderson, P. (2005). Stress granules and processing bodies are dynamically linked sites of mRNP remodeling. *J. Cell Biol.* **169,** 871–884.

Knop, M., Siegers, K., Pereira, G., Zachariae, W., Winsor, B., Nasmyth, K., and Schiebel, E. (1999). Epitope tagging of yeast genes using a PCR-based strategy: More tags and improved practical routines. *Yeast* **15,** 963–972.

Longtine, M. S., McKenzie, A., III, Demarini, D. J., Shah, N. G., Wach, A., Brachat, A., Philippsen, P., and Pringle, J. R. (1998). Additional modules for versatile and economical PCR-based gene deletion and modification in *Saccharomyces cerevisiae. Yeast* **14,** 953–961.

Patterson, G. H., and Lippincott-Schwarz, J. (2002). A photoactivatable GFP for selective photolabeling of proteins and cells. *Science* **297,** 1873–1877.

Rabut, Q., and Ellenberg, J. (2005). "Photobleaching Techniques to Study Mobility and Molecular Dynamics of Proteins in Living Cells: FRAP, iFRAP, and FLIP." Cold Spring Harbor Laboratory Press, Cold Spring Harbor, NY.

Rodriguez, A. J., Shenoy, S. M., Singer, R. H., and Condeelis, J. (2006). Visualization of mRNA translation in living cells. *J. Cell Biol.* **175,** 67–76.

Sarkar, S., and Hopper, A. K. (1998). tRNA nuclear export in *Saccharomyces cerevisiae*: *In situ* hybridization analysis. *Mol. Biol. Cell* **9,** 3041–3055.

Sheth, U., and Parker, R. (2003). Decapping and decay of messenger RNA occur in cytoplasmic processing bodies. *Science* **300,** 805–808.

St Johnson, D. (2005). Moving messages: The intracellular localization of mRNAs. *Nature Reviews Mol. Cell Biol.* **6,** 363–375.

Stotz, A., and Linder, P. (1990). The ADE2 gene from *Saccharomyces cerevisiae*: Sequence and new vectors. *Gene* **95,** 91–98.

Sung, H., Chul Han, K., Chul Kim, J., Wan Oh, K., Su Yoo, H., Tae Hong, J., Bok Chung, Y., Lee, C.-k., Lee, K. S., and Song, S. (2005). A set of epitope-tagging integration vectors for functional analysis in *Saccharomyces cerevisiae. FEMS Yeast Research* **5,** 943–950.

White, J., and Stelzer, E. (1999). Photobleaching GFP reveals protein dynamics inside live cells. *Trends Cell Biol.* **9,** 61–65.

Zhang, J., Campbell, R. E., Ting, A. Y., and Tsien, R. Y. (2002). Creating new fluorescent probes for cell biology. *Nat. Rev. Mol. Cell Biol.* **3,** 906–918.

CHAPTER FOUR

In Vitro and Tissue Culture Methods for Analysis of Translation Initiation on the Endoplasmic Reticulum

Samuel B. Stephens *and* Christopher V. Nicchitta

Contents

Abstract

For mRNAs encoding secretory and integral membrane proteins, translation initiation is thought to begin a process of mRNA localization where mRNA/ribosome/nascent chain complexes (RNCs) are trafficked from the cytosol compartment to the endoplasmic reticulum (ER). At the ER membrane, RNCs bind to a protein-conducting channel via the large ribosomal subunit and protein translocation ensues through coupling of the ribosomal nascent protein exit site with the protein-conducting channel. At the termination of translation, ribosomal subunits are thought to dissociate from the ER to return to a common

Department of Cell Biology, Duke University Medical Center, Durham, North Carolina

Methods in Enzymology, Volume 431
ISSN 0076-6879, DOI: 10.1016/S0076-6879(07)31004-5

cytoplasmic pool and participate in additional cycles of initiation, translation, targeting, termination, and ER membrane release. Experimental evidence has demonstrated that ER-membrane ribosomes are capable of *de novo* initiation, that mRNA partitioning to the ER membrane does not, *per se*, require translation of an encoded signal sequence, and that ribosomal subunit dissociation from the ER membrane is not obligatorily coupled to protein synthesis termination. These findings suggest that the cycle of protein synthesis—initiation, elongation, and termination—can occur on the two-dimensional plane of the ER membrane and challenge current views on the subcellular restriction of translation initiation to the cytosol, the role of the ribosome cycle in partitioning mRNA between the cytosol and ER, and the *in vivo* basis for termination-induced ribosomal subunit dissociation. In the following chapter, we provide detailed experimental methods to study protein synthesis initiation on the ER membrane.

1. BACKGROUND

The complex subcellular organization of the endoplasmic reticulum (ER), with its highly ordered spiral arrays of bound ribosomes and, in terminally differentiated secretory cells, the densely packed arrays of ER cisternae, has fascinated the cell biology community for over 5 decades. This fervent attention has yielded remarkable insights into the processes that segregate the protein synthesis machinery of the cytoplasmic and ER compartments. Beginning with the pioneering studies of George Palade and colleagues in the 1950s and extending to the current era, remarkable molecular insights into the role of the ER in the synthesis, folding, and assembly of secretory and integral membrane proteins have been achieved. A central mystery in these studies concerned the mechanism by which mRNA encoding secretory and integral membrane proteins were segregated to the ER and their translation coupled to the process of protein translocation. In large part, this mystery was solved by Blobel and colleagues with the discovery of the signal recognition particle (SRP) (Blobel and Dobberstein, 1975a,b; Blobel *et al.*, 1979). SRP is well established to direct the ER–directed trafficking of cytoplasmic ribosomes engaged in the translation of secretory or integral membrane protein-encoding mRNAs (Walter and Blobel, 1981a,b; Walter *et al.*, 1981). This targeting process defines a trafficking cycle, where SRP intersects with translationally active ribosomes in the cytosol, directs their trafficking to the ER, and then dissociates to return to the cytosol. This cycle is thought to be mirrored by the ribosomes themselves, with the exception being that ribosome dissociation from the ER to the cytosol is coupled to the protein synthesis termination reaction rather than to the targeting reaction itself (Blobel, 2000). In this view, the cytosol is the sole subcellular site of translation initiation, and translationally active ribosomes either reside in the cytosol (default pathway) or are trafficked to the ER (SRP pathway). Indeed, well-characterized *in vitro* experimental systems have demonstrated that SRP can direct such a cytosol–ER RNC trafficking itinerary, and that the trafficking process is dependent on translation and a functional signal sequence (Blobel, 2000; Blobel and Dobberstein, 1975a,b; Blobel *et al.*, 1979; Walter and Blobel, 1981a,b; Walter and Johnson, 1994; Walter *et al.*, 1981). Nonetheless, and as summarized in a 2005 review on this topic, studies of ribosome trafficking and mRNA partitioning between the cytosol and ER compartments have demonstrated that mRNAs encoding cytosolic/nucleoplasmic proteins can be highly partitioned to ER ribosomes and that the predominant fate of ribosomes following termination is continued association with the ER membrane (Nicchitta *et al.*, 2005). These findings have raised important new questions regarding the subcellular location(s) of the protein synthesis initiation reaction as well as the mechanisms governing mRNA partitioning between the cytosol and ER compartments. For example, if mRNAs that do not encode signal sequences or transmembrane domains can undergo initiation/translation on ER-bound ribosomes, there must be a mechanism for translation-independent targeting of the mRNA to the ER compartment. Extrapolating from this premise, it would seem necessary that ER-bound ribosomes be capable of de novo initiation, so that mRNAs encoding cytoplasmic/nucleoplasmic proteins can undergo translation on the ER. Ironically, it was George Palade who, in his 1975 Nobel Prize lecture, noted the need for insights into "the relationship between free and attached ribosomes, the position of polysomes at the time of initiation, and the duration of polysome attachment to the endoplasmic reticulum." (Palade, 1975). In light of the findings reported previously, these questions represent a fertile area for study. In the following chapter, we present methods for studying the initiation of protein synthesis on the ER membrane, the post-termination structural state of ER-bound ribosomes, and the cycle of ER initiation, elongation, and termination in tissue culture cell.

2. Methods

2.1. *In vitro* analysis of protein synthesis initiation on membrane-bound ribosomes

2.1.1. Preparation of rough microsomes and ribosome-free reticulocyte lysate

Canine pancreas rough microsomes (RM) are a well-characterized form of ER membranes and have been in use for decades. Currently, RM are available from a commercial source (Promega, Madison, WI), though detailed methods for preparing highly active RM fractions are described in an early volume of *Methods in Enzymology* (Walter and Blobel, 1983). This method provides a cost-effective means for preparing this reagent in large quantities. We recommend the use of canine pancreas as a source tissue; though RM can be obtained from alternative tissue sources, such as porcine and bovine pancreas, we have observed that tissues other than canine pancreas display relatively high levels of ribonuclease activity. As isolated, the canine pancreas RM fraction contains significant quantities of bound mRNAs, which will contribute substantial background signals to assays of total protein synthesis. To eliminate this background signal, RM are treated with *S. aureus* nuclease (Calbiochem) as described (Walter and Blobel, 1983). This treatment is known to decrease the overall translational capacity of the RM, presumably by stranding ribosomes on cleaved mRNA fragments (Jackson and Hunt, 1983). Commercial RM preparations are pretreated with nuclease and require no further processing.

Though the protocols for purifying RM yield a highly enriched RM fraction, it is rare that they are devoid of free ribosomes and/or loosely attached ribosomes. To ensure that the translation initiation reactions under investigation reflect solely the activity of ER-bound ribosomes, it is necessary to remove contaminating free ribosomes. Commonly, this is achieved by diluting 100 equivalents (eq; as defined in Walter and Blobel, 1983) of RM 5-fold with 2.5 M sucrose, 25 mM K-HEPES (pH 7.2), 350 mM KOAc, 5 mM Mg (OAc)$_2$, and 1 mM DTT. This solution is then overlaid with 750 μl of 1.9 M sucrose, 25 mM K-HEPES (pH 7.2), 350 mM KOAc, 5 mM Mg(OAc)$_2$, and 1 mM DTT, and 200 μl of the same buffer lacking sucrose. The sample is then subjected to centrifugation in an SW55 rotor (Beckman, Palo Alto, CA) at 55,000 rpm for 4 h at 4°. Under these conditions, the RM, with tightly associated ribosomes, float and free ribosomes are recovered in the pellet. The upper 400 μl of the step gradients are collected, diluted 3-fold with RM buffer (250 mM sucrose, 25 mM K-HEPES (pH 7.2), 50 mM KOAc, 5 mM Mg(OAc)$_2$, and 1 mM DTT), and centrifuged for 10 min at 60,000 rpm (4°) in a TLA100.2 rotor (Beckman, Palo Alto, CA). The RM pellet is resuspended in RM buffer, the A$_{280}$ in 1% SDS determined, and aliquots snap-frozen in liquid nitrogen. Frozen aliquots can be stored for at least 6 months at $-80°$.

Translation-competent ER membrane fractions can also be prepared from tissue culture cells. We recommend a terminally differentiated secretory suspension cell line, such as a plasmacytoma (e.g., J558L), which contains abundant levels of ER membrane. In this protocol, cells are collected by centrifugation (5 min, 500×g) and resuspended in a homogenization buffer containing 10 mM KOAc, 10 mM K-HEPES, pH 7.5, 1.5 mM Mg(OAc)$_2$, 2 mM DTT, 1 mM PMSF, and 200 U/ml RNAsin. The cell suspension is incubated on ice and homogenized with a Dounce homogenizer (tight-fitting pestle) until >95% of cells are ruptured, as assayed by vital dye staining. The homogenate is then mixed with 2 volumes of 2.5 M sucrose in 150 mM KOAc, 50 mM K-HEPES, pH 7.5, and 5 mM Mg(OAc)$_2$ (HKM), and 2 ml fractions are overlaid with 0.75 ml of 1.3 M sucrose-HKM and

0.5 ml of 0.25 M sucrose-HKM. Gradients are centrifuged for 45 min at 500,000×g in a TLA-100.3 rotor (Beckman). After centrifugation, the rough endoplasmic reticulum (ER) membrane fraction, banding at the 1.3 M/0.25 M sucrose interface, is removed, diluted in HKM buffer to a total volume of 3 ml, and the membranes recovered by centrifugation (85,000×g, 20 min). The ER membrane pellet is resuspended in 0.25 M sucrose-HKM, 1 mM DTT, the A_{280} determined in 1% SDS, and the membranes snap-frozen in small aliquots.

Ribosome-free reticulocyte lysate is prepared directly from standard rabbit reticulocyte lysate preparations. Commercial sources for reticulocyte lysate include Promega (Madison, WI), Ambion (Austin, TX), Novagen (Madison, WI), and Green Hectares (Oregon, WI). Alternatively, rabbit reticulocyte lysate can be prepared directly, as described (Jackson and Hunt, 1983). Ribosome/ribosomal subunit-depleted depleted reticulocyte lysate is prepared by centrifugation at 100,000 rpm for 30 min at 4° in a TLA100 rotor (Beckman). The supernatant is removed and the centrifugation step repeated to ensure that ribosomes and ribosomal subunits are completely depleted. PRS prepared in this manner is devoid of translation activity, but fully supports translation when an exogenous source of ribosomes is provided. PRS can be aliquoted, snap-frozen, and stored in liquid nitrogen.

2.2. *In vitro* translation

Cell-free translations are performed in 20 μl final volume reactions, containing 8 μl of nuclease-treated rabbit reticulocyte PRS fraction, 25 μCI of [^{35}S] methionine/cysteine, 0.05 U/ml RNasin, 1 mM DTT, 80 μM (-) methionine amino acid mix, and from 1–4 equivalents of purified RM. Final reaction conditions are adjusted to 110 mM KOAc, 2.5 mM Mg(OAc)$_2$. Assays are programmed with 0.4–2 μg of *in vitro* transcribed mRNAs and translations are performed for 30 min at 25°. In general, though not always, optimal translation *in vitro* is observed with uncapped transcripts lacking a polyA tail and with relatively small 5′ and 3′ UTRs. A Kozak consensus site commonly stimulates translation significantly as does inclusion of a small region of the β-globin mRNA. These features have been previously assembled in commercially available transcription vectors, such as pSPUTK (Strategene). For any given transcript, it is useful to perform preliminary experiments to assess the optimal mRNA and RM concentrations.

In a typical experiment, paired reactions will be performed where either mRNA or RM are absent, to confirm that all translation products derive from the intended mRNA and reflect the activity of ER-bound ribosomes. As an additional control, mock translations can be performed and the membrane fraction removed by centrifugation (10 min, 60K, TLA100.2 rotor). The supernatant is then recovered, programmed with mRNA, and reaction incubated at 25° for 30 min. In this case, any ribosomes/ribosomal subunits released in the soluble fraction will be recovered and their relative translation activity determined. We observe very little to no activity in the supernatant fractions derived from canine pancreas RM/reticulocyte lysate PRS incubations. This protocol can be used to evaluate models for translation initiation on ER-bound ribosomes that invoke membrane dissociation, initiation in the soluble fraction and re-targeting to the ER membrane.

2.3. Ribosomal assembly status—ER membrane-restricted initiation

As detailed in companion chapters, methodologies exist for the study of translation initiation *in vitro*, using purified ribosomal subunits and initiation factors. Through such experimental approaches, past research has identified an ordered series of biochemical reactions that resolve in selection of the translation initiation codon (reviewed in Kapp and Lorsch, 2004). In this model, the translation initiation ternary complex binds to the 40S ribosomal subunit, to form the 43S complex. This complex, in combination with eIF4F, is thought to undergo $5'$-$3'$ scanning of the $5'$ UTR, to yield assembly of the 80S ribosome at the initiator AUG codon (Kapp and Lorsch, 2004). This model is rather difficult to examine in the RM system because the bound ribosomes represent a heterogeneous population of ribosomes engaged in initiation, elongation, or termination. To surmount this problem, it is possible to synchronize the cellular translation status of the ribosomes prior to isolation of the RM fraction and thereby provide a more biochemically uniform ribosome population. In past experiments, we have accomplished this by treating J558L cells with dithiothreitol (DTT), which elicits the ER unfolded protein response (UPR) (see later). UPR induction yields activation of an ER-restricted eIF2α kinase, PERK, the inhibition of initiation, and subsequent polyribosome breakdown (Marciniak and Ron, 2006; Schroder and Kaufman, 2005). In this experimental system, J558L cell suspensions are treated with 10 mM DTT for 30 min at 37°, chilled on ice, and processed to yield the ER membrane fraction, as described previously. A typical ER-polysome profile from such an experiment is shown later (Fig. 4.1).

Of particular interest, the data depicted in Fig. 4.1 demonstrate that ribosomes remain membrane-bound after termination and, surprisingly, are recovered predominantly as 80S rather than the predicted post-termination 40S + 60S subunits. It would be of high interest to determine if 80S post-termination, ER-bound ribosomes are competent for initiation, to identify the cohort of initiation factors necessary for initiation with 80S ribosomes, and also to determine whether initiation and scanning require dissociation of the 80S couplet. Methods to address these questions are under development but unavailable at the time of preparation of this contribution.

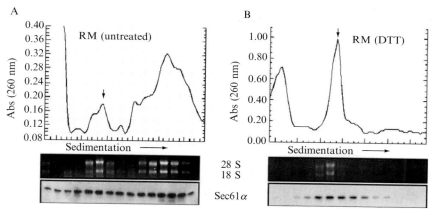

Figure 4.1 UPR induction elicits polyribosome breakdown and yields continued association of 80S ribosomes with the ER membrane. J558L cells were treated with 10 m*M* DTT for 30 min and an RM fraction was prepared from the DTT treated (B) and paired mock treated (A) cells. RM were detergent solubilized and the ribosome fraction resolved on 15–50% sucrose gradients. The position of the ribosomes in the gradient was assessed by UV spectrometry (A260 nm) and denaturing agarose gel electrophoresis (rRNA). In parallel, samples were assayed for the ER ribosome binding and translocation channel component Sec61p. Note difference in y-axis scale in panel A vs panel B. The position of the 80 S ribosome is indicated by the downward arrow.

 ## 3. ANALYSIS OF PROTEIN SYNTHESIS INITIATION ON ER MEMBRANE-BOUND RIBOSOMES OF TISSUE CULTURE CELLS

3.1. Background

The field of translation initiation has focused on the initial round of ribosomal subunit recruitment to an mRNA. Presumably, these events are mirrored in the subsequent rounds of initiation necessary for polyribosome formation. Importantly, because mRNAs are typically present in large polyribosomes (averaging 9–13 ribosomes per mRNA), the initiation events that govern ribosome recruitment to preexisting polyribosomes constitute the majority of translation initiation cycles occurring in an mRNA's lifetime. Whether or not these initiation events mimic the first round of initiation is not yet known. Since eukaryotic cells divide ribosomes between two subcellular compartments, the cytosol and ER membrane, it is also important to know if the mechanism of translation initiation on ER-bound ribosomes is similar to that occurring on soluble ribosomes and, importantly, whether ER-bound ribosomes can directly (re)initiate translation on bound polyribosmes.

The methods described in this section can be used to examine the *in vivo* dynamics of polyribosome assembly on the ER membrane. Here, we will examine polyribosome structure via sucrose gradient velocity sedimentation following translational recovery from a physiological suppression of translation initiation. For this system, we use activation of the unfolded protein response (UPR) to elicit eukaryotic initiation factor 2 (eIF2) phosphorylation by the ER resident kinase, PERK (Harding *et al.*, 1999, 2000). Phosphorylation of eIF2α effectively diminishes recharging of the eIF2 complex with GTP and the subsequent reloading of the initiator methionyl-tRNA (Brostrom and Brostrom, 1998; Hinnebusch, 2000; Scheuner *et al.*, 2001). This results in attenuation of the formation of competent translation initiation complexes, yielding significant polyribosome breakdown and suppression of protein synthesis. In this section, we use the membrane-permeable reducing agent dithiothreitol (DTT) to induce ER stress by disrupting disulfide bond formation in the ER lumen (Braakman *et al.*, 1992; Lodish and Kong, 1993). A critical aspect of this technique is that DTT-elicited eIF2α phosphorylation is rapidly reversible by simply replacing the culture media with fresh media not containing DTT. Thus, this method provides a physiologically relevant scenario for the examination of the dynamics of ER membrane initiation.

The isolation of distinct subcellular fractions can be achieved using a selective sequential detergent extraction that takes advantage of the distinct molecular composition of the plasma membrane, relative to the intracellular organelle membranes (Adam *et al.*, 1990; Lerner *et al.*, 2003; Seiser and Nicchitta, 2000; Stephens *et al.*, 2005, 2008). In this protocol, the β-sterol binding detergent, digitonin (or saponin), is used to preferentially solubilize the cholesterol-rich plasma membrane, without affecting the integrity of the organelle membranes. This allows release of soluble cytosolic contents, such as free ribosomes, without loss of lumenal and membrane (i.e., membrane-bound) components. Subsequently, a second more broadly acting detergent (NP-40/deoxycholate) is used to solubilize the remaining intracellular membranes, including ER membrane-bound ribosomes. A number of nonionic detergents can be used at this step with varying efficiencies. While NP40/DOC (or Triton X-100/DOC) are effective, some protein interactions can be disrupted in these detergent admixtures, including the ribosome-translocon (Sec61) complex. In cases in which this is important, detergents such as 2% digitonin (Calbiochem) or 2% dodecylmaltoside (Calbiochem) can be used to preserve these interactions (Potter and Nicchitta, 2002). The ribosome component of this fraction can then be resolved by sucrose gradient velocity sedimentation.

3.2. Modulating polyribosome loading

For the analysis of ribosome loading in the cytosol or ER by velocity sedimentation, it is necessary to utilize a minimum of 1 to 2 $\times$ 10^7 cells. The techniques described here are optimized for a 75 cm^2 culture flask and may be altered to accommodate other cell culture flask sizes. For time

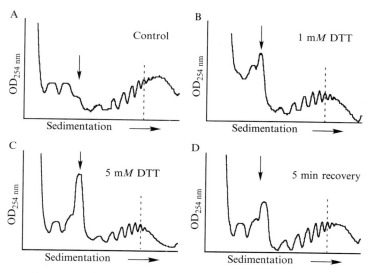

Figure 4.2 Recovery of ER-bound polyribosomes following DTT-elicited ER stress. Cos7 cells were either untreated (UT) (A), treated with 1 m*M* DTT (B) or 5 m*M* DTT (C) for 30 min. Culture media was replaced with fresh media not containing DTT. Cycloheximide (0.2 m*M*) was added to DTT-treated cultures at 0(B, C) and 5(D) min recovery. Cycloheximide was also added to the untreated flask and incubated an additional 5 min. ER-bound fractions were harvested by sequential detergent extraction and resolved by velocity sedimentation through continuous 15–50% sucrose gradients at 151,000×g for 3 hrs. Gradients were collected using an Isco automated gradient fractionator equipped with a continuous UV (254 nm) flow cell. The downward facing arrow denotes the 80S monosome peak. The dotted line denotes the "septasome" (7-mer polyribosome).

course experiments, it is therefore necessary to prepare multiple flasks, containing similar cell numbers. In the experiments depicted in Fig. 4.2, for example, four 75 cm² flasks were used. One flask should be the untreated control. As noted previously, in our current protocols, we exploit intracellular signaling pathways to modulate steady-state protein synthesis initiation rates. In the example depicted in Fig. 4.2, DTT was used to elicit UPR activation. Add 10 to 50 μl of 1 *M* DTT (freshly made) to the cell culture flasks and incubate at 37° for 30 min (final concentration 1–5 m*M* DTT). Different cell lines display different sensitivities to ER stress agents. While 1 to 5 m*M* DTT is commonly used with cell lines such as Cos7, HeLa, HEK293, the exact concentration to achieve maximal polyribosome breakdown (protein synthesis inhibition) may need to be empirically defined. Thus, it may be necessary to perform a DTT concentration titration and examine the effects on protein synthesis via [35S]-methionine incorporation to ascertain the most effective concentration. Importantly, there are distinct thresholds for activating the UPR translational program (e.g., PERK) and the transcriptional program (XBP-1 and ATF6) and thus the transcriptional response cannot be used as surrogate assay in this case. Replace the culture media of all flasks, with fresh prewarmed media. Treat cultures with 0.1 ml 20 mM cycloheximide at various recovery times (t = 0–20 min) to follow reassembly of polyribosomes (final concentration is 0.2 mM cycloheximide). Incubate for 5 min following cycloheximide addition. Place flasks on ice.

The suppression and recovery of protein synthesis from DTT treatment (without cycloheximide treatment) can be monitored via metabolic pulse radiolabeling of cell cultures using [^{35}S]-methionine and subsequent determination of radiolabeled protein content either by SDS-PAGE/ phosphorimager analysis or liquid scintillation of tricholoroacetic acid insoluble material (Stephens et al., 2005).

3.3. Sequential detergent extraction

Some cell lines, such as HEK293, may detach during the permeabilization step due to a strong Ca^{2+} dependence for attachment. While it is critical that cells are permeabilized as a monolayer, detachment does not seem to hinder cytosolic ribosome release, as they tend to detach as a (partial) monolayer. Following the permeabilization step, cells can simply be separated from the soluble cytosol phase by centrifugation at 750–1000×g for 5 min. Transfer the supernatant (cytosol) to a new tube. Remove any remaining cells attached to the flask via the wash buffer, combine with the cell pellet, and recover by centrifugation. Proceed with the membrane solubilization step.

The method described here uses volumes adjusted for a confluent cell monolayer in a 75 cm^2 tissue culture flask. As noted previously, the volumes used may be adjusted to accommodate other culture flask sizes; however, the upper limit for loading cell extract atop a 10-ml sucrose gradient is approximately 2 ml (maximum volume is 12 ml total). The volume used to release the cytosol fraction need only be enough to effectively cover the cell monolayer. It is essential that cell permeabilization occur on the monolayer rather than in suspension (scraped or trypsinized adherent cells). For reasons that are not entirely clear, release of supramolecular structures such as polyribosomes is partially inhibited when adherent cells are permeabilized in suspension, due perhaps to the dramatic rounding and contraction that accompanies lifting of adherent cells (Stephens et al., 2008). Remove media and wash cells with 5–10 ml ice-cold PBS (containing 0.2 mM cycloheximide) to remove any excess culture media. Gently add 1 ml permeabilization buffer and place on ice for 5 min or rock in a 4° cold room. Stand the flask upright and allow the buffer to drain to the bottom for 1 min. Transfer the soluble material (cytosol) to a 1.5-ml microcentrifuge tube and ice and gently wash the flask with 5 ml wash buffer. Discard the wash buffer, add 1 ml lysis buffer, and place on ice for 5 min, or rock in a 4° cold room. Again, stand the flask upright and allow the buffer to drain to the bottom. Transfer the soluble material (membrane-bound fraction) to a 1.5-ml microcentrifuge tube, centrifuge for 10 min at 7500×*g* (4°), remove supernatant, and place on ice.

4. VELOCITY SEDIMENTATION

The sedimentation protocol described here provides sufficient resolution of the small and large ribosomal subunits, the 80S monosome, and heavy polyribosomes (averaging 9–13 ribosomes per mRNA). One of the simplest methods is described here for the formation of continuous sucrose gradients. Nevertheless, numerous methods exist to form linear sucrose gradients; it is necessary only that the method be reliable and consistent. Additionally, alternative gradient conditions (such as sucrose concentrations) and centrifugation times can certainly be used to increase the resolution of specific sizes of ribosome-containing translation complexes as needed. Generate a continuous 10-ml 15–50% sucrose gradient(s) as follows. Carefully layer 5 ml of 15% sucrose solution atop 5 ml of 50% sucrose solution in a SW41 centrifuge tube, taking care not to disturb the interface of the two solutions. Cap the tube with a silicon rubber stopper (or seal with parafilm) and gently lay it on its side for 1–3 hrs to allow formation of the gradient. We typically use 2 hrs. Carefully return the tube to the upright position. Gradients may be chilled on ice for 15 min or stored overnight without loss of resolution. Remove the cap and gently layer the soluble lysate (2 ml maximum) atop the sucrose gradient so as not to disturb the gradient. Load into the rotor buckets and centrifuge for 180 min at 35,000 rpm (151,000×*g*) using a SW41 Beckman rotor. Collect gradient fractions and determine ultraviolet (UV) absorbance using an automated gradient fractionator equipped with a UV (254 nm) flow cell (such as Teledyne Isco) according to the manufacturer's instructions. Fractions are typically collected by bottom displacement using 60% sucrose at a flow rate of 1.0 ml/min, collecting approximately 30–35 fractions at 18–20 s intervals. Additionally, for the Teledyne Isco instrument, the analog voltage output (1–10 V) can be converted to a digital signal using a standard analog–digital converter (Measurement Computing 12-bit USB 1208LS; Norton, MA) and the information saved in ASCII file format for import into any number of graphical software applications, such as Microsoft Excel or GraphPad Prizm. Gradient fractions may be analyzed by any number of techniques to evaluate mRNA (RT-PCR, Northern blot; Stephens *et al.*, 2008), ribosome, or protein content (SDS-PAGE, immunoblot). An alternative method for collecting fractions is to manually puncture the bottom of the centrifuge tube with a 16-gauge needle and dropwise (1 drop per second), collect ∼0.2–0.35 ml fractions. A ribosome trace can be subsequently generated by manually measuring the UV absorbance (254 nm) with a spectrophotometer.

This system can be as reproducible and reliable as an automated system (Stephens *et al.*, 2005).

Figure 2 demonstrates the changes in the polyribosome profiles from control (A), DTT treated (B, C), and DTT-treated cells after DTT wash-out (D). Two different concentrations of DTT (1 and 5 mM) were used to demonstrate the disassembly of polyribosomes with complementary increases in 80S monosome peaks (downward arrow; note the different degrees of polyribosome loss between the two different DTT concentrations). In panel D, recovery of large polyribosomes (>"septasomes"; dotted line) sedimenting in the heavy sucrose was clearly evident within 5 min of replacing the culture media following a 1 mM DTT treatment.

4.1. Reagents

1. Digitonin (Calbiochem; San Diego, CA) is prepared as a 1% (w/v) stock in DMSO (freeze in 0.2-ml aliquots).
2. RNase Out ribonuclease recombinant inhibitor 40 U/μl (Invitrogen) or equivalent.
3. Diethyl pyrocarbonate (DEPC)-treated water. Prepare as a 0.1% (v/v) solution and incubate at 37° for 18 hrs. Autoclave for 15 min to destroy excess DEPC.
3. Stock solutions (in DEPC water): 4 M potassium acetate (KOAc), 2 M potassium 4-(2-hydroxyethyl)-1-piperazineethanesulfonate (KHEPES), pH 7.2, 1 M magnesium acetate (Mg(OAc)$_2$), 20–50 mM cycloheximide (freeze in 0.5–1.0 ml stocks), 1 M dithiothreitol (DTT) (freeze in 50-μl aliquots), 100 mM phenylmethylsulfonate fluoride (PMSF) (prepared in isopropanol; stable at 4° or at -20° for long-term storage), 0.2 M ethylene-glycol bis (2-aminoethylether)-N,N,N′,N′-tetraacetic acid (EGTA), pH 8.0, 10% (v/v) Nonidet P-40 (NP-40), 10% (w/v) sodium deoxycholate (DOC).
4. Permeabilization buffer: 110 mM KOAc, 25 mM KHEPES, pH 7.2, 2.5 mM Mg(OAc)$_2$, 1 mM EGTA, 0.015% digitonin, 1 mM DTT, 1mM PMSF, 0.2 mM cycloheximide, 40 U/ml RNase Out. Digitonin, cycloheximide, DTT, PMSF, and RNase Out must be added fresh.
5. Wash buffer: 110 mM KOAc, 25 mM KHEPES, pH 7.2, 2.5 mM Mg(OAc)$_2$, 1 mM EGTA, 0.004% digitonin, 1 mM DTT, 1 mM PMSF. 0.2 mM cycloheximide. Digitonin, cycloheximide, DTT, and PMSF must be added fresh.
6. Lysis buffer: 400 mM KOAc, 25 mM KHEPES, pH 7.2, 15 mM Mg(OAc)$_2$, 1% (v/v) NP-40, 0.5% (w/v) DOC, 1 mM DTT, 1mM PMSF, 0.2 mM cycloheximide, 40 U/ml RNase Out. NP40 must be added before DOC to prevent DOC from precipitating. Long-term storage of this buffer containing detergent is not recommended. DTT, PMSF, cycloheximide, and RNase Out must be added fresh.

4.2. Velocity sedimentation

1. 15% (w/v) sucrose solution containing (400 mM KOAc, 25 mM KHEPES, pH 7.2, 15 mM Mg(OAc)$_2$, 0.2 mM cycloheximide, and 10 U/ml RNase Out. Cycloheximide and RNase Out must be added fresh.

2. 50% (w/v) sucrose solution containing (400 mM KOAc, 25 mM KHEPES, pH 7.2, 15 mM Mg(OAc)$_2$, 0.2 mM cycloheximide, and 10 U/ml RNase Out. Cycloheximide and RNase Out must be added fresh.

3. 60% (w/v) sucrose

4. SW41 centrifuge tubes (Beckman)

5. SW41 rotor and buckets

6. Ultracentrifuge

7. Automated gradient fractionator with continuous UV (260 nm) flow cell (Teledyne Isco Lincoln, NE)

ACKNOWLEDGMENTS

This work was supported by NIH grant GM077382 to C.V.N. and American Heart Association predoctoral fellowship 0515333U to S.B.S.

REFERENCES

Adam, SA., Marr, R. S., and Gerace, L. (1990). Nuclear protein import in permeabilized mammalian cells requires soluble cytoplasmic factors. *J. Cell Biol.* **111,** 807–816.

Blobel, G. (2000). Protein targeting. *Biosci. Rep.* **20,** 303–344.

Blobel, G., and Dobberstein, B. (1975a). Transfer of proteins across membranes. I. Presence of proteolytically processed and unprocessed nascent immunoglobulin light chains on membrane-bound ribosomes of murine myeloma. *J. Cell Biol.* **67,** 835–851.

Blobel, G., and Dobberstein, B. (1975b). Transfer of proteins across membranes. II. Reconstitution of functional rough microsomes from heterologous components. *J. Cell Biol.* **67,** 852–862.

Blobel, G., Walter, P., Chang, C. N., Goldman, B. M., Erickson, A. H., and Lingappa, V. R. (1979). Translocation of proteins across membranes: The signal hypothesis and beyond. *Symp. Soc. Exp. Biol.* **33,** 9–36.

Braakman, I., Helenius, J., and Helenius, A. (1992). Manipulating disulfide bond formation and protein folding in the endoplasmic reticulum. *EMBO J.* **11,** 1717–1722.

Brostrom, C. O., and Brostrom, M. A. (1998). Regulation of translational initiation during cellular responses to stress. *Prog. Nucleic Acid Res. Mol. Biol.* **58,** 79–125.

Harding, H. P., Zhang, Y., Bertolotti, A., Zeng, H., and Ron, D. (2000). Perk is essential for translational regulation and cell survival during the unfolded protein response. *Mol. Cell* **5,** 897–904.

Harding, H. P., Zhang, Y., and Ron, D. (1999). Protein translation and folding are coupled by an endoplasmic–reticulum–resident kinase. *Nature* **397,** 271–274.

Hinnebusch, A. G. (2000). Mechanism and regulation of initiator methionyl-tRNA binding to ribosomes. *In* "Translational Control of Gene Expression" (N. Sonenberg,

J. W. B. Hershey, and M. B. Mathews, eds.), pp. 185–243. Cold Spring Harbor Press, Cold Spring Harbor, NY.

Jackson, R. J., and Hunt, T. (1983). Preparation and use of nuclease-treated rabbit reticulo-cyte lysates for the translation of eukaryotic messenger RNA. *Methods Enzymol.* **96,** 50–74.

Kapp, L. D., and Lorsch, J. R. (2004). The molecular mechanics of eukaryotic translation. *Annu. Rev. Biochem.* **73,** 657–704.

Lerner, R. S., Seiser, R. M., Zheng, T., Lager, P. J., Reedy, M. C., Keene, J. D., and Nicchitta, C. V. (2003). Partitioning and translation of mRNAs encoding soluble proteins on membrane-bound ribosomes. *RNA* **9,** 1123–1137.

Lodish, H. F., and Kong, N. (1993). The secretory pathway is normal in dithiothreitol-treated cells, but disulfide-bonded proteins are reduced and reversibly retained in the endoplasmic reticulum. *J. Biol. Chem.* **268,** 20598–20605.

Marciniak, S. J., and Ron, D. (2006). Endoplasmic reticulum stress signaling in disease. *Physiol. Rev.* **86,** 1133–1149.

Nicchitta, C. V., Lerner, R. S., Stephens, S. B., Dodd, R. D., and Pyhtila, B. (2005). Pathways for compartmentalizing protein synthesis in eukaryotic cells: The template-partitioning model. *Biochem. Cell Biol.* **83,** 687–695.

Palade, G. (1975). Intracellular aspects of the process of protein synthesis. *Science* **189,** 347–358.

Potter, M. D., and Nicchitta, C. V. (2002). Endoplasmic reticulum-bound ribosomes reside in stable association with the translocon following termination of protein synthesis. *J. Biol. Chem.* **277,** 23314–23320.

Scheuner, D., Song, B., McEwen, E., Liu, C., Laybutt, R., Gillespie, P., Saunders, T., Bonner-Weir, S., and Kaufman, R. J. (2001). Translational control is required for the unfolded protein response and *in vivo* glucose homeostasis. *Mol. Cell* **7,** 1165–1176.

Schroder, M., and Kaufman, R. J. (2005). The mammalian unfolded protein response. *Annu. Rev. Biochem.* **74,** 739–789.

Seiser, R. M., and Nicchitta, C. V. (2000). The fate of membrane-bound ribosomes following the termination of protein synthesis. *J. Biol. Chem.* **275,** 33820–33827.

Stephens, S. B., Dodd, R. D., Brewer, J. W., Lager, P. J., Keene, J. D., and Nicchitta, C. V. (2005). Stable ribosome binding to the endoplasmic reticulum enables compartment-specific regulation of mRNA translation. *Mol. Biol. Cell* **16,** 5819–5831.

Stephens, S. B., Dodd, R. D., Lerner, R. S., Pyhtila, B. M., and Nicchita, C. V. (2008). Analysis of the partitioning of mRNAs between cytosolic and ER membrane-bound compartments in mammalian cells. *Methods Mol. Biol.* **419,** 197–214.

Walter, P., and Blobel, G. (1981a). Translocation of proteins across the endoplasmic reticu-lum III. Signal recognition protein (SRP) causes signal sequence-dependent and site-specific arrest of chain elongation that is released by microsomal membranes. *J. Cell Biol.* **91,** 557–561.

Walter, P., and Blobel, G. (1981b). Translocation of proteins across the endoplasmic reticulum. II. Signal recognition protein (SRP) mediates the selective binding to micro-somal membranes of *in vitro*-assembled polysomes synthesizing secretory protein. *J. Cell Biol.* **91,** 551–556.

Walter, P., and Blobel, G. (1983). Preparation of microsomal membranes for cotranslational protein translocation. *Methods Enzymol.* **96,** 84–93.

Walter, P., Ibrahimi, I., and Blobel, G. (1981). Translocation of proteins across the endo-plasmic reticulum. I. Signal recognition protein (SRP) binds to *in vitro*-assembled polysomes synthesizing secretory protein. *J. Cell Biol.* **91,** 545–550.

Walter, P., and Johnson, A. E. (1994). Signal sequence recognition and protein targeting to the endoplasmic reticulum membrane. *Annu. Rev. Cell Biol.* **10,** 87–119.

CHAPTER FIVE

MAMMALIAN STRESS GRANULES AND PROCESSING BODIES

Nancy Kedersha *and* Paul Anderson

Contents

Abstract

The packaging of cytoplasmic mRNA into discrete RNA granules regulates gene expression by delaying the translation of specific transcripts. Specialized RNA granules found in germ cells direct the timing of maternal mRNA translation to promote germ cell development in the early embryo and establish the germ line for the next generation. Similarly, select neuronal mRNA transcripts are packaged into translationally inert RNA granules, transported to sites where their protein products are required, and only then activated and translated. Following translation, however, newly inactivated mRNAs released from polysomes can also be packaged into dynamic, transient structures known as stress granules (SGs) and processing bodies (PBs). Stress granules are composed largely of stalled preinitiation complexes, and contain mRNA, small ribosomal subunits, eIF3, eIF4E, eIF4G, and PABP, as their core components. PBs are associated with mRNA decay and contain the decapping enzymes DCP1/2, the $5'$ to $3'$ exonuclease Xrn1, the Lsm proteins (1–7), and the scaffolding proteins hedls/GE-1 and GW182. Both SGs and PBs contain mRNA, eIF4E, microRNAs and argonaute proteins, and various regulators of mRNA stability and translation (TTP, RCK/p54, and CPEB). Thus, SGs and PBs share some protein and mRNA components,

Division of Rheumatology, Immunology and Allergy, Brigham and Women's Hospital, Boston, Massachusetts

Methods in Enzymology, Volume 431
ISSN 0076-6879, DOI: 10.1016/S0076-6879(07)31005-7

but also contain a number of unique markers specific to each structure. We describe markers and staining procedures used to identify these distinct types of RNA granules, describe conditions that promote their assembly and disassembly, and establish YB-1 as a useful marker of SGs and PBs.

1. INTRODUCTION

Stress granules are cytoplasmic phase-dense structures that occur in eukaryotic cells exposed to environmental stress (heat, viral infection, oxidative conditions, ultraviolet [UV] irradiation, hypoxia). SGs were first described in cells cultured at supra–ambient temperatures and characterized by their inclusion of low molecular weight heat shock proteins (Collier and Schlesinger, 1986; Nover et al., 1989). Subsequent analysis has identified a large number of proteins and mRNAs that are components of SGs (Anderson and Kedersha, 2006). The core constituents of SGs are components of a noncanonical, translationally silent 48S pre-initiation complex that includes the small ribosomal subunit and early initiation factors eIF4E, eIF3, eIF4A, eIFG, and PABP (Kedersha et al., 2002). SGs also include RNA-binding proteins that regulate mRNA translation and decay, as well as proteins involved in various aspects of mRNA metabolism. SGs also contain proteins that regulate diverse cell signaling pathways (e.g., TRAF2; Hofmann et al., 2006; Kim et al., 2005), the functional significance of which remains to be determined.

SG assembly is usually initiated by the phosphorylation of translation initiation factor eIF2α, a component of the eIF2–GTP–tRNA^{Met} ternary complex (Kedersha et al., 1999). Different types of environmental stress activate distinct eIF2α kinases (PKR, PERK, HRI, and GCN2) to initiate SG assembly. Phosphorylation of eIF2α reduces the availability of the eIF2–GTP–tRNA^{Met} ternary complex, thereby preventing the assembly of the 48S pre-initiation complex (Srivastava et al., 1998). When this stalled initiation complex is assembled at the 5$'$ end of polysomal mRNA, actively translating ribosomes "run-off" the transcript, resulting in polysome disassembly. These nonpolysomal transcripts are then actively organized into stress granules, a process mediated by a number of RNA-binding proteins (Kedersha et al., 2002).

Many fundamental studies of PBs have been performed in yeast (Sheth and Parker, 2003; Teixeira et al., 2005). Mammalian PBs contain many core proteins found in yeast, but also contain a number of metazoan-specific proteins; hence, some functions of mammalian PBs are likely lacking in yeast. Mammalian PBs (also known as GWBs [Eystathioy et al., 2002, 2003] or DCP foci [Ingelfinger et al., 2002; van Dijk et al., 2002], contain components of the 5$'$ to 3$'$ decay machinery (Xrn1, DCP1/DCP2,

Hedls/GE-1, Lsm1), nonsense-mediated decay pathway (SMG5, SMG7, UPF1), and RNA-induced silencing machinery (GW182, microRNA, argonaute; for review, see Anderson and Kedersha, 2006). Activators of mRNA decay pathways (4E-T, p54/RCK, CBEP) are also found in PBs; many of these proteins are translational regulators and also present in SGs. Actively growing, unstressed cells display PBs, but more PBs are assembled in response to arsenite; this response is independent of eIF2α phosphorylation (Kedersha *et al.*, 2005). Arsenite treatment induces PB assembly concurrent with SG assembly, and results in physically associated SG-PB structures (Kedersha *et al.*, 2005; Wilczynska *et al.*, 2005). Dissecting the physical and functional links between SGs and PBs is currently a very active field of research in many labs.

Our laboratory is interested in the biology of mammalian SGs and PBs, and in understanding their roles in the spatial regulation of mRNA translation and decay. The methods and procedures presented here describe our present knowledge of SG and PB assembly and composition. We indicate some commercially available antibodies useful as SG and PB markers, describe some immunocytochemical protocols that we employ, and offer some caveats regarding data interpretation.

2. Detection and Identification of SGs and PBs

Despite the efforts of several groups, no one has yet succeeded in purifying SGs, so their contents are morphologically defined using immunostaining. SG contents can be broadly categorized as follows: (1) preinitiation and translation-related factors (Anderson and Kedersha, 2002; Kedersha *et al.*, 2002), (2) mRNAs, (3) mRNA-binding proteins with known functions in translational control or mRNA stability (e.g., TIA, TTP, PABP, HuR, FXR1, CPEB, eIF-4E) (Anderson and Kedersha, 2006), (4) proteins linked to mRNA metabolism (e.g., G3BP, p54/rck, PMR1, SMN, Staufen), and (5) signaling proteins with no direct known links to RNA metabolism (e.g., TRAF2, roquin, plakophilins 1 and 3, Disrupted-in-Schizophrenia). PolyA(+) mRNA, preinitiation factors eIF4E, eIF4G, eIF3, PABP-1, and small but not large ribosomal subunits are translational components that appear universal to SGs induced by multiple stimuli in all cells examined, whereas the SG association of PMR1, eIF2B, phospho-eIF2α, HSP70, and HSP27 varies with cell type and specific stress (Anderson and Kedersha, 2002; Yang *et al.*, 2006). In addition to noncanonical 48S complexes, SG assembly is promoted by any one of several RNA-binding proteins, many of which are able to oligomerize as well as bind RNA. Such proteins include TIA-1/TIAR (Gilks *et al.*, 2004), Fragile X Mental Retardation Protein (FMRP/FXR1; Mazroui *et al.*, 2006), ras-gap SH3-binding protein (G3BP1; Tourriere *et al.*, 2003), cytoplasmic polyadenylation-binding protein (CPEB; Wilczynska *et al.*, 2005), Survival of Motor Neurons protein (SMN; Hua and Zhou, 2004), p54/RCK (Wilczynska *et al.*, 2005), smaug (Baez and Boccaccio, 2005), and tristetraprolin (TTP; Stoecklin *et al.*, 2004). Overexpression of any of these proteins is sufficient to induce SG assembly in the absence of stress, unless protein kinase R (PKR) activity is blocked (Kedersha *et al.*, 1999), indicative of a requirement for eIF2α phosphorylation. A prionlike aggregation has been demonstrated for TIA-1 (Gilks *et al.*, 2004) and CPEB (Wilczynska *et al.*, 2005) whereas G3BP possesses an oligomerization domain that is regulated by phosphorylation(Tourriere *et al.*, 2003). SMN self-aggregation appears mediated by its Y–G box, whereas the aggregation of FMRP family members appears mediated by a coiled-coil domain. Dominant–negative mutants of TIA-1 (e.g., delta RRM; Gilks *et al.*, 2004; Kedersha *et al.*, 1999), G3BP1 (Tourriere *et al.*, 2003), and SMN (Hua and Zhou, 2004) block SG formation in response to stress, suggesting that these proteins are important to SG assembly.

Additional SG proteins not known to directly bind to RNA include fas-activated serine–threonine phosphoprotein (FASTK), TRAF-2, plakophilins 1 and 3, and roquin. FASTK interacts with TIA-1, likely explaining its recruitment to SGs. TRAF2 contains no obvious RNA-binding domains, and its recruitment to SGs appears mediated by its interactions with eIF4G (Kim *et al.*, 2005). Its recruitment to SGs correlates with altered solubility and its inactivation, suggesting that its recruitment to SGs serves to inactivate TRAF2-mediated signaling events. Interestingly, the endonuclease PMR1 exhibits variable recruitment to SGs, mediated by a region in its N-terminal region that promotes arsenite-regulated binding to TIA-1 (Yang *et al.*, 2006).

Table 5.1 lists a number of SG and PB markers, and indicates the relative specificity of each reagent for SGs or PBs. This is only a partial list of known SG and PB markers, focusing on some that are broadly useful and for which antibodies are readily available. The staining patterns obtained using some of these antibodies are shown in Fig. 5.1.

Proteins exclusively found in SGs include eIF3 (Fig. 5.1A, B, C, and D, blue), PABP-1 (not shown), eIF4G (Fig. 5.1C, red), FXR1 (not shown; Kedersha *et al.*, 2005); proteins exclusively found in PBs include DCP1a, DCP2, hedls/GE-1, 4E-T, and GW182. A number of useful SG markers (TIA-1, TIAR, HuR) are not entirely SG specific and may associate with PBs in unstressed cells. At present, we are unaware of good commercial antibodies against DCP1a, DCP2, or GW182, and must refer the reader to those investigators who have kindly shared their reagents. Regarding hedls/GE-1, we accidentally discovered that a monoclonal antibody recognizing p70 S6kinase exhibits a very strong reactivity to hedls/GE-1. We have confirmed the dual specificity of this antibody using recombinant hedls,

Table 5.1 Protein markers and antibodies to detect SGs and PBs

Detects	Protein	Catalog number/ Commercial source	Marker for	Host species dilution	Comments
SG	p-eIF2α	KAP-P130 StressGen Biotech	SGs only	Rabbit 1/1000	Phospho-protein, causes SG assembly
	eIF3, p116	N20 sc-16377 Santa Cruz Biotech	SGs only	Goat 1/200	Strong signal, robust SG marker
	eIF4AI	sc-14211	SGs	Goat 1/200	Helicase, impairment causes SG assembly
	eIF4B	Santa Cruz Biotech 3592-Cell Signaling Technologies	SGs	Rabbit 1/100	Weak signal
	eIF4G	sc-11373 Santa Cruz Biotech		Rabbit 1/200	May not work well in mouse cells
	FXR1	sc-10544 Santa Cruz Biotech		Goat 1/100	Signal weak in U2OS; stronger when in SGs
	G3BP-1	BD-TL 611126 BD Transduction Labs		Mouse 1/200	Robust SG marker
	PABP-1	sc-32318 Santa Cruz Biotech	SGs only	Mouse 1/200	Strong signal human cells, does not detect mouse PABP.
SG>>PB	HuR	sc-5261 Santa Cruz Biotech goat sc-1751		Mouse 1/200	Rarely detectable at PBs in cells lacking SGs
	TIA-1	Santa Cruz Biotech goat sc-1749		Goat 1/200	Rarely detectable at PBs in cells lacking SGs
	TIAR	Santa Cruz Biotech		Goat 1/200	Rarely detectable at PBs in cells lacking SGs

(continued)

Table 5.1 (continued)

Detects	Protein	Catalog number/Commercial source	Marker for	Host species dilution	Comments
PB	DCP1a	none known	PBs only		
	4E-T	sc-13453, sc-13454 Santa Cruz Biotech	PBs only	Goat 1/200	Weak signal in U2OS cells
	GE-1/hedls	sc- 8418, Santa Cruz Biotech	PBs only	Mouse 1/1000	Robust PB marker but crossreacts with nuclear S6 kinase; use at 1/1000.
	P54/RCK	BL2142 A300-461A Bethyl Labs	PB>>SG	Rabbit 1/1000	Robust. Sees C-terminus
	Xrn1	Abcam A300-443A-1	PB>>SG	Rabbit 1/500	Use at 1/500
PB/SG, Stress-dependent		These proteins are weak markers of PBs in unstressed cells, but vary in their SG/PB ratio during stress. Distribution here refers to relative proportion of signal during arsenite stress.			
	eIF4E	sc-13963 Santa Cruz Biotech	SGs/PBs	Rabbit 1/200	Not always detectable at PBs; may not detect mouse 4E
	eIF4E	sc 9976 Santa Cruz Biotech	SGs/PBs	Mouse 1/200	Human, mouse; not always present in PBs
	Lsm1	15-288-22100F Genway Biotech	PBs>SGs	Chicken 1/1000	Protein localizes to the cytoplasm; minor fraction at PBs
	YB-1	Ab12148 Abcam	SGs>PBs	Rabbit 1/500	Weakly detectable in PBs; predominates in SGs upon stress

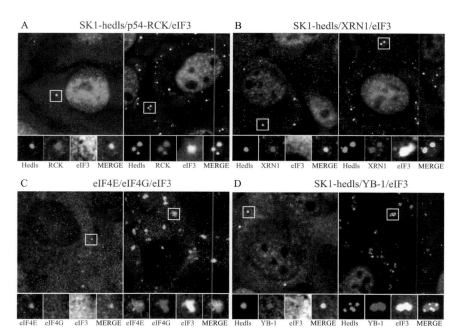

Figure 5.1   SGs and PBs detected using different antibodies. U2OS cells were untreated (left panels) or arsenite treated (0.5 m*M*, 45 min) (right panels), then fixed and stained using the described protocol. Enlarged views of boxed regions are displayed underneath the corresponding panels, showing the separate views of the same field. (A) Untreated (left panels) or arsenite-treated (right panels) cells stained for Hedls/S6K1 in green, p54-RCK in red, and eIF3b in blue. Hedls and p54-RCK stain PBs exclusively, whereas eIF3 is specific for SGs. (B) Untreated (left panels) or arsenite-treated (right panels) cells stained for Hedls/S6K1 in green, XRN1 in red, and eIF3b in blue. Hedls and XRN1 stain PBs exclusively, whereas eIF3 is specific for SGs. (C) Untreated (left panels) or arsenite-treated (right panels) cells stained for eIF4E in green, eIF4G in red, and eIF3b in blue. Note that eIF4E (green) is present in PBs in unstressed cells (left panel), but also is detectable in SGs in arsenite-treated cells (right panels). In contrast, eIF4G and eIF3 remain exclusively associated with SGs. (D) Untreated (left panels) or arsenite-treated (right panels) cells stained for Hedls/S6K1 in green, YB-1 in red, and eIF3b in blue. Note that YB-1 is detectable in PBs in unstressed cells, but is largely relocalized to SGs upon this particular stress. (See color insert.)

and also have used it to visualize PBs in p70 S6 kinase knockout cells (not shown). This antibody stains PBs with remarkable clarity, works well by IP and western blot, and its use has been previously documented (Stoecklin *et al.*, 2006, Fig. S2, supplemental). While this antibody also reacts with its intended target p70 S6 kinase, most of the non–hedls signal is nuclear and does not interfere with the antibody's ability to recognize PBs (shown green in Fig. 5.1A, B, D, and Fig. 5.3). Other useful PB markers include p54/RCK (Fig. 5.1A, red), XRN1 (Fig. 5.1B, red), and Lsm1 (not shown). These are largely PB-specific in the absence of stress (Fig. 5.1), although

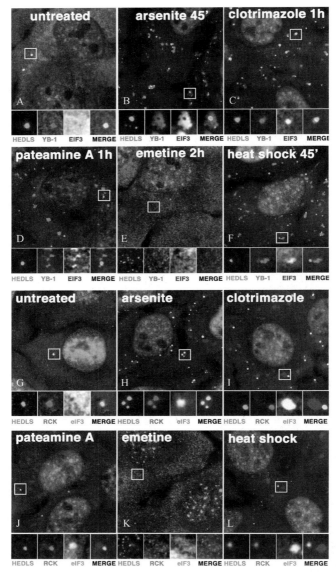

Figure 5.2   SG and PB formation in response to different stresses. U2OS cells were subjected to some of the stresses described in Table 5.1, then stained as indicated. Untreated cells display few PBs and no SGs, whereas arsenite treatment (0.5 mM, 45 min, panels B and H) strongly induces assembly of both PBs and SGs. Clotrimazole (C, I), pateamine A (D, J), and heat shock (F, L) induce more SGs than PBs, whereas emitine treatment (E, K) abolishes PBs and does not induce SGs. Note that YB-1 is present in PBs in unstressed cells (A, red), but relocalizes to SGs upon most stress conditions (C, D, F, whereas p54/RCK (red, G, H, I, J, L) remains predominantly associated with PBs regardless of stress. (See color insert.)

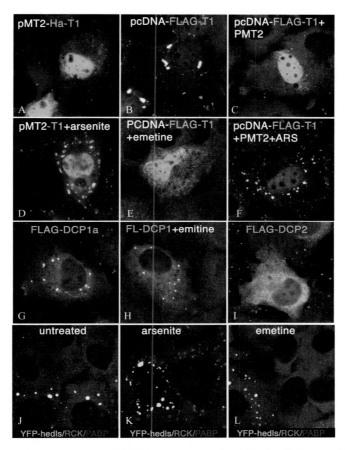

Figure 5.3 Overexpression of SG/PB marker proteins. COS7 cells (panels A–I) were transiently transfected with different vectors containing the same coding region of TIA-1, treated, and stained as indicated in the individual panels. Overexpressed PB markers such as DCP1a (G, H) and hedls/GE-1 (panels J, K, L) induce formation of abnormally large PBs (compare yellow foci to red foci in nontransfected cells) that are resistant to emetine, which dissolves normal PBs (absence of red foci in panel H). In contrast, overexpressed DCP2 (I) is not detected in PBs but appears diffuse. (J–L) U2OS cells stably expressing YFP-Hedls/GE-1 (green, panels J, K, L) were co-cultured with wild -type U2OS cells (lacking green), treated as indicated, and counterstained for the PB marker RCK (red) and the SG marker PABP-1 (blue). YFP-hedls PBs appear yellow, endogenous PBs in untransfected cells appear red. As with overexpressed DCP1, the YFP-hedls cells exhibit huge PB-like structures that are not disassembled upon emitine treatment (panel L, comparing large yellow foci to absent red foci). (See color insert.)

small amounts of each exhibit limited association with some types of SGs in stressed cells. Of interest are eIF4E (Fig. 5.1C, green) and YB-1 (Fig. 5.1D, red), both of which can associate with PBs in unstressed cells, but which partially (eIF4E, Fig. 5.1C green, right panel) or predominantly (YB-1, Fig. 5.1D, right panel, red) relocalize to SGs during stress.

3. Immunostaining Protocol

This method employs an unconventional fixation procedure, using conventional paraformaldehyde fixation followed by a methanol permeabilization/postfixation step, rather than the more typical detergent permeabilization (0.5% TX-100, saponin, digitonin, etc.). This technique has several advantages for SG-PB visualization: (1) It flattens the cells, compressing SGs and PBs into a thinner specimen amenable to wide-field microscopy, useful for those without ready access to a confocal system; (2) it enhances the adhesion of the cells and results in better retention of cells throughout the staining process; and (3) it preserves more total cytoplasmic signal, allowing for a more accurate comparison of diffuse versus granule-associated signal. Using mild paraformaldehyde fixation followed by detergent permeabilization (rather than methanol) typically removes a variable percentage of SG- or PB-associated proteins, leaving the viewer with the erroneous impression that all the protein is in SGs or PBs, when, in fact, only a small fraction of it may actually be localized within SGs or PBs prior to fixation. Using GFP-tagged constructs is important in testing one's fixation conditions, but care must be taken to ensure that the protein is not grossly overexpressed (see Fig. 5.2), especially when overexpressing proteins associated with SGs/PBs. All of the images shown here were taken on a conventional widefield Eclipse E800 microscope with epifluorescence optics and a CCD-SPOT RT digital camera. The images were compiled using Adobe Photoshop software (v7.0).

3.1. Buffers and solutions

PBS: phosphate-buffered saline

Dissolve the following in 800 ml of deionized (milliQ) H_2O
80 g NaCl
2.0 g KCL
14.4 g Na_2HPO_4
2.4 g KH_2PO_4
Adjust pH to 7.4
QS to 1 l with deionized H_2O

4% paraformaldehyde in PBS:

(Perform all procedures in chemical fume hood, wear gloves, and be careful!)

1. Weigh a capped 50-ml tube and record the tube weight. Transfer tube to fume hood. Open bottle containing paraformaldehyde and transfer ~20 ml of powder into the 50-ml tube. Cap securely, then remove from hood and weigh on balance. Subtract tube weight from total weight and calculate the amount of paraformaldehyde in the tube. Divide number of grams by 4 and multiple by 100: this is the total volume needed to obtain a 4% solution.

2. In the fume hood, add PBS (prewarmed in microwave to save time!) to a flask containing a stirbar. With constant stirring, add the weighed pow-dered paraformaldehyde. Heat to hot-not-boiling (~60°) on a heater/stir plate, until solution clears. Check that pH is around 7.5 using pH paper.

3. Cover and let cool until just warm, and aliquot immediately into 50-ml polypropylene tubes (or smaller, if appropriate for your needs) in the fume hood. Let cool to room temperature in the fume hood and make sure there is no spilled solution on the tubes before removing from the hood. Store aliquots at −20°F.

For use, thaw in hot water and use at room temperature. Store at 4° and use within one week, or else refreeze at −20°F.

Blocking solution:
5% normal horse serum (Sigma H1270), made up in PBS containing 0.02% sodium azide, and filtered through a 0.4 microfilter to remove precipitated serum proteins.

3.2. Staining procedure

All steps should be performed at room temperature, optimally on a gently moving rotary shaker. If the cells are very delicate (e.g., neurons) and cannot withstand gentle shaking, incubations may be performed without shaking, but times or antibody concentrations may need to be increased.

1. **Grow cells on glass coverslips to desired confluence**. Circular glass coverslips are available (Fisher catalog #12-545-80) which fit nicely into 24-well tissue culture plates. These coverslips may be autoclaved in a petri dish; in our hands, no prerinsing is necessary. To sterilely transfer the autoclaved coverslips, use a sterile Pasteur pipette attached to a vacuum and pick up each coverslip using aspiration: this is much easier than fumbling around with sterile forceps! Incubate the glass coverslips in serum-containing media for 5 h to overnight before adding the cells. This will greatly improve spreading and adhesion. Plate cells at a reason-able density to allow 70–90% confluence in 1–2 days (e.g., U2OS cells at

1×10^5 cells per well for use 2 days after plating). Administer stresses (Table 5.1) as desired prior to fixation.

2. **Fixation.** Rinse the cells with PBS (phosphate-buffered saline, pH 7.4). Remove PBS and immediately (do *not* allow cells to dry out) add 4% paraformaldehyde in PBS buffer. Incubate samples for 15 min at room temp, gently shaking. Remove paraformaldehyde solution (aspirate into a collection flask) and transfer into an appropriate hazardous waste container for proper disposal. Immediately add $-20°F$ methanol and incubate cells for 10 min (on a shaker at room temperature). Make sure all additions are done gently, by pipetting each liquid into the plate slowly by running the stream along the wall of the well. Do not bomb the cells! Rinse cells once in PBS. If necessary, cells can usually be stored in the cold room overnight prior to staining.

3. **Block** by incubation in 5% normal horse serum in PBS containing 0.02% sodium azide. This assumes that the secondary antibody to be used is donkey; if using another species, blocking should be done using the sera of the species used. Blocking should be performed for at least 45 min at room temperature with gentle shaking.

4. **Primary antibody incubation.** Prepare primary antibody cocktail in the blocking buffer (5% normal horse serum, buffer + azide), and incubate ~1 h at room temperature. We typically use a mix of mouse, rabbit, goat, and/or chicken primary antibodies together; see Table 5.2 for appropriate dilutions of the recommended antibodies. At this point (or during secondary antibody incubation), it is usually safe to incubate the samples overnight at 4°.

5. **Wash** cells thoroughly twice (5–10 min each in PBS), aspirating off each solution and gently adding ~1 ml PBS to the well. We use a squirt bottle containing PBS for the washes, with the tip cut off to avoid a jet stream, and add gently by letting the buffer flow along the side of each well upon addition.

6. **Secondary antibody incubation.** Prepare secondary antibody cocktail in blocking buffer. We routinely triple stain using species–specific secondary antibodies using "ML" grade fluorescent conjugates from Jackson Immunoresearch Labs. Note that Cy2 and Cy5 conjugates are diluted 1/200, whereas Cy3 dye conjugates require a 1/2000 dilution. We routinely add Hoechst dye (Sigma, catalog #HOE 33258, at 50 ng/ml) to the secondary cocktail, which allows us to visualize DNA using a UV/DAPI filter set. SG and PB assembly are largely blocked in mitotic cells; thus, DNA visualization is useful when counting cells with SGs/PBs.

7. **Wash** cells thoroughly (3×, 5–10 min each).

8. **Mount** in polyvinyl–based mounting media. Store in the dark; slides are usually stable for weeks to months.

Table 5.2 Treatments to induce SGs and PBs

Treatment/reagent	Stock solution, if applicable	Conditions	Effects	Comments
Arsenite Sigma S-7400, NaAsO2	65 mg/ml makes stock of 0.5M; soluble in PBS; stable frozen	0.5–1.0 mM, 30 min–1 h	Induces SGs and PBs	Also causes active export of rare SG-associated proteins (e.g., TTP) from SGs Reversible.
Clotrimazole Sigma C6019	6.9 mg/ml makes 20 mM stock inDMSO	10–20 ml for 30 min–1h: requires serum-free media	Induces SGs	Requires serum-free media, works on all cells tested
Cycloheximide	Stock = 5.0 mg/ml in DMEM, sterile	1–2 h treatment, 20–50 μg/ml	Disassembles both SGs and PBs	
(Edeine)			Induces SGs	Wilczynska et al., 2005 Prevents 60S joining
(eIF2αS51D mutant)	Transfection		Assembles SGs, not PBs	Induces SGs in COS cells but does not localize to SGs
Emetine Sigma E2375	Stock solution 10 mg/ml in H20; use at 100 μg/ml	1–2 h treatment, 20–50 μg/ml	Disassembles both SGs and PBs	Useful SG/PB diagnostic

(continued)

Table 5.2 (*continued*)

Treatment/reagent	Stock solution, if applicable	Conditions	Effects	Comments
FCCP Sigma C2920	Stock of 1.0 mM/ DMSO	1–2 h Dilute 1/1000 in glucose-free media	Induces SGs without increasing PBs, promotes SG–PB fusion	Can cause detachment of fibroblasts from substrate without SG induction
Glucose starvation	GIBCO 1196-025 Glucose-free DMEM	30 min to 1 h		Most effective on tumor cell lines such as HeLa or DU145; does not work on fibroblasts
(G3BP overexpression)	Transfection		Causes large SGs, does not affect PBs	Unusual in that G3BP is not a translational silencer
Heat shock	Preheated incubator	42–44°, 30–45 min. Overlay media with mineral oil if using non-CO$_2$ incubator	Induces SGs early, may induce PBs later	Cells adapt and SGs can disappear within 1 h; need time course to determine optimum time

Osmotic shock	Media containing 1.0 M sorbitol	Expose cells to sorbitol-containing media for 30 min; replace with regular media for 30–60 min	Induces SGs	Duration of exposure more important that degree of hypertonicity
(Pateamine A)	DMDA-PatA 2.0 mM/DMSO	50 nM	Induces SGs without PBs	SGs often linear, possibly associated with cytoskeleton
Puromycin	Stock 100 mg/ml	20 μg/ml for several hours	Induces very large SGs, few per cell	Only ~20% cells respond; effects vary with cell line
Thapsigargin Calbiochem # 586005	1.0 mM stock in DMSO	1.0 μM 30–90 min	Induces SGs without increasing PBs	Does not work on HeLa, DU145 cells; effective on fibroblasts and U2OS cells
UV	Statolinker set at 100 using "energy" setting, ($\times$100 μJ/cm^2)	Remove media, expose cells, and replace media; incubate cells 30 min to 24 h	Induces SGs	SGs are small; may persist for 24 h

Note: Reagents in parentheses are not commercially available.

4. STIMULI FOR THE INDUCTION OF SGS AND PBS

Several conditions must be kept in mind when studying SGs: (1) SGs result from sudden translational arrest, which requires that the cells be actively translating at the application of the stress stimulus, (2) SGs are transient and may only be present for a limited time following the application of stress, and (3) different types of cells/cell lines respond differently to the same stress. These issues are described in detail.

Stalled translational initiation, usually a consequence of stress-induced phosphorylation of eIF2α, initiates SG assembly. Thus, untranslated mRNA is the substrate from which SGs are assembled. Cells that are sick, overgrown, mycoplasma-infected, or otherwise mistreated have low levels of translating polysomes, rendering them resistant to SG assembly following the application of exogenous stress. In a similar fashion, viral infection is a potent inducer of SGs (Iseni *et al.*, 2002; McInerney *et al.*, 2005), but in productively infected cells, SG assembly may be blocked in response to virus-induced translational shutdown (McInerney *et al.*, 2005). Table 5.2 describes various conditions used to induce SG assembly in tissue culture cells, and describes some of their limitations and uses. Among the many stimuli that we have tried that generally do NOT induce SGs are the following: DNA-damaging agents such as cisplatin or etoposide, treatment with PMA, H_2O_2, ceramide, actinomycin D, DRB (dichlorobenzimidizole riboside), hydoxyurea, lipopolysaccharide, interferons (α or γ), ricin, or staurosporin. Note that many of these agents would be expected to affect translation, and might induce SGs in cell lines other than the ones we have tested (DU145, U2OS, HeLa, COS, MEFs).

Generally, at least some phospho-eIF2α is essential for the generation of SGs (Kedersha *et al.*, 1999; McEwen *et al.*, 2005). This requirement for minimal phospho-eIF2α suggests that polysome disassembly is required for SG assembly; in other words, the ribosomes must run off a transcript before it is routed to a SG (reviewed in Anderson and Kedersha, 2002). Two studies have shown that different toxins (pateamine A and hippuristanol) can induce SGs without inducing phosphorylation of eIF2α (Dang *et al.*, 2006; Mazroui *et al.*, 2006). Both these agents functionally inactivate the RNA helicase eIF4A and prevent ribosome scanning, which effectively aborts translation initiation, promotes ribosome runoff, and causes polysome disassembly. Thus, disassembled polysomes provide the mRNA substrate from which SGs are assembled. In contrast, SGs are disassembled when polysomes are stabilized using pharmacological inhibitors of translational elongation such as emetine or cycloheximide. These drugs disperse SGs even in the continuous presence of phospho-eIF2α, providing further indication that SGs are in equilibrium with polysomes. They will also eventually reduce the number of arsenite-induced PBs, although with slower kinetics than SGs.

SGs can be induced in all mammalian cells examined to date. Ideally, to ensure reproducible SG induction, cells should be in log phase growth prior to application of stress. Cells should be plated at a known density and cultured in abundant media for ~1–3 days until 50–90% confluent. Whenever feasible, drugs should be diluted in the same conditioned media in which the cells are growing, to avoid addition of fresh nutrients and glucose along with the drug. Only when the drugs themselves require dilution in serum-free (e.g., clotrimazole; see Table 5.2) or glucose-free (e.g., FCCP) media should the medium be changed when stressing the cells. Figure 5.2 shows U2OS cells exposed to a series of stresses, and the effects of these on SG and PBs. Untreated U2OS display no SGs and few PBs (panel A), whereas a 45-min treatment with 0.5 mM arsenite (panel B) induces the assembly of both SGs (blue or purple) and PBs (green), usually juxtaposed or fused (Fig. 5.2B inset). Clotrimazole treatment (Fig. 5.2C) induces SGs and fewer PBs, as does heat shock (Fig. 5.2F). Pateamine A causes SG assembly without increasing PBs (Fig. 5.2D), whereas treatment with emetine completely eliminates PBs (Fig. 5.2E).

5. TRANSFECTION-INDUCED SGS

Overexpression of many SG-associated proteins (G3BP; Tourriere *et al.*, 2003), TIA-1/TIAR (Gilks *et al.*, 2004), FXR1/FMRP (Mazroui *et al.*, 2002), TTP/BRF-1 (Stoecklin *et al.*, 2004), SMN (Hua and Zhou, 2004), and mutant eIF2αS51D (Kedersha *et al.*, 1999)) will induce "spontaneous" SGs in the absence of additional stress. The vector, choice of transfection agent, and interpretation are important when using transfection to examine SGs. Certain vectors designed for high protein expression (notably, pMT2) may express inhibitors of the eIF2α kinase PKR, which appears induced and activated during transfection and which is required for spontaneous SG formation (Kedersha *et al.*, 1999). For example, when TIA-1 is expressed in the pMT2 vector (Fig. 5.3A and D), its behavior is similar to that of endogenous TIA-1; the recombinant TIA-1 appears predominantly nuclear in unstressed cells (Fig. 5.3A), but moves to SGs upon arsenite treatment (Fig. 5.3D). In contrast, when TIA-1 is expressed in the commonly used pcDNA3 (Invitrogen) or pEGFP vectors (Clontech), the tagged proteins are predominantly cytoplasmic and found in self-nucleated "spontaneous" SGs (Fig. 5.3, panel B). These artificial SGs are dispersed upon emetine treatment as are normal SGs (Fig. 5.3, panel E) and appear to contain all the same components as normal SGs (Kedersha *et al.*, 2005). This apparent paradox is due to the fact that pMT2 contains VA-I, an adenoviral noncoding small RNA inactivator of the eIF2α kinase PKR (adenoviral VAI), which is necessary to induce spontaneous SG assembly. Cotransfection of empty pMT2 vector with pcDNA3-TIA-1 prevents spontaneous SGs (Fig. 5.3C), while allowing arsenite treatment (which activates a different eIF2α kinase; (McEwen *et al.*, 2005) to induce SGs normally (Fig. 5.3F). Cotransfection of pcDNA3-TIA-1 with a plasmid encoding only VA-I (pAdVAntage, Promega) similarly prevents spontaneous SG assembly (not shown). Thus, the vector influences the experimental outcome, and this must be considered in the experimental design and data interpretation. Furthermore, the specific transfection agent can affect the results. Some cationic agents used for transfection (notably, LipoFectamine) can induce SGs that persist for hours or days, while others (SuperFect) are less prone to do so. Both vector and agent must be considered in the context of any experimental protocol—transfecting an empty vector control sample should be routine practice when assessing novel candidate SG proteins in any specific system.

6. INDUCTION OF PBS

Much less is known about stimuli required for the induction of PBs, and the molecular mechanisms that affect PB assembly. Serum-starvation and cell-cycle arrest (Eystathioy *et al.*, 2002) will ablate PBs, which are reassembled hours after serum addition or rapidly upon arsenite treatment (30 min). Cells not amenable to cell cycle arrest (such as most strains of HeLa, COS7, and DU145 cells) constitutively display many PBs per cell, while growing U2OS cells only display PBs in 20–50% of the cells. Arsenite strongly induces PBs, as does overexpression of DCP1, RCK, hedls/GE-1, etc. (see later for caveats about overexpression studies). However, many stresses that induce SGs, such as mild heat shock, short clotrimazole treatment, thaipsigargin, pateamine A (Fig. 5.2D), and glucose starvation, do not induce PBs. Some of these treatments, such as arsenite treatment (Fig. 5.1 and Fig. 5.2), overexpression of TTP/BRF1 (Kedersha *et al.*, 2005), or overexpression of CPEB (Wilczynska *et al.*, 2005) induce both SGs and PBs, and promote physical interactions among these structures, similar to results using arsenite. One consideration in interpreting data obtained using overexpression studies is that PBs induced by overexpression of a single PB-component may not be functional, because large PBs are assembled when decay functions are blocked (Cougot *et al.*, 2004). Overexpression of DCP1a has widely been used as a marker of PBs, but its overexpression can generate abnormal PBs that are likely functionally aberrant because they do not exhibit normal dispersal upon emetine treatment (Fig. 5.3H; compare yellow foci to absent red foci). Moreover, overexpression of DCP1a enhances the expression of other PB components such as hedls/GE-1 (Fenger-Gron *et al.*, 2005; Fillman and Lykke-Andersen, 2005). Conversely, overexpression of GE-1/hedls generates huge PBs that do not disperse upon emetine or cycloheximide treatment, as do endogenous PBs. Figure 5.2 J–L shows a mixture of wild-type U2OS cells and cells stably overexpressing YFP-hedls (green). Overexpressed hedls induces huge PBs (yellow foci), much larger than those in neighboring wild-type cells (Fig. 5.2 J–L, red foci). These cells exhibit normal SG response upon arsenite treatment (Fig. 5.2J; SGs appear blue), but the giant hedls PBs are completely resistant to the effects of emetine (Fig. 5.2K), even though PBs in adjoining wild-type cells are completely dispersed (Fig. 5.2K, red). Conversely, knockdown experiments indicate that reduced levels of many individual PB components (e.g., DCP1, Lsm1, GW182, hedls/GE-1 (Cougot *et al.*, 2004; Stoecklin *et al.*, 2005; Yu *et al.*, 2005)) will decrease the numbers and size of PBs.

7. Conclusions

As the complex and ever-expanding list of SG and PB components indicates, mammalian SGs and PBs appear integrated with many aspects of cell metabolism beyond translational control and mRNA decay. Therefore, the state of the cells is critical to reproducible results. We cannot overemphasize the importance of controlling cell density, feeding schedule, CO_2 and pH regulation, temperature control, etc., so as to minimize conflicting environmental cues. A complex integration of kinase cascades, energy levels, redox potential, and other factors regulates SG/PB assembly and composition, and our understanding of these interactions is far from complete. Study of these fascinating and dynamic structures has just begun—more surprises and connections will doubtless emerge.

REFERENCES

Anderson, P., and Kedersha, N. (2002). Stressful initiations. *J. Cell Sci.* **115**, 3227–3234.

Anderson, P., and Kedersha, N. (2006). RNA granules. *J. Cell Biol.* **172**, 803–808.

Baez, M. V., and Boccaccio, G. L. (2005). Mammalian Smaug is a translational repressor that forms cytoplasmic foci similar to stress granules. *J. Biol. Chem.* **280**, 43131–43140.

Collier, N., and Schelsinger, M. J. (1986). The dynamic state of heat shock proteins in chicken embryo fibroblasts. *J. Cell Biol.* **103**, 1495–1507.

Cougot, N., Babajko, S., and Seraphin, B. (2004). Cytoplasmic foci are sites of mRNA decay in human cells. *J. Cell Biol.* **165**, 31–40.

Dang, Y., Kedersha, N., Low, W. K., Romo, D., Gorospe, M., Kaufman, R., Anderson, P., and Liu, J. O. (2006). Eukaryotic initiation factor 2alpha-independent pathway of stress granule induction by the natural product pateamine A. *J. Biol. Chem.* **281**, 32870–32878.

Eystathioy, T., Chan, E. K., Tenenbaum, S. A., Keene, J. D., Griffith, K., and Fritzler, M. J. (2002). A phosphorylated cytoplasmic autoantigen, GW182, associates with a unique population of human mRNAs within novel cytoplasmic speckles. *Mol. Biol. Cell* **13**, 1338–1351.

Eystathioy, T., Jakymiw, A., Chan, E. K., Seraphin, B., Cougot, N., and Fritzler, M. J. (2003). The GW182 protein colocalizes with mRNA degradation associated proteins hDcp1 and hLSm4 in cytoplasmic GW bodies. *RNA* **9**, 1171–1173.

Fenger-Gron, M., Fillman, C., Norrild, B., and Lykke-Andersen, J. (2005). Multiple processing body factors and the ARE binding protein TTP activate mRNA decapping. *Mol. Cell* **20**, 905–915.

Fillman, C., and Lykke-Andersen, J. (2005). RNA decapping inside and outside of processing bodies. *Curr. Opin. Cell Biol.* **17**, 326–331.

Gilks, N., Kedersha, N., Ayodele, M., Shen, L., Stoecklin, G., Dember, L. M., and Anderson, P. (2004). Stress granule assembly is mediated by prion-like aggregation of TIA-1. *Mol. Biol. Cell* **15**, 5383–5398.

Hofmann, I., Casella, M., Schnolzer, M., Schlechter, T., Spring, H., and Franke, W. W. (2006). Identification of the junctional plaque protein plakophilin 3 in cytoplasmic particles containing RNA-binding proteins and the recruitment of plakophilins 1 and 3 to stress granules. *Mol. Biol. Cell* **17**, 1388–1398.

Hua, Y., and Zhou, J. (2004). Survival motor neuron protein facilitates assembly of stress granules. *FEBS Lett.* **572**, 69–74.

Ingelfinger, D., Arndt-Jovin, D. J., Luhrmann, R., and Achsel, T. (2002). The human LSm1-7 proteins colocalize with the mRNA-degrading enzymes Dcp1/2 and Xrnl in distinct cytoplasmic foci. *RNA* **8**, 1489–1501.

Iseni, F., Garcin, D., Nishio, M., Kedersha, N., Anderson, P., and Kolakofsky, D. (2002). Sendai virus trailer RNA binds TIAR, a cellular protein involved in virus-induced apoptosis. *EMBO J.* **21**, 5141–5150.

Kedersha, N., Chen, S., Gilks, N., Li, W., Miller, I. J., Stahl, J., and Anderson, P. (2002). Evidence that ternary complex (eIF2-GTP-tRNA(i)(Met))-deficient preinitiation complexes are core constituents of mammalian stress granules. *Mol. Biol. Cell* **13**, 195–210.

Kedersha, N., Stoecklin, G., Ayodele, M., Yacono, P., Lykke-Andersen, J., Fitzler, M., Scheuner, D., Kaufman, R., Golan, D. E., and Anderson, P. (2005). Stress granules and processing bodies are dynamically linked sites of mRNP remodeling. *J. Cell Biol.* **169**, 871–884.

Kedersha, N. L., Gupta, M., Li, W., Miller, I., and Anderson, P. (1999). RNA-binding proteins TIA-1 and TIAR link the phosphorylation of eIF-2a to the assembly of mammalian stress granules. *J. Cell Biol.* **147**, 1431–1441.

Kim, W. J., Back, S. H., Kim, V., Ryu, I., and Jang, S. K. (2005). Sequestration of TRAF2 into stress granules interrupts tumor necrosis factor signaling under stress conditions. *Mol. Cell Biol.* **25**, 2450–2462.

Mazroui, R., Huot, M. E., Tremblay, S., Filion, C., Labelle, Y., and Khandjian, E. W. (2002). Trapping of messenger RNA by Fragile X Mental Retardation protein into cytoplasmic granules induces translation repression. *Hum. Mol. Genet.* **11**, 3007–3017.

Mazroui, R., Sukarieh, R., Bordeleau, M. E., Kaufman, R. J., Northcote, P., Tanaka, J., Gallouzi, I., and Pelletier, J. (2006). Inhibition of ribosome recruitment induces stress granule formation independently of eukaryotic initiation factor 2alpha phosphorylation. *Mol. Biol. Cell* **17**, 4212–4219.

McEwen, E., Kedersha, N., Song, B., Scheuner, D., Gilks, N., Han, A., Chen, J. J., Anderson, P., and Kaufman, R. J. (2005). Heme-regulated inhibitor (HRI) kinase-mediated phosphorylation of eukaryotic translation initiation factor 2 (eIF2) inhibits translation, induces stress granule formation, and mediates survival upon arsenite exposure. *J. Biol. Chem.* **280**, 16925–16933.

McInerney, G. M., Kedersha, N. L., Kaufman, R. J., Anderson, P., and Liljestrom, P. (2005). Importance of eIF2alpha phosphorylation and stress granule assembly in alpha-virus translation regulation. *Mol. Biol. Cell* **16**, 3753–3763.

Nover, L., Scharf, K. D., and Neumann, D. (1989). Cytoplasmic heat shock granules are formed from precursor particles and are associated with a specific set of mRNAs. *Mol. Cell Biol.* **9,** 1298–1308.

Sheth, U., and Parker, R. (2003). Decapping and decay of messenger RNA occur in cytoplasmic processing bodies. *Science* **300,** 805–808.

Srivastava, S. P., Kumar, K. U., and Kaufman, R. J. (1998). Phosphorylation of eukaryotic translation initiation factor 2 mediates apoptosis in response to activation of the double-stranded RNA-dependent protein kinase. *J. Biol. Chem.* **273,** 2416–2423.

Stoecklin, G., Mayo, T., and Anderson, P. (2006). ARE-mRNA degradation requires the 5′–3′ decay pathway. *EMBO Rep.* **7,** 72–77.

Stoecklin, G., Stubbs, T., Kedersha, N., Blackwell, T. K., and Anderson, P. (2004). MK2-induced tristetraprolin:14-3-3 complexes prevent stress granule association and ARE-mRNA decay. *EMBO J.* **23,** 1313–1324.

Teixeira, D., Sheth, U., Valencia-Sanchez, M. A., Brengues, M., and Parker, R. (2005). Processing bodies require RNA for assembly and contain nontranslating mRNAs. *RNA* **11,** 371–382.

Tourriere, H., Chebli, K., Zekri, L., Courselaud, B., Blanchard, J. M., Bertrand, E., and Tazi, J. (2003). The RasGAP-associated endoribonuclease G3BP assembles stress granules. *J. Cell Biol.* **160,** 823–831.

van Dijk, E., Cougot, N., Meyer, S., Babajko, S., Wahle, E., and Seraphin, B. (2002). Human Dcp2: A catalytically active mRNA decapping enzyme located in specific cytoplasmic structures. *EMBO J.* **21,** 6915–6924.

Wilczynska, A., Aigueperse, C., Kress, M., Dautry, F., and Weil, D. (2005). The translational regulator CPEB1 provides a link between dcp1 bodies and stress granules. *J. Cell Sci.* **118,** 981–992.

Yang, F., Peng, Y., Murray, E. L., Otsuka, Y., Kedersha, N., and Schoenberg, D. R. (2006). Polysome-bound endonuclease PMR1 is targeted to stress granules via stress-specific binding to TIA-1. *Mol. Cell Biol.* **26,** 8803–8813.

Yu, J. H., Yang, W. H., Gulick, T., Bloch, K. D., and Bloch, D. B. (2005). Ge-1 is a central component of the mammalian cytoplasmic mRNA processing body. *RNA* **11,** 1795–1802.

CHAPTER SIX

METHODS TO ANALYZE MICRORNA-MEDIATED CONTROL OF MRNA TRANSLATION

Jennifer L. Clancy,* Marco Nousch,* David T. Humphreys,*
Belinda J. Westman,[1] Traude H. Beilharz,*,[†] *and*
Thomas Preiss*,[†],[‡]

Contents

* Molecular Genetics Program, Victor Chang Cardiac Research Institute (VCCRI), Sydney, Australia
† St. Vincent's Clinical School, University of New South Wales, Sydney, Australia
‡ School of Biotechnology & Biomolecular Sciences, University of New South Wales, Sydney, Australia
1 Division of Gene Regulation & Expression, School of Life Sciences, Wellcome Trust Biocentre, University of Dundee, Dundee, United Kingdom.

Methods in Enzymology, Volume 431
ISSN 0076-6879, DOI: 10.1016/S0076-6879(07)31006-9

Abstract

MicroRNAs (miRs) are an important class of gene regulators that affect a wide range of biological processes. Despite the early recognition of miRs as translational regulators and intense interest in studying this phenomenon, it has so far not been possible to derive a consensus model for the underlying molecular mechanism(s). The potential of miRs to act in a combinatorial manner and to also promote mRNA decay creates conceptual and technical challenges in their study. Here, we discuss critical parameters in design and analysis of experiments used to study miR function including creation of synthetic miR and mRNA partners for assay of translational inhibition using luciferase reporters; measurement of mRNA stability after miR action; defining poly(A) tail length in miR target mRNA; determining the distribution of miRs and their target mRNAs in polysome profiles; and visualization of P-body components. We describe protocols for each of these procedures.

1. INTRODUCTION

MicroRNAs (miRs) have quickly become known as a large class of versatile gene regulators. For example, the miRBase Sequence Database Release 9.0 (http://microrna.sanger.ac.uk/) lists 474 distinct miR species encoded by the human genome, and bioinformatic analyses predict that around 1/3 of all human protein-coding genes are targets of miR regulation (Bartel and Chen, 2004; Lewis *et al.*, 2003). miRs are ~22nt sized noncoding RNAs derived from longer, stem-loop forming precursors. They can affect gene transcription (Kim *et al.*, 2006) but so far, their role in post-transcriptional gene regulation has received most attention. miRs recruit ribonucleoprotein complexes (termed miRNP or RISC) to target mRNAs by virtue of base complementarity. A precise match triggers endonucleolytic cleavage of the mRNA, while imperfect base-pairing typically results in repression of protein synthesis (for a review, see Engels and Hutvagner, 2006; Humphreys *et al.*, 2007; Pillai, 2005; Valencia-Sanchez *et al.*, 2006). At present, the expectation is that most animal miRs will exhibit partial matches to sequences in the 3′ untranslated region of their target mRNAs. Several studies have addressed the mechanism of miR action in such a configuration, yielding data to support conflicting models. Thus, miRs have been proposed to inhibit initiation (Humphreys *et al.*, 2005; Pillai *et al.*, 2005) or post-initiation stages of translation (Maroney *et al.*, 2006; Petersen *et al.*, 2006). Other evidence suggests that they stimulate degradation of the nascent polypeptide (Nottrott *et al.*, 2006;

Olsen and Ambros, 1999). Furthermore, miRs can stimulate deadenylation (Giraldez *et al.*, 2006; Wu *et al.*, 2006) and exonucleolytic mRNA decay (Bagga *et al.*, 2005; Giraldez *et al.*, 2006; Lim *et al.*, 2005). Finally, miRs, components of the miRNP, and their mRNA targets are found in processing (P-) bodies (Liu *et al.*, 2005a,b; Pillai, 2005; Sen and Blau, 2005) and stress granules (Leung *et al.*, 2006), cytoplasmic foci related to mRNA degradation and/or translational inhibition.

Careful experimentation will be required for a consensus model to emerge and this challenging subject matter is catching the attention of a growing number of investigators who apply techniques from translational control research in model systems from diverse species including *D. melanogaster*, *C. elegans*, *A. thaliana*, and *H. sapiens*. Here, we describe methods to study miR action in mammalian cells, including use of reporter constructs to measure translational inhibition, methods for measuring target mRNA stability and/or deadenylation, localization of miRs and miRNP components in polysome profiles, and visualization of P-bodies.

2. Systems to Measure miR-Mediated Repression

At present, mRNA targets for most miRs have been assigned only on the basis of bioinformatic predictions. Thus, a common experimental task is to verify predicted miR/mRNA pairings, which is typically done by the use of luciferase reporter assays. It involves cloning either the full-length 3′ UTR of an endogenous mRNA, or (repeated) fragments thereof, into the 3′ region of a luciferase reporter plasmid and measuring the expression of the construct in transfected cells. Tests for effects of the miR under study can be done either in a cell line that expresses this miR [in this case, a wild-type UTR plasmid is compared to a mutated companion construct (Lewis *et al.*, 2003; Pillai *et al.*, 2005)] or in a cell line that does not express the miR. In the latter case, the miR is chemically synthesized, prepared as a duplex with its passenger strand, and transfected into cells together with the reporter plasmid (Doench *et al.*, 2003; Humphreys *et al.*, 2005; Petersen *et al.*, 2006). Many studies addressing the miR mechanism have employed this strategy, often using a completely synthetic pairing of an artificial miR and a reporter plasmid harboring multiple, concatenated binding sites designed to mimic the mismatched base pairing typical of known endogenous miR/mRNA pairings (Doench *et al.*, 2003; Pillai *et al.*, 2005). A related approach, with certain advantages as outlined later, is to synthesize target mRNAs by *in vitro* transcription and transfect cells directly with an mRNA/miR mix.

2.1. Co-transfection of synthetic miRs and target plasmids into HeLa cells

The following procedure works well for the synthetic system developed by Sharp and colleagues (Doench *et al.*, 2003). It uses a Renilla luciferase plasmid (based on pRL-TK, Promega, WI) harboring 4 concatenated, partially mismatched target sites for the artificial CXCR4 miR in its 3' UTR (p-RL-TK-4 sites).

Procedure:

1. Seed HeLa cells into 24 well plates at least 6 h prior to transfection. The aim is to have cells at 70% confluency at the time of transfection.
2. Prepare transfection mix A that contains 1 μg of Renilla luciferase target plasmid (e.g., p-R-luc-4 sites) and 100 μl of optiMEM (Invitrogen, CA). This is enough transfection mixture for 1 well, and can be scaled up for multiple transfections. For cultures that are to receive miR duplex, add 2 nM (calculated for a final volume of 500 μL for each well).
3. For each well, prepare transfection mix B, containing 45 ng of the control Firefly luciferase (F-luc) plasmid pGL3 (Promega), 1 μl Lipofectamine 2000 (Invitrogen), and 100 μl of optiMEM.
4. Mix equal volumes of transfection mixes A and B and leave at room temperature for at least 20 min.
5. Wash plated cells with PBS and replace with 300 μl of optiMEM.
6. Add 200 μl of transfection A+B mix to each well and return to incubator for 24 h before harvest.
7. Cells are lysed for Firefly and Renilla luciferase assays using the Dual-Luciferase Reporter Assay system (Promega), following the manufacturer's instructions. We use a multimode microplate reader with automatic injectors (FLUOROstar Optima from BMG Labtech, Offenburg, Germany) for luminescence measurements.

Comments: The typical variables of transient transfection experiments may need to be adjusted, particularly if other cell lines are used. A variety of dedicated RNA transfection reagents are also on the market; these may be tried if results are not satisfactory. Critical parameters for repression are the dosage of, and ratio between, miR and target plasmid; these should be systematically optimized. As the target plasmid used here contains the relatively weak TK promoter, plasmids with stronger promoters may need to be dosed at lower amounts. There is a nonlinear relationship between the number of miR target sites and the magnitude of repression (Doench *et al.*, 2003). Increasing the number of target sites may therefore give much improved repression. We also recommend the routine use of specificity controls (e.g., a plasmid without target sites and an unrelated miR duplex).

Variations: If a cell line is available that expresses the miR under study, then transfections with a plasmid exhibiting wild-type 3' target sites may be compared to parallel transfections with a plasmid in which these sites have been mutated (e.g., in the "seed" region; Lewis et al., 2003; Pillai et al., 2005). Alternatively, inhibition of endogenous miR activity can be gained by using antisense 2′-O-methyl oligonucleotides (Bhattacharyya et al., 2006; Hutvagner et al., 2004). The advantage of these approaches is that repression by an endogenous miR is measured. Results can, however, be complicated by problems with varying plasmid preparation quality or unexpected effects of the target site mutations on mRNA metabolism.

2.2. mRNA transfection into HeLa cells

The advantage of mRNA over plasmid transfection is the ability of *in vitro* transcription to allow precise control over features contained within the mRNA (Humphreys et al., 2005; Pillai et al., 2005; Westman et al., 2005). For example, mRNA can be prepared either with or without the physiological $m^7G(5′)ppp(5′)G$ cap structure and 3′ poly(A) tail, which are important mediators of canonical translation initiation (Gallie, 1991; Hentze et al., 2006; Iizuka et al., 1994; Kahvejian et al., 2005; Tarun and Sachs, 1995).

The first step in creating a specific mRNA species is to construct a plasmid encoding the mRNA behind a T7, T3, or SP6 RNA polymerase promoter sequence. The 5′ boundary of the mRNA is defined by the position of the promoter transcription start site and the 3′ end is defined by restriction digest of the plasmid prior to *in vitro* transcription.

Procedure:

1. Linearize plasmid by restriction digest.
2. Check linearization by gel electrophoresis.
3. Purify linearized plasmid by excision of the band from the gel (optional) and phenol/chloroform extraction.
4. Assemble a 20 μl *in vitro* transcription reaction using the T7, T3, or SP6 megascript kit (Ambion, Cambridgeshire , UK): 2 μl ATP 75 mM, 2 μl CTP 75 mM, 2 μl UTP 75 mM, 2 μl GTP 15 mM, 2 μl of a 40 mM cap analogue solution (Kedar Inc, Warsaw, Poland), 2 μl 10× reaction buffer, 1 μg linearized pDNA, 2 μl RNA polymerase.
5. Incubate the reaction 2 to 4 h at 37°.
6. Add 1 μl of DNAse 1 (2 U/μl) and incubate the reaction at 37° for 15 min to remove template DNA.
7. At this stage, mRNA can be polyadenylated using the poly(A) tailing kit (Ambion), according to the manufacturer's instructions.
8. RNA is then purified using column purification (we use the Ambion MegaClear kit, following the manufacturer's instructions).
9. mRNA size is confirmed and integrity is measured by gel electrophoresis or microfluidics using the RNA 6000 Nano LabChip® Kit on an Agilent 2100 Bioanalyzer (Agilent Technologies, CA).

10. The transfection of mRNA uses the same procedure as for that of plasmid (see previously), except that for a 24 well plate, 20 ng of R–luc-4 sites mRNA and 80 ng of F-luc (control) mRNA are used per well in place of plasmid (the amount of miR remains the same, 2 n*M* in 500 μl).

2.2.1. Normalization of data

Co-transfection of an unrelated reference construct is a common practice to compensate for variation in transfection efficiency between cultures. In the systems described here, we use Renilla luciferase as a reporter for miR-mediated repression, and co-transfect Firefly luciferase as a reference for transfection efficiency. Renilla luciferase activity from each transfection is normalized to the corresponding Firefly luciferase measurement. Repression by a miR is then calculated by dividing the normalized Renilla luciferase activity without miR by the normalized Renilla luciferase activity in the presence of miR.

The commercial dual luciferase kit protocol requires that the Firefly enzyme be measured first, followed by a quenching step and recording of the Renilla enzyme activity. Carryover of residual Firefly signal into the Renilla measurements is not normally a problem but may require attention with Renilla luciferase constructs giving very low levels of enzymatic activity (e.g., IRES-driven mRNAs; see later). It is thus desirable to transfect only the minimal amount of Firefly reference construct required for accurate measurements. Each transfection series should include a negative control experiment where cells are transfected only with the Firefly luciferase construct. As a compromise, any significant Renilla signal from this negative control can then be deducted from all other Renilla luciferase measurements in the series.

Example experiments using the previous methodologies are shown in Fig. 6.1. The major mRNA constructs described in this chapter are diagrammatically represented in Fig. 6.1B and an example of *in vitro* transcribed and polyadenylated R–luc-4 sites mRNA is shown in Fig. 6.1A. In these experiments, translation of R–luc-4 sites mRNA is synergistically promoted by the physiological cap structure and the poly(A) tail (Fig. 6.1C), and full miR-dependent translational repression requires the presence of both modifications (Fig. 6.1D, Humphreys *et al.*, 2005). (The EMCV IRES-containing constructs are discussed later.)

Comment: It is critical to remove excess cap analogue from *in vitro* transcribed transcripts prior to transfection, because the cap analogue will compete with transcripts for the cellular translational machinery. Also, even after DNase treatment, the RNA sample may still contain traces of functional pDNA, which may interfere with subsequent detection by RT-PCR. Furthermore, plasmids containing a mammalian promoter may even give rise to *de novo* transcription in transfected cells.

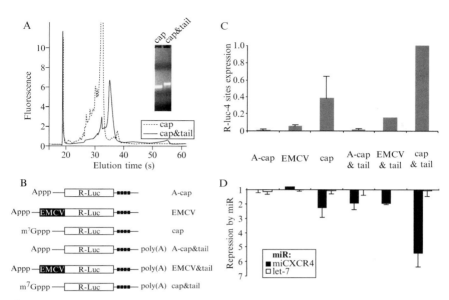

Figure 6.1   mRNA transfection of HeLa cells to investigate miR-mediated repression of translation (as detailed in text). (A) Quality control of *in vitro* transcribed R-luc mRNAs. Aliquots of "cap" and "cap&tail" R-luc-4 sites mRNA were analyzed on a Nano LabChip® using the Agilent 2100 Bioanalyzer (Agilent Technologies). mRNA were also analyzed by 1% agarose gel electrophoresis and stained with ethidium bromide (insert). Both analysis methods indicate good quality mRNA preparations (one major band/peak; size shift after polyadenylation). (B) Schematic of the *in vitro* transcribed R-luc-4 sites mRNAs. The 5′ end of the mRNAs was modified with either a physiological cap (m⁷Gppp) or with a "blocked" A(5′)ppp(5′)G cap (Appp). Addition of a poly(A) tail at the 3′ end is indicated, as is presence of a EMCV IRES sequence in the 5′ UTR. (C) Expression of each R-luc-4 sites mRNA in the absence of a miR after normalization against the signal from the co-transfected Firefly luciferase mRNA and arbitrarily setting the expression level of R-luc-4 sites cap&tail mRNA to 1.0. (D) Repression of translation from the different R-luc-4 sites mRNAs by the specific miCXCR4 (filled bars) or nonspecific let-7 miR (open bars). Repression was calculated as the fold difference in R-luc activity upon addition of miR. Averaged results from three to five experiments are shown with standard deviation. Data in panels C and D were previously published (Humphreys *et al.*, 2005). Copyright *PNAS*, reprinted with permission.

Variations: The poly(A) tailing kit (Ambion) produces a mRNA population with varying lengths of poly(A) tails, controlled by altering poly(A) polymerase concentrations and incubation times. An alternate method to incorporate a poly(A) tail is to clone a defined stretch of adenosines/thymidines into the 3′ UTR of the template pDNA. To allow transcripts to finish on an adenosine, the insert should be followed by a restriction site for an enzyme that cleaves 5′ of the last antisense strand thymidine, such as Nsi I. In this way, the poly(A) tail can be incorporated directly into the mRNA in one step via the *in vitro* transcription reaction. A limitation of this method is the length (<150 nt) of adenosines/thymidines stretch that can be reliably propagated in plasmids (Preiss *et al.*, 1998). Another potential option for studying the biochemical mechanism of miR action are *in vitro* translation reactions. Many researchers have pursued this option, largely without published success thus far. A 2006 publication using the classical rabbit reticulocyte lysate system reports successful reconstitution of the process *in vitro*, suggesting that the ice has now been broken (Wang *et al.*, 2006).

2.3. Systematic differences between plasmid and mRNA transfections

The time-course of reporter enzyme accumulation differs considerably between DNA and mRNA transfections. Plasmid DNA (pDNA) transfections generate a sustained production of mRNA molecules in the cell for the lifetime of the plasmid. In contrast, mRNA transfections supply only a single burst of mRNA into the cell. It is important to identify the amount of transfected mRNA that will give robust signal without saturating the translational machinery (Fig. 6.2A). We transfect 20 ng of R-luc-4 sites mRNA per well (24 well plate). In reporter systems where exogenous miR is to be co-transfected, it is also important to pre-determine the optimal concentration of miR in the transfection mix to use. For mRNA transfections, we suggest testing miRs in a final concentration range of 0.2–10 nM. For the CXCR4 miR, we find 2 nM to be the optimal concentration (Fig. 6.2B).

As pDNA and mRNA transfection differ in both the timing of mRNA expression and the gross amount of mRNA delivered to the cell, it is important to identify a suitable time point to measure miR-mediated repression. We observe that at any time point after transfection, pDNA transfections have higher measurable levels of miR-mediated repression compared to mRNA transfections (Fig. 6.2C). This difference may, in part, reflect a time lag of active miR-protein-complex formation relative to the onset of translation of the transfected Renilla luciferase mRNA. For single time point experiments, we decided to measure miR-mediated repression in mRNA and DNA transfections at 16 and 24 h, respectively.

2.4. Internal ribosome entry sites

Viral IRES elements can be useful tools to identify the sub-step in translation targeted by a given regulator (Jackson, 2005; Ostareck *et al.*, 2001). The IRES elements listed in Table 6.1 are of particular interest, because careful biochemical and structural analyses have defined different initiation factor requirements for translation in each case (Borman and Kean, 1997; Fraser and Doudna, 2006; Hellen and Sarnow, 2001; Poyry *et al.*, 2004). We and others have employed viral IRES-containing reporter constructs to

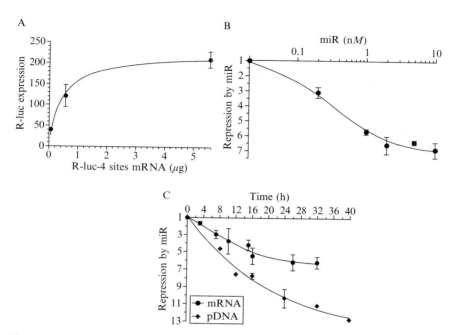

Figure 6.2 Critical parameters of the miR/mRNA co-transfection method. (A) Titration of mRNA amount. HeLa cells were transfected with increasing amounts of cap & tail R–luc-4 sites mRNA and a fixed amount of firefly (F-luc) mRNA. R-luc expression (luciferase activity) was measured 5 h after transfection. (B) Titration of miCXCR4 concentration. HeLa cells were transfected with cap&tail R–luc-4 sites mRNA, F-luc mRNA, and varying concentrations of miCXCR4. Luciferase activity was measured 16 h after transfection and fold-repression by the miR was calculated as in Fig. 6.1D (C) Time-course of miR-mediated repression. HeLa cells were co-transfected with cap&tail R–luc-4 sites and F-luc (control) mRNAs, either with or without miCXCR4, and harvested at different time points. Repression was calculated as detailed in Fig. 6.1 and plotted against time (mRNA transfection data series depicted by the circles). Analogous plasmid DNA transfections are shown for reference (pDNA, diamonds). Averaged results from several experiments are shown with standard deviation. Data were previously published (Humphreys *et al.*, 2005). Copyright *PNAS*, reprinted with permission.

investigate which sub-step of translation is targeted by a miR (Humphreys *et al.*, 2005; Petersen *et al.*, 2006; Pillai *et al.*, 2005), reaching, in part, dramatically different conclusions. The basic premise, however, was similar in each case, namely, that if the miR targets a component of the translation machinery not required for IRES-mediated translation, then translation driven by the IRES should no longer be repressed by the miR.

In our work, we opted to deploy the IRES sequences within the 5′ UTR of mono-cistronic reporter mRNA (Humphreys *et al.*, 2005; Fig. 6.1B), which were directly transfected into HeLa cells. IRES-containing transcripts were further capped with the nonphysiological A(5′)ppp(5′)G cap structure,

Table 6.1 Translation initiation factor requirements of internal ribosome entry site (IRES) elements

IRES	Mode of initiation
Hepatitis A Virus (HAV)	All eIFs required
Encephalomyocarditis Virus (EMCV)	Independent of eIF4E
Classical Swine Fever Virus (CSFV) or Hepatitis C Virus (HCV)	Independent of eIF4E, -4G, -4A, -4B
Cricket Paralysis Virus (CrPV), intergenic region	Factorless initiation from a CGU codon placed in the aminoacyl site of the small ribosomal subunit

which is inactive in translation but protects against accelerated decay (Bergamini *et al.*, 2000). This approach avoids potential problems due to interference of canonical, cap-dependent translation with genuine IRES-mediated translation on the same mRNA template. It is also free of most complications that have been observed with the common bi-cistronic reporter approach (Bert *et al.*, 2006).

The efficiency of translation driven by IRES elements is often significantly lower than that of canonical translation (e.g., Fig. 6.1C) and, thus, control experiments may be needed to demonstrate that genuine IRES activity is measured. This can be done by inserting a stable stem-loop structure into the mRNA 5′ UTR, upstream of the IRES element, or by comparing the translational output of a reporter mRNAs with the wild-type IRES to that of the equivalent transcript harboring a mutated, inactive IRES element (which should be much lower; Humphreys *et al.*, 2005; Wilson *et al.*, 2000).

In the example experiment shown in Fig. 6.1D, R–luc-4 sites mRNA with an EMCV IRES encoded in the 5′ UTR (and a nonphysiological A–cap) were synthesized with and without a poly(A) tail. In HeLa cells, miCXCR4 did not repress translation from the EMCV construct lacking a poly(A) tail and repressed translation from the EMCV&tail construct to the same degree as a the A–cap&tail construct. This suggests that EMCV-driven translation is not repressed by miRs and any repression observed in the EMCV&tail construct was due to the function of the poly(A) tail alone. Previously, we have also shown that miRs cannot repress translation driven by the CrPV IRES (Humphreys *et al.*, 2005). In this way, we have obtained evidence to suggest that a miR can affect translation initiation by inhibiting the roles of eIF4E, the mRNA cap structure, and poly(A) tail (Humphreys *et al.*, 2005; Fig. 6.1).

3. DETECTION OF miR-MEDIATED CHANGES IN mRNA STABILITY

A further aspect of miR function is that, depending on the model system utilized, effects on target mRNA stability can range from minor to severe (Bagga *et al.*, 2005; Behm-Ansmant *et al.*, 2006; Giraldez *et al.*, 2006; Humphreys *et al.*, 2005; Lim *et al.*, 2005; Wu *et al.*, 2006). Most studies of the mechanism by which miRs affect translation have gone to some length to show that in their model system, miR-mediated acceleration of mRNA decay is either absent or quantitatively minor compared to the observed reduction in protein synthesis (Bhattacharyya *et al.*, 2006; Humphreys *et al.*, 2005; Petersen *et al.*, 2006; Wang *et al.*, 2006). Commonly used methods are northern blotting, RNAse protection assay (RPA), conventional RT-PCR, and quantitative (q)PCR. Here, we describe measurement of mRNA levels using either RPA, or a method that infers the functional stability of mRNA from the time course of protein accumulation after direct mRNA transfection. A qPCR method is described in the section on polysome profiles.

3.1. RNA extraction

We typically use RNA purification by the Trizol method (Invitrogen, following the manufacturer's instructions), which has the advantage over column-based methods that it can purify small amounts of RNA and retain miRs. Purified RNA is dissolved in RNAse-free water and stored at $-80°$. RNA quality is assessed on an Agilent Bioanalyzer (Agilent Technologies) or by gel electrophoresis.

DNAse 1 treatment may be required to reduce both genomic and plasmid DNA contamination in the RNA sample. We suggest using the Turbo DNA-free kit (Ambion, following the manufacturer's protocol in the stringent version).

3.2. RNase protection assay

The RPA is a sensitive method for quantifying specific RNAs from a mixture of RNAs. This is achieved using a small-volume hybridization of an RNA probe to the RNA under study. Unhybridized probe and sample is then digested with RNAses and the protected probe fragment is visualized after denaturing gel electrophoresis. Commonly, the probe is radiolabeled for maximum sensitivity. Following is a method for RPA detection of R-luc-4 sites and F-luc mRNA.

Procedure:

a. Preparation of riboprobes: Probes for RPA should be between 100 and 900 bases and are commonly created by *in vitro* transcription in

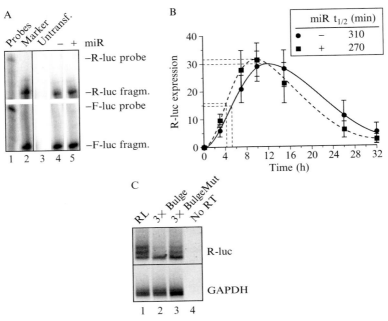

Figure 6.3 Measurement of miR-dependent changes to mRNA stability and deadenylation. (A) HeLa cells were co-transfected with an R-luc-4 sites mRNA, miCXCR4 (as specified above the lanes), and the F-luc mRNA. Total RNA was isolated 7 h after transfection, and 1 μg aliquots were subjected to dual-probe ribonuclease protection assays. Undigested probe mix (lane 1), a model reaction with input R-luc and F-luc mRNAs (marker, lane 2), and total RNA from untransfected HeLa cells (lane 3) are shown for comparison. Positions of probes and protected fragments are indicated on the right. (B) Functional half-life measurements of R-luc-4 sites mRNA. HeLa cells were transfected with cap&tail R-luc-4 sites mRNA, F-luc mRNA with or without miCXCR4, and cells were harvested 0 to 32 h post-transfection. R-luc activity was not normalized to F-luc but made relative to the total R-luc activity measured in each time series. The functional half-life ($t_{1/2}$) of the mRNA was calculated as the time taken to produce 50% of the maximal amount of luciferase activity. (C) Determination of polyadenylation state of miR-targeted transcripts by PAT assay. HeLa cells were transfected with plasmids encoding R-luc with no (RL), three functional (3× Bulge), or three mutant (3× BulgeMut) let-7 binding sites in the 3′ UTR (Pillai, 2005) and harvested after 24 h. Shown are LM-PAT products for both R-luc and GAPDH (control) derived from these transfections (including a RT negative control, final lane). Data in panels A and B have been previously published (Humphreys et al., 2005). Copyright PNAS, reprinted with permission.

the presence of α-^{32}P–UTP (Amersham Pharmacia Biosciences). To generate probes to detect R–luc-4 sites and F-luc mRNA (as shown in Fig. 6.3A), regions of F-luc and R–luc-4 sites sequences corresponding to nucleotides 504 to 762 and 35 to 264 (position one refers to the start of the luciferase coding region), respectively, were amplified with the primers TAAGACGACTCGAAATCCA (F-lucF), CACATCTCAT-CTACCTCC (F-lucR), AACGGATGATAACTGGTCC (R-lucF), and ACCAGATTTGCCTGATTTG (R-lucR). PCR products were cloned into the pGEM-T Easy Vector (Promega) and confirmed by sequencing. The sizes of RNA probes and protected fragments after RNase digestion were as follows: F-luc, 200/136 nt; R-luc-4 sites RNA, 324/230 nt. Probes for R-luc-4 sites and F-luc mRNA were transcribed by T7 RNA polymerase (see section on mRNA transfection into HeLa cells). The probe must be full-length; therefore, probe size should be confirmed by gel electrophoresis (gel purification is optional).

b. Protection assay: Dual probe RPA assay was performed using the RPA III kit (Ambion) and is outlined below (also see product insert).

1. Incubate ≥ 1 μg of RNA with a 3- to 10-fold molar excess of radio-labeled probe (about 150 to 600 pg per 10 μg total RNA) in hybridization buffer.
2. Co-precipitate probe and RNA with 1/10th volume 5 M NH_4OAc and 2.5 volumes ethanol. Precipitate at $-20°$ for at least 15 min.
3. Centrifuge samples at maximum speed for 15 min at 4° and remove supernatant.
4. Air dry pellet and resuspend in 10 μl hybridization buffer.
5. Denature samples at 95° for 3 min, vortex, and briefly centrifuge to collect sample at bottom of tube.
6. Hybridize sample at 42° overnight (preferably in an oven to eliminate condensation). The optimal hybridization temperature must be empirically determined and is usually between 40 and 60°; however, most probes will hybridize at 42 to 45°.
7. Create an RNAse working buffer. RNAse T1 and/or RNAse A can be used at various concentrations. In Fig. 6.3A, RNAse T1 was diluted 1:10 in RNAse digestion buffer.
8. Add 150 μl of RNAse solution to each sample and incubate at 37° for 30 min.
9. Inactivate by precipitating with 225 μl of inactivation solution and incubate at $-20°$ for 15 min.
10. Centrifuge samples at maximum speed for 15 min at 4° and remove supernatant.
11. Resuspend sample and detect protected probe fragments by denaturing polyacrylamide gel electrophoresis and autoradiography.

An example experiment is shown in Fig. 6.3A. Here, a dual probe RPA was performed on 1 μg of total RNA purified from cells transfected with cap&tail R-luc-4 sites and F-luc mRNA, with and without miCXCR4. The assay was performed on total RNA from untransfected cells to test for specificity. The level of R-luc-4 sites mRNA was not significantly different between cells transfected with miCXCR4 (+) and those not (−).

Comments: There are several suggested controls for this assay, including use of yeast total RNA as a negative control (check for probe species specificity) and a no RNAse control to determine probe stability. In Fig. 6.3A, the positive control marker lane was produced by addition of R-luc-4 sites or F-luc mRNA only to the assay. Also, optimal times for RNAse digestion will vary from probe to probe. In addition, for maximum sensitivity a probe with high specific activity is preferable (yet still in molar excess to the mRNA).

3.3. Measurement of mRNA functional half-life

A potential complication of direct reporter mRNA transfection experiments is that a proportion of the mRNA that was added to the cell culture may remain sequestered in an inert state and contaminate RNA samples extracted from the transfected cells. In principle, inert subpopulations of reporter mRNA may stick to the outside of cells, the plastic surfaces of culture dishes, or may even be sequestered within vesicles inside transfected cells (Barreau *et al.*, 2006). This may lead to erroneous results with any method (Northern blotting, RPA, or RT-PCR) that relies on detection of the physical stability of mRNA. In our work, we have therefore additionally employed an indirect method that infers the functional stability of transfected mRNA (Gallie, 1991) to complement our physical stability measurements (Humphreys *et al.*, 2005). It takes advantage of the fact that mRNA transfection delivers a single cohort of transcripts to the cells, which become engaged in translation and decay in a synchronous manner. Thus, the accumulation of reporter protein over time is a function of reporter protein stability, as well as the efficiency of translation and the stability of only those mRNA molecules that entered the cells in a functional state. Assuming that the stability of the completed reporter protein molecules is unchanged between different experimental conditions, it is then possible to generate a measure for the stability of the reporter mRNA by determining the time required for half-maximal accumulation of reporter protein in each condition (referred to as the functional half-life of the mRNA). In our experimental model, we found no evidence of miR-mediated changes in physical stability of the R-luc-4 sites reporter mRNA as assayed by RPA (Fig. 6.3A) or semi-quantitative RT-PCR (Humphreys *et al.*, 2005), and a minor reduction of ~15% in functional half-life (Fig. 6.3B), indicating that the miR had only minor effects on target mRNA stability and, thus, predominantly acted by repressing mRNA translation (Fig. 6.1D).

Comment: Application of this method makes the further assumption that miRNA-mediated repression results from a graded reduction in translation of all functional mRNA molecules, rather than a complete translational shutdown of a subpopulation thereof. It also critically depends on reproducible transfection conditions, because the reporter enzyme activity cannot be normalized against a co-transfected control. Several independent repeat time courses will be necessary for each experimental condition to establish the reproducibility of the measurements.

Variations: There are pros and cons associated with each method used to measure mRNA levels. Experiments involving plasmid transfection or measuring miR effects on an endogenous mRNA cannot normally use the functional half-life methodology. For these situations, Northern blotting is an accurate, if somewhat laborious method. If higher sensitivity is required, then RT-PCR/qPCR or RPA are good choices. For experiments involving direct mRNA transfection, Northern blotting, RT-PCR/qPCR, or RPA may be applied but any sequestered, inert subpopulations of the mRNA may affect the results. Thus, we recommend complementing them with measurements of functional half-life. RT-PCR/qPCR approaches can be significantly affected by plasmid DNA contamination. This is because it can be difficult to implement the typical strategy of choosing an amplicon that stretches across a substantial intron with common plasmid constructs, and complete removal of plasmid DNA by DNAse treatment is difficult to achieve with cellular material from transient plasmid transfections. For plasmids with small introns, it may be possible to design one primer to overlap the exon–exon junction, effectively making the PCR specific for the transcript. A distinct advantage of Northern blotting is that it provides a direct measure the mRNA size and thus controls for (unexpected) effects of mRNA fragmentation. RT-PCR/qPCR or RPA can be adapted to address this issue, for instance, by choosing the amplicon or protected fragment to stretch across the miR–target sites.

4. MEASUREMENT OF MIR TARGET mRNA DEADENYLATION

We and others have previously shown a contribution of the poly(A) tail to miR-mediated translational repression (Humphreys *et al.*, 2005; Wang *et al.*, 2006). Several studies further demonstrate that miRs can stimulate deadenylation of their mRNA targets (Giraldez *et al.*, 2006; Wu *et al.*, 2006). Here, we describe the Ligation-Mediated Poly(A) Test (LM-PAT), a reliable, high-resolution method developed for analysis of transcript adenylation-state (Sallés and Strickland, 1995; Woolstencroft *et al.*, 2006). This assay, being PCR based, has the advantage of being very sensitive, requiring just 1 to 2 μg of input total RNA. In brief, the assay consists of ligating hybridized oligo(dT)$_{12-18}$ primers as well as a specific (dT)$_{12}$-Anchor primer to cover the full poly(A) tail of mRNAs, followed by first strand cDNA synthesis. Aliquots of the cDNA samples are then used in PCR reactions, which amplify a region between a gene-specific primer site in the 3′ UTR and the anchor at the 3′ end of the poly(A) tail. The range of PCR product sizes thus generated reflects the range of poly(A) tail lengths that were present on that mRNA in the original RNA sample. Following is a description of the PAT assay method.

Procedure:

1. DNase1 treat total RNA using the Turbo DNA-free reagent (Ambion).
2. Dilute 1 to 2 μg RNA to a total of 6 μl with dH$_2$O and add 1 μl of 20 ng/μl linker pd(T) 12–18 (GE Healthcare, MA). Incubate at 65° for 5 min to melt any secondary structure.
3. Meanwhile prepare and pre-warm to 42°, a master mix containing for each reaction: 4 μl dH$_2$O; 4 μl 5× superscript reaction buffer; 2 μl 0.1 M DTT; 1 μl dNTP mix (10 mM each); 1 μl 10 mM ATP; 1 μl T4 DNA ligase (New England Biolabs, MA); 0.5 μl RNAsin (Promega, optional).
4. Transfer the denatured RNA to 42° and add 13 μl pre-warmed master mix. Incubate for 30 min.
5. Add 1 μl PAT primer GCGAGCTCCGCGGCCGCGTTTTTTT-TTTTT (200 ng/μl) and transfer to 12° for 2 to 3 h.
6. Return to 42° for 2 min before addition of 1 μl Superscript II (Invitrogen). Incubate at 42° for at least 2 h or overnight.
7. Inactivate the reverse transcriptase by incubation at 70° for 10 min.
8. Dilute reaction 5 fold in dH$_2$O in preparation for PCR.
9. Prepare 25 μl PCR reactions using 5 μl diluted PAT assay as template. Use standard PCR components and buffers supplied with the Fast-start Taq (Roche, Basel, Switzerland). Cycling conditions are: 2 min at 95°; then 30 s each at 95, 60, and 72° for 25 to 35 cycles (template-dependent); 2 min 72°.
10. Run 1/3rd PCR product on 2% high-resolution agarose (Agarose 1000, Invitrogen) containing ethidium bromide for visualization; (we use a FLA-5100 Fluorescence imager and MultiGauge software (Fuji, Tokyo, Japan).

An example of miR-dependent deadenylation of mRNA measured by this method is shown in Fig. 6.3C. In this case, HeLa cells (which express let-7) were transfected with a pDNA R-luc construct encoding three let-7 binding sites in the 3′ UTR (3× bulge), or with control constructs encoding either no sites (plasmid) or three mutated let-7 sites (3× bulge mut) (constructs described in Pillai *et al.*, 2005). Cells were harvested 24 h after transfection, RNA was purified for the PAT assay, and luciferase activity was measured from cell lysates. As reported previously, the presence of functional let-7 target sites results in specific repression of luciferase expression with very minor effects on mRNA stability (Pillai *et al.*, 2005). The experiment in Fig. 6.3C demonstrates that the let-7 targeted reporter mRNA is selectively deadenylated.

Variations: Another high–resolution method used to measure mRNA adenylation state is by a modified Northern blot, where size resolution is increased by DNA-oligo directed RNase H cleavage of the target 3′ UTR followed by Urea-PAGE for high-level size discrimination. The cleaved polyadenylated UTR fragment is detected using a radioactive probe. The advantage of this method is that the length of the 3′ UTR is measured directly, the disadvantage is the amount of input RNA required, especially if more than one or two probes are to be analyzed. (For detailed comparison of the methods, see Salles *et al.*, 1999.) The most informative deadenylation data may be gained from a pulse chase regimen, where a regulatable promoter is employed to generate a brief transcriptional pulse of reporter mRNA, the adenylation state of which is then followed over time (Wu *et al.*, 2006).

Comment: Limitations to the LM-PAT assay can arise where a gene contains alternate 3′ polyadenylation sites generating 3′ UTRs of different lengths. This becomes particularly important if these alternate amplicons all fall within a similar size range. Sequencing amplicons arising from both short- and long-tailed transcripts give confirmation of the primer specificity and that a single 3′ UTR of differing poly(A) tail length is under investigation. Finally, problems of nonspecific PCR products can arise from adenosine tracts within the open reading frame or 3′ UTR, allowing annealing of the PAT primer and 1st-strand synthesis. Finally, we suggest using pDNA transfection to measure miR-mediated adenylation-state changes, because of the potential for inert subpopulations of directly transfected mRNA to complicate the analysis (see previously and Barreau *et al.*, 2006).

5. MEASURING THE DISTRIBUTION OF MiRNP COMPONENTS IN POLYSOME PROFILES

Sucrose density gradient centrifugation is commonly used in translational control research because it can separate polysomes from free 80S ribosomes and ribosomal subunits (for examples, see Beilharz and Preiss, 2004; Preiss *et al.*, 2003). A change in the association of an mRNA with polysomes is indicative of changes in its translation state. For instance, a block in translation initiation would result in reduced ribosome density on the affected mRNA and a shift toward the lower-density fractions of the gradient. A block during the elongation phase of translation would have the opposite effect. The miR literature provides conflicting observations in that regard, with several studies reporting essentially unchanged polysome association of an mRNA when it becomes targeted by a miR (Nottrott *et al.*, 2006; Olsen and Ambros, 1999; Petersen *et al.*, 2006; Seggerson *et al.*, 2002), while a 2005 publication presents evidence for a clear shift of the targeted mRNA away from polysomes (Pillai *et al.*, 2005). Further evidence exists for the presence of both miR and targeted mRNA in polysomal fractions (Kim *et al.*, 2004; Maroney *et al.*, 2006; Nelson *et al.*, 2004). It is thus clear that further careful experimentation is required to resolve this matter. Here, we describe the core methodology required to investigate the distribution of miR, associated miRNP components, and target mRNA within sucrose density gradients.

5.1. Preparation of polysome gradients

The following procedure works well for resolving polysomal complexes in lysates from cultured mammalian cells.

Procedure:

1. We prepare the sucrose density gradients by a simple and reproducible freeze thawing method (Luthe, 1983). In short, dissolve 50, 41.9, 33.8, 25.6, or 17.5% (w/v) sucrose in gradient buffer, pass through a 0.2 μm filter, and layer 2 ml of each solution into centrifugation tubes, starting with the highest sucrose concentration. We use Beckman polyallomer centrifugation tubes, 14 × 89 mm (Beckman Coulter, CA). Freeze each liquid layer for 15 min at −80° before applying the next layer. Multiple gradients can be made in one batch and stored at −80° for several months.
2. Before use, thaw the required number of gradient tubes overnight at 4°, to establish continuous sucrose density gradients.
3. For making cellular extracts, wash 1 × 10^6 HeLa cells twice with ice-cold PBS containing 100 μg/ml cycloheximide (CHM), scrape into 1 ml PBS-CHM, and pellet by centrifugation at 1000×g for 2 min at 4°.
4. Lyse the cell pellet by suspension in 450 μl lysis buffer [20 mM Hepes (pH 7.5), 125 mM KCl, 5 mM MgCl$_2$, 2 mM DTT, 0.5% NP-40, 100 μg/ml cyclohexamide, 100 U/ml RNAse inhibitor, 1× complete protein inhibitor cocktail (Roche)] for 10 min on ice, then centrifuge at 16,000×g for 10 min at 4°.
5. Remove the supernatant and layer on top of the thawed sucrose gradient.
6. Centrifuge gradients at 35,000 rpm for 150 min at 4° in a Beckman L-80 ultracentrifuge using the SW 41 Ti rotor (Beckman Coulter, CA).
7. Collect fractions starting from the bottom of the gradient, while recording an absorbance profile at 254 nm (which is dominated by the very abundant ribosomal RNA). We use a setup consisting of a peristaltic pump, ultraviolet (UV) detector, and fraction collector commonly used for chromatography experiments (GE Healthcare Life Sciences).
8. To precipitate proteins from fractions, add 1/10 volume 50% TCA and 10 mM Na-deoxycholate (carrier) and incubate on ice for 20 min.

Centrifuge sample at maximum speed for 15 min at 4°. Remove supernatant and wash with ice-cold ethanol. Remove ethanol, air dry sample, and resuspend in sample buffer.

9. Analyze proteins by gel electrophoresis and western blotting.

Example polysome profiles from sucrose gradient fractionation of HeLa cell lysates, either untreated or treated with EDTA, are shown in Fig. 6.4A and B. The polysome profile of untreated HeLa lysates shows three defined peaks in less dense fractions (6 to 9), which correspond to the 80S, 60S, and 40S peaks (Fig. 6.4A). Treatment of lysates with 30 μM EDTA results in ribosome dissociation leaving predominantly free 60S and 40S subunits

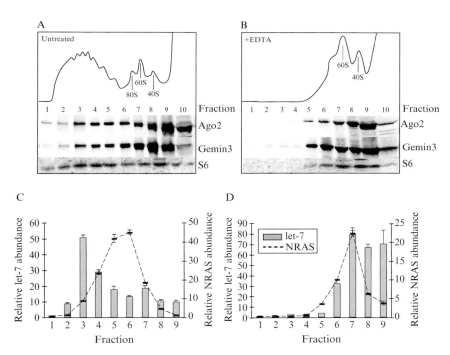

Figure 6.4 Distribution of miR, miRNP components, and a target mRNA in polysome fractions. (A, B) HeLa cell lysates, prepared in the presence of cycloheximide, were fractionated on a sucrose gradient and polysome trace was gained by measuring the absorbance at 254 nm (top). The positions of monosomes (80S) and ribosomal subunits 60S and 40S in the fractions are indicated. The distribution of the miRNP components Ago2 and Gemin3, as well as of the S6 ribosomal protein (S6), were measured by western blotting (bottom). Lysates were either (A) untreated or (B) treated with 30 μM EDTA to dissociate polysomes. (C, D) The relative levels of let-7 miR (bars) and one of its mRNA targets NRAS (dashed line) in each gradient fraction were determined by qPCR. RNA levels were normalized against those of a control mRNA spiked into each fraction before RNA extraction. Error bars indicate standard error in multiple qPCR reactions.

(Fig. 6.4B). Polysome fractions with increasing numbers of associated ribosomes can be seen in the bottom (denser) fractions (1 to 5). Western blotting was performed for the miRNP constituent proteins Ago 2 and Gemin 3 and for S6 ribosomal protein as a positive control for the presence of ribosomes. A proportion of Ago 2 and Gemin 3 sediments into the fractions containing polysomes in untreated lysates and moves to upper fractions after EDTA treatment. Antibodies used in this experiment (dilution in brackets): Argonaute 2 clone 7C3 (1:1000, gift from Tom Hobman), Gemin 3 clone 12H12 (1:1000, gift from Z. Mourelatos), S6 ribosomal protein, (1:500, Cell Signalling Technologies, MA), secondary antibodies (1:20000, Invitrogen).

Comment: A simple test that is often done is to treat the lysate with 30 μM EDTA prior to loading on the gradient, which leads to the disassembly of polysomes due to chelation of Mg^{2+} (Fig. 6.4B). EDTA is, however, not a very specific reagent and instead more specific drugs such as puromycin, which causes polypeptide chain termination and disaggregates polysomes, may be used.

Variations: Dedicated density gradient forming and fractionating instruments are available (e.g., from Biocomp, Fredericton, NB, Canada), which offer convenient operation and increased accuracy. The range of sucrose concentration and centrifugation conditions (time or speed) can be adjusted to obtain optimal resolution. The exact composition of the lysis and gradient buffer (e.g., K^+ or Mg^{2+} concentration) may be varied to help preserve or dissociate particular complexes. A gentle formaldehyde (FA) cross-linking can help to preserve labile complexes during centrifugation (0.25% FA added directly to the lysate, incubated for 5 min on ice, and stopped by adding 250 μM glycine; (Niranjanakumari *et al.*, 2002).

5.2. Extraction of RNA from sucrose gradients

A technical challenge with this step is to achieve RNA extraction of uniform quality and efficiency for each fraction. This is because the amount of RNA in each sucrose gradient fraction varies considerably and the high concentration of sucrose in the bottom fractions interferes with phase separation in typical phenol-based extraction steps. To address these problems, we "spike" each fraction with an aliquot of a foreign (control) RNA, which can be used later to correct for differences in RNA recovery (and reverse transcription efficiency) between samples. We then remove sucrose from the samples by precipitation of total nucleic acid and protein with ethanol. To purify RNA, a standard Trizol (Invitrogen) extraction is performed as outlined later (also see product insert).

Procedure:

1. Add three volumes 100% ethanol to fraction, mix well, and precipitate lysate at $-80°$ overnight.

2. Spin precipitate at $16,000 \times g$ for 20 min and remove supernatant. Dry pellet under vacuum.
3. Add 1 ml Trizol to the pellet and vortex to dissolve pellet. Let the nucleoprotein complexes dissociate for 5 min. Add 0.2 ml of chloroform, mix vigorously for 15 s and incubate for 2 to 3 min at RT.
4. Continue with standard Trizol extraction (see product insert).

Variations: Other protocols use high-salt solutions in combination with isopropanol to precipitate RNA from sucrose fractions. Standard proteinase K digestion in the presence of SDS may be used to extract RNA from the precipitated material.

5.3. Detection of miRs by reverse transcription (RT) and quantitative PCR (qPCR)

Detection of miRs presents technical challenges, particularly when the source material is limiting. Northern blotting requires a large amount of isolated RNA and a robust strategy for cross-linking small RNA species to a membrane. Even so, detection of rare miRs will be difficult. Using PCR to detect miRs has the advantage of detecting miRs at significantly lower abundance, and there are several published methods (Johnson *et al.*, 2005; Raymond *et al.*, 2005; Shi and Chiang, 2005). The method described later relies on a hairpin RT primer and was adapted from Chen *et al.*, 2005. The hairpin primer binds to the last six bases of the miR and then folds back on itself, protected from interaction with other nucleic acids. The use of a hairpin RT primer introduces an initial level of specificity by targeting nucleic acids ending in a specific 6-base sequence, while generating an miR-specific cDNA template of increased size. Subsequent qPCR utilizes the remaining miR sequence as the forward primer and a reverse primer recognizing the hairpin RT primer sequence. This results in specific amplification of a single miR species, theoretically distinguishing among miR family members (although in practice there is some nonspecific amplification of closely related family members). This method has the additional advantage that the $3'$ end of the miR is not modified, allowing detection of exogenously $3'$ labeled miR species. Here, we describe the methodology for detecting the let-7 target mRNA, NRAS (Johnson *et al.*, 2005), and the spike-in control RNA (in this case, an *in vitro* transcribed R-luc mRNA) by RT-qPCR, followed by adaptations required for the specific detection of miRs.

5.3.1. mRNA reverse transcription and qPCR

Although random priming, in general, is less efficient, it may be preferable to oligo dT priming in this instance, because the miR-target mRNA may be significantly deadenylated by miR action (see Fig. 6.3C).

Procedure:

1. For a 20-μl reaction add 200 to 1000 ng of total RNA; 1 pmole random hexanucleotide primer (Promega); 1μl of dNTP mix containing 10 mM each of dATP, dGTP, dCTP, and dTTP; and water up to 12 μl.
2. Heat sample at 65° for 5 min and incubate on ice.
3. Add 4 μl of 5× first strand buffer, 2 μl of 0.1 M DTT, 200 U of superscript II reverse transcriptase (Roche), and 40 U of RNAse inhibitor (RNAsin, Promega), making the reaction volume up to 20 μl.
4. Incubate reaction at 42° for 2 to 3 h and 75° for 10 to 15 min.
5. Dilute cDNA 1/10 and add 2 μl to the following 20 μl PCR reaction (using the Lightcycler FastStart DNA Master Plus SYBR green 1 kit, Roche): 5 pmole each primer, 4 μl 5× reaction master mix (containing FastStart Taq DNA polymerase, reaction buffer, dNTP/dUTP mix, SYBR green 1 dye, and $MgCl_2$), and 1 to 5 mM total $MgCl_2$.
6. Using a Rotorgene thermocyler (Corbett Research, Mortlake, NSW, Australia), perform the following reaction: 10 min 95° to activate the enzyme; 40 to 50 cycles of 10 s 95°, 10 s annealing (approximately 2° less than primer melting temperature), and 72° extension. The extension is usually 1 s/25 bp of product. This is followed by product melting from 65 to 95° at 0.1°/s to create a melting curve.
7. Visualize the PCR product on agarose gel to confirm correct size of a single product and analyze melting curve to further confirm only one product was amplified.

Comment: Unique primer pairs are, where possible, designed to amplify over exon–exon borders to discriminate between genomic DNA- (or plasmid, see previously) and mRNA–derived sequence, and to have a difference in melting temperatures of less than 4°. Reactions are optimized for Mg^{2+} concentration and efficiency of the reaction is optimized using a template dilution series. All reactions are checked for their specificity by using no template controls. Sometimes, specificity is lost as the template concentrations fall, allowing primers to bind less specific targets. Thus, it is important to take the expected concentration range of samples into consideration when selecting a primer pair.

5.3.2. miR reverse transcription and qPCR
Procedure:

1. For a 20-μl reaction, add 200 to 1000 ng of total RNA; 1 pmole miR-specific hairpin primer (Table 6.2); 1μl of dNTP mix containing 10 mM each of dATP, dGTP, dCTP, and dTTP; and water up to 12 μl.
2. Heat sample at 65° for 5 min and incubate on ice.

Table 6.2　Sequence of let-7c miR and the primers used to detect its expression

Primer/miR	Sequence
Let-7c miR	UGAGGUAGUAGGUUGUAUGGUU
RT primer	GTCGTATCCAGTGCAGGGTCCGAGGTATTC-GCACTGGATACGAC AACCAT
qPCR for	GCCGCTGAGGTAGTAGGTTGT
qPCR rev	GTGCAGGGTCCGAGGT

3. For a 20-μl reaction, add 4 μl of 5× first strand buffer, 2 μl of 0.1 M DTT, 200 U of superscript II reverse transcriptase (Invitrogen), and 40 U of RNAse inhibitor (RNAsin, Promega).
4. Incubate reaction at 16° for 30 min, 42° for 30 min, and 75° for 15 min.
5. qPCR is performed as for mRNA, with a 4 s extension.

Example data showing the levels of let-7c miR and its target NRAS mRNA in sucrose gradient fractions of untreated and EDTA-treated HeLa lysates is displayed in Fig. 6.4C and D. The relative levels of each was normalized against the spike-in control mRNA (R-luc). Let-7c is predominantly detected in the polysome fractions in untreated cells (similar to experiments with miR-124a reported in Nelson *et al.*, 2004) and moves to upper fractions upon treatment of lysates with EDTA. The primers used to detect let-7c are shown in Table 6.2. NRAS mRNA peaks in the lower polysomal region of the gradient. Thus, all miR-related components analyzed in Fig. 6.4 sediment, at least partly, into the polysomal region of the gradient, although to differing extents.

Comment: For the hairpin method, the forward qPCR primer and the RT hairpin primer cannot overlap otherwise a primer product will be formed.

Variations: If the sequence difference between miR family members resides in the last 6 base pairs, utilization of fluorescent probes (such as Taqman technology, as shown in Chen *et al.*, 2005) will give optimal specificity. In the example in Fig. 6.4A and B, both let-7c and let-7a are detected, but because both target the RAS family (Johnson *et al.*, 2005), there was no need to distinguish between these family members in this experiment.

6. LOCALIZATION OF miRNP COMPONENTS TO P-BODIES

Components of the miR complex are concentrated in cytoplasmic foci, called P-bodies (Liu *et al.*, 2005a,b; Pillai *et al.*, 2005; Sen and Blau, 2005). P-bodies are enriched for factors involved in mRNA degradation

(Cougot *et al.*, 2004; Teixeira *et al.*, 2005). Co–localization of these factors is most commonly shown by confocal fluorescence microscopy, a method using antibodies or fluorescent-tagged proteins to assess the subcellular localization of proteins (see Allan, 2000, for detailed methodologies). Usually, antibodies against endogenous proteins are preferred because tagged proteins can be mislocalized, either due to the presence of a tag or through excessive expression. The following protocol works well for visualization of P-bodies in adherent HeLa cells, using an antibody raised against GW182 as a P-body marker (Liu *et al.*, 2005a).

Procedure:

1. Place a coverslip in a 6-well plate and seed cells in 2 ml DMEM medium (plus 10% FCS, glutamate, penicillin, and streptomycin).
2. For the detection of Flag-tagged Argonaute proteins, transfect cells at 60 to 80% density with 1 to 2 μg pIRESneo Flag/HA-Ago 1, 2, or 3 plasmid (kind gift from Lana Dinic and Thomas Tuschl) per well using Lipofectamine 2000 (Invitrogen), following the supplier's instructions.
3. 24 h after transfection, wash cells 3× with PBS, fix with sterile filtered 4% paraformaldehyde/PIPES-buffer [60 mM PIPES, 30mM HEPES, 10 mM EGTA, 4 mM MgSO$_4$ (pH 7.0, KOH)] for 10 min at RT, and permeabilize with ice-cold methanol for 10 min at RT (Kedersha and Anderson, 2002).
4. Wash coverslips 3× with PBS and place them on drops of 5% normal donkey serum (NDS)/PBS for 10 min to block unspecific epitopes.
5. Incubate with primary and secondary antibodies in 3% NDS/PBS for 1 h each at RT. Wash coverslips 3× with PBS between incubations.
6. After washing 3× with PBS, place coverslips on slides with a drop of Vectashield mounting medium for fluorescence (Vector Laboratories, CA).
7. Take images using a confocal microscope. We use a Leica DM RBE TCS SP1 microscope (Leica, Wetzlar, Germany) and process images in Adobe Photoshop^{TM}.

The experiment depicted in Fig. 6.5A demonstrates co-localization of Flag-tagged Ago1, 2, and 3 proteins (detected with anti–Flag M2 antibody, Sigma-Aldrich, MO) with the P-body marker protein GW182 (detected with anti-GW182 antibody, Santa Cruz Biotechnology, CA). As reported in Leung *et al.* (2006), the Argonautes were observed to be predominantly diffuse in the cytoplasm while being localized to a proportion of P-bodies. As a negative control, 4E–BP (antibody from Cell Signalling) was shown not to co–localize with the Argonautes (Fig. 6.5B; data for Ago2 and Ago3 not shown). Secondary antibodies used were anti-mouse antibody conjugated to FITC and anti-rabbit conjugated to Cy3 (Jackson ImmunoResearch Laboratories, Suffolk, UK). All primary antibodies were used in a 1:100 dilution and all secondary antibodies in a 1:200 dilution.

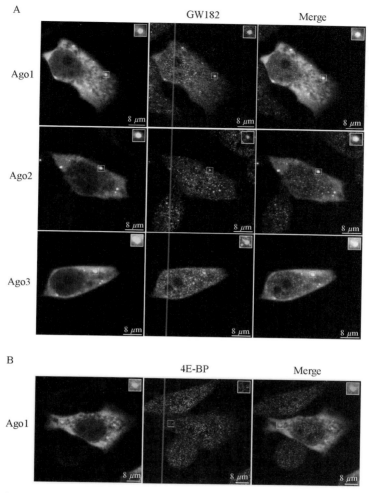

Figure 6.5 Enrichment of Argonaute proteins in P-bodies. HeLa cells were transfected with plasmids expressing Flag-tagged Argonaute 1, 2, or 3 protein and analyzed by confocal microscopy after 24 h. (A) The Argonaute proteins were detected using an anti-Flag antibody (green), while GW182 was detected with anti-GW182 antibody (red). Co-localization is indicated by the yellow color in the merged panels. Inserts show regions of co-localization at higher magnification. (B) The protein 4E-BP (red) was used as a specificity control and did not co-localize with Ago1 foci (green) (insert shows higher magnification of an Ago1 foci). (See color insert.)

Comments: Permeabilization can also be done with 0.1 to 0.3% Triton X-100 or NP-40 for 5 min at RT. Permeabilization methods should be optimized for different antibodies and preservation of different subcellular structures (Allan, 2000).

Variations: An alternative marker of P-bodies is the decapping protein Dcp1a (Fenger-Gron *et al.*, 2005). Currently, there is also interest in localization of miRNP components to stress granules, which can be visualized by detecting TIA-1 protein (Leung *et al.*, 2006). Visualization of miRs themselves requires *in vitro* fluorescent labeling (which can be performed with labeling kits or directly during synthesis); however, analysis of localization may be complicated if the transfection method sequesters miRs away from the cytoplasm (Barreau *et al.*, 2006). To specifically detect single-stranded miR, presumably incorporated into RISC, the passenger strand can be labeled with a quencher (such as BHQ-2) (Leung *et al.*, 2006). An important specificity control for all localization experiments is to show a protein that does not localize to the body of interest and to show that there is no signal from the secondary antibody alone.

7. Concluding Remarks

As indicated in passing throughout this chapter, there is currently no consensus model for the mode by which miR affect the process of mRNA translation. It may well be that miR act through different molecular mechanisms, perhaps in a context-dependent manner. Few studies have, so far, addressed this possibility. Detailed biochemical characterization of the miR mechanism will depend on the development and application of tractable cell-free translation systems that preserve the key physiological features of their action. Further, it may be just as instructive to determine "where" in the cell miR-dependent translational repression takes place, as to investigate "how" it happens. An important task will also be to attempt independent verification of key findings that led the different investigators to propose their divergent models. We hope that the collection of methods described here will be helpful to some in these endeavors.

ACKNOWLEDGMENTS

We thank David I. K. Martin for ongoing support of this work. Research in the authors' lab is supported by grants from the National Health and Medical Research Council, the Australian Research Council, the Sylvia & Charles Viertel Charitable Foundation, and by the Victor Chang Cardiac Research Institute.

REFERENCES

Allan, V. (2000). Protein localization by fluorescence microscopy: A practical approach. Oxford University Press, Oxford, UK.

Bagga, S., Bracht, J., Hunter, S., Massirer, K., Holtz, J., Eachus, R., and Pasquinelli, A. E. (2005). Regulation by let-7 and lin-4 miRNAs results in target mRNA degradation. *Cell* **122,** 553–563.

Barreau, C., Dutertre, S., Paillard, L., and Osborne, H. B. (2006). Liposome-mediated RNA transfection should be used with caution. *RNA* **12**, 1790–1793.

Bartel, D. P., and Chen, C. Z. (2004). Micromanagers of gene expression: The potentially widespread influence of metazoan microRNAs. *Nat. Rev. Genet.* **5**, 396–400.

Behm-Ansmant, I., Rehwinkel, J., Doerks, T., Stark, A., Bork, P., and Izaurralde, E. (2006). mRNA degradation by miRNAs and GW182 requires both CCR4:NOT deadenylase and DCP1:DCP2 decapping complexes. *Genes Dev.* **20**, 1885–1898.

Beilharz, T. H., and Preiss, T. (2004). Translational profiling: The genome-wide measure of the nascent proteome. *Brief Funct. Genomic Proteomic* **3**, 103–111.

Bergamini, G., Preiss, T., and Hentze, M. W. (2000). Picornavirus IRESes and the poly(A) tail jointly promote cap-independent translation in a mammalian cell-free system. *RNA* **6**, 1781–1790.

Bert, A. G., Grepin, R., Vadas, M. A., and Goodall, G. J. (2006). Assessing IRES activity in the HIF-1{alpha} and other cellular 5′ UTRs. *RNA* **12**, 1074–1083.

Bhattacharyya, S. N., Habermacher, R., Martine, U., Closs, E. I., and Filipowicz, W. (2006). Relief of microRNA-mediated translational repression in human cells subjected to stress. *Cell* **125**, 1111–1124.

Borman, A. M., and Kean, K. M. (1997). Intact eukaryotic initiation factor 4G is required for hepatitis A virus internal initiation of translation. *Virology* **237**, 129–136.

Chen, C., Ridzon, D. A., Broomer, A. J., Zhou, Z., Lee, D. H., Nguyen, J. T., Barbisin, M., Xu, N. L., Mahuvakar, V. R., Andersen, M. R., Lao, K. Q., Livak, K. J., *et al.* (2005). Real-time quantification of microRNAs by stem-loop RT-PCR. *Nucleic Acids Res.* **33**, e179.

Cougot, N., Babajko, S., and Seraphin, B. (2004). Cytoplasmic foci are sites of mRNA decay in human cells. *J. Cell Biol.* **165**, 31–40.

Doench, J. G., Petersen, C. P., and Sharp, P. A. (2003). siRNAs can function as miRNAs. *Genes Dev.* **17**, 438–442.

Engels, B. M., and Hutvagner, G. (2006). Principles and effects of microRNA-mediated post-transcriptional gene regulation. *Oncogene* **25**, 6163–6169.

Fenger-Gron, M., Fillman, C., Norrild, B., and Lykke-Andersen, J. (2005). Multiple processing body factors and the ARE binding protein TTP activate mRNA decapping. *Mol. Cell.* **20**, 905–915.

Fraser, C. S., and Doudna, J. A. (2006). Structural and mechanistic insights into hepatitis C viral translation initiation. *Nat. Rev. Microbiol.* **5**(1), 29–38.

Gallie, D. R. (1991). The cap and poly(A) tail function synergistically to regulate mRNA translational efficiency. *Genes Dev.* **5**, 2108–2116.

Giraldez, A. J., Mishima, Y., Rihel, J., Grocock, R. J., Van Dongen, S., Inoue, K., Enright, A. J., and Schier, A. F. (2006). Zebrafish MiR-430 promotes deadenylation and clearance of maternal mRNAs. *Science* **312**, 75–79.

Hellen, C. U., and Sarnow, P. (2001). Internal ribosome entry sites in eukaryotic mRNA molecules. *Genes Dev.* **15**, 1593–1612.

Hentze, M. W., Gebauer, F., and Preiss, T. (2006). Cis-regulatory sequences and trans-acting factors in translational control. *In* "Translational Control in Biology and Medicine" (M. B. Mathews, *et al.*, eds.), pp. 273–299. Cold Spring Harbor Laboratory Press, Cold Spring Harbor, NY.

Humphreys, D. T., Westman, B., Martin, D. I., and Preiss, T. (2008). Inhibition of translation initiation by a microRNA. *In* "MicroRNAs: From Basic Science to Disease Biology" (K. Appasani, ed.), pp. 83–99. Cambridge University Press, Cambridge UK.

Humphreys, D. T., Westman, B. J., Martin, D. I., and Preiss, T. (2005). MicroRNAs control translation initiation by inhibiting eukaryotic initiation factor 4E/cap and poly(A) tail function. *Proc. Natl. Acad. Sci. USA* **102**, 16961–16966.

Hutvagner, G., Simard, M. J., Mello, C. C., and Zamore, P. D. (2004). Sequence-specific inhibition of small RNA function. *PLoS Biol.* **2**, E98.

Iizuka, N., Najita, L., Franzusoff, A., and Sarnow, P. (1994). Cap-dependent and cap-independent translation by internal initiation of mRNAs in cell extracts prepared from *Saccharomyces cerevisiae*. *Mol. Cell Biol.* **14,** 7322–7330.

Jackson, R. J. (2005). Alternative mechanisms of initiating translation of mammalian mRNAs. *Biochem. Soc. Trans.* **33,** 1231–1241.

Johnson, S. M., Grosshans, H., Shingara, J., Byrom, M., Jarvis, R., Cheng, A., Labourier, E., Reinert, K. L., Brown, D., and Slack, F. J. (2005). RAS is regulated by the let-7 microRNA family. *Cell* **120,** 635–647.

Kahvejian, A., Svitkin, Y. V., Sukarieh, R., M'Boutchou, M. N., and Sonenberg, N. (2005). Mammalian poly(A)-binding protein is a eukaryotic translation initiation factor, which acts via multiple mechanisms. *Genes Dev.* **19,** 104–113.

Kedersha, N., and Anderson, P. (2002). Stress granules: Sites of mRNA triage that regulate mRNA stability and translatability. *Biochem. Soc. Trans.* **30,** 963–969.

Kim, D. H., Villeneuve, L. M., Morris, K. V., and Rossi, J. J. (2006). Argonaute-1 directs siRNA-mediated transcriptional gene silencing in human cells. *Nat. Struct. Mol. Biol.* **13,** 793–797.

Kim, J., Krichevsky, A., Grad, Y., Hayes, G. D., Kosik, K. S., Church, G. M., and Ruvkun, G. (2004). Identification of many microRNAs that copurify with polyribosomes in mammalian neurons. *Proc. Natl. Acad. Sci. USA* **101,** 360–365.

Leung, A. K., Calabrese, J. M., and Sharp, P. A. (2006). Quantitative analysis of Argonaute protein reveals microRNA-dependent localization to stress granules. *Proc. Natl. Acad. Sci. USA* **103,** 18125–18130.

Lewis, B. P., Shih, I. H., Jones-Rhoades, M. W., Bartel, D. P., and Burge, C. B. (2003). Prediction of mammalian microRNA targets. *Cell* **115,** 787–798.

Lim, L. P., Lau, N. C., Garrett-Engele, P., Grimson, A., Schelter, J. M., Castle, J., Bartel, D. P., Linsley, P. S., and Johnson, J. M. (2005). Microarray analysis shows that some microRNAs downregulate large numbers of target mRNAs. *Nature* **433,** 769–773.

Liu, J., Rivas, F. V., Wohlschlegel, J., Yates, J. R., 3rd, Parker, R., and Hannon, G. J. (2005a). A role for the P-body component GW182 in microRNA function. *Nat. Cell Biol.* **7,** 1161–1166.

Liu, J., Valencia-Sanchez, M. A., Hannon, G. J., and Parker, R. (2005b). MicroRNA-dependent localization of targeted mRNAs to mammalian P-bodies. *Nat. Cell Biol.* **7,** 719–723.

Luthe, D. S. (1983). A simple technique for the preparation and storage of sucrose gradients. *Anal. Biochem.* **135,** 230–232.

Maroney, P. A., Yu, Y., Fisher, J., and Nilsen, T. W. (2006). Evidence that microRNAs are associated with translating messenger RNAs in human cells. *Nat. Struct. Mol. Biol.* **13,** 1102–1107.

Nelson, P. T., Hatzigeorgiou, A. G., and Mourelatos, Z. (2004). miRNP:mRNA association in polyribosomes in a human neuronal cell line. *RNA* **10,** 387–394.

Niranjanakumari, S., Lasda, E., Brazas, R., and Garcia-Blanco, M. A. (2002). Reversible cross-linking combined with immunoprecipitation to study RNA-protein interactions *in vivo*. *Methods* **26,** 182–190.

Nottrott, S., Simard, M. J., and Richter, J. D. (2006). Human let-7a miRNA blocks protein production on actively translating polyribosomes. *Nat. Struct. Mol. Biol.* **13,** 1108–1114.

Olsen, P. H., and Ambros, V. (1999). The lin-4 regulatory RNA controls developmental timing in *Caenorhabditis elegans* by blocking LIN-14 protein synthesis after the initiation of translation. *Dev. Biol.* **216,** 671–680.

Ostareck, D. H., Ostareck-Lederer, A., Shatsky, I. N., and Hentze, M. W. (2001). Lipoxygenase mRNA silencing in erythroid differentiation: The 3′ UTR regulatory complex controls 60S ribosomal subunit joining. *Cell* **104,** 281–290.

Petersen, C. P., Bordeleau, M. E., Pelletier, J., and Sharp, P. A. (2006). Short RNAs repress translation after initiation in mammalian cells. *Mol. Cell* **21**, 533–542.

Pillai, R. S. (2005). MicroRNA function: Multiple mechanisms for a tiny RNA? *RNA* **11**, 1753–1761.

Pillai, R. S., Bhattacharyya, S. N., Artus, C. G., Zoller, T., Cougot, N., Basyuk, E., Bertrand, E., and Filipowicz, W. (2005). Inhibition of translational initiation by let-7 microRNA in human cells. *Science* **309**, 1573–1576.

Poyry, T. A., Kaminski, A., and Jackson, R. J. (2004). What determines whether mammalian ribosomes resume scanning after translation of a short upstream open reading frame? *Genes Dev.* **18**, 62–75.

Preiss, T., Baron-Benhamou, J., Ansorge, W., and Hentze, M. W. (2003). Homodirectional changes in transcriptome composition and mRNA translation induced by rapamycin and heat shock. *Nat. Struct. Biol.* **10**, 1039–1047.

Preiss, T., Muckenthaler, M., and Hentze, M. W. (1998). Poly(A)-tail-promoted translation in yeast: Implications for translational control. *RNA* **4**, 1321–1331.

Raymond, C. K., Roberts, B. S., Garrett-Engele, P., Lim, L. P., and Johnson, J. M. (2005). Simple, quantitative primer-extension PCR assay for direct monitoring of microRNAs and short-interfering RNAs. *RNA* **11**, 1737–1744.

Salles, F. J., Richards, W. G., and Strickland, S. (1999). Assaying the polyadenylation state of mRNAs. *Methods* **17**, 38–45.

Sallés, F. J., and Strickland, S. (1995). Rapid and sensitive analysis of mRNA polyadenylation states by PCR. *PCR Methods and Applications* **4**, 317–321.

Seggerson, K., Tang, L., and Moss, E. G. (2002). Two genetic circuits repress the *Caenorhabditis elegans* heterochronic gene lin-28 after translation initiation. *Dev. Biol.* **243**, 215–225.

Sen, G. L., and Blau, H. M. (2005). Argonaute 2/RISC resides in sites of mammalian mRNA decay known as cytoplasmic bodies. *Nat. Cell Biol.* n, 633–636.

Shi, R., and Chiang, V. L. (2005). Facile means for quantifying microRNA expression by real-time PCR. *Biotechniques* **39**, 519–525.

Tarun, S. Z., Jr., and Sachs, A. B. (1995). A common function for mRNA 5′ and 3′ ends in translation initiation in yeast. *Genes & Dev.* **9**, 2997–3007.

Teixeira, D., Sheth, U., Valencia-Sanchez, M. A., Brengues, M., and Parker, R. (2005). Processing bodies require RNA for assembly and contain nontranslating mRNAs. *RNA* **11**, 371–382.

Valencia-Sanchez, M. A., Liu, J., Hannon, G. J., and Parker, R. (2006). Control of translation and mRNA degradation by miRNAs and siRNAs. *Genes Dev.* **20**, 515–524.

Wang, B., Love, T. M., Call, M. E., Doench, J. G., and Novina, C. D. (2006). Recapitulation of short RNA-directed translational gene silencing *in vitro*. *Mol. Cell.* **22**, 553–560.

Westman, B., Beeren, L., Grudzien, E., Stepinski, J., Worch, R., Zujberek, J., Jemielity, J., Stolarski, R., Darzynkiewicz, E., Rhoads, R. E., and Preiss, T. (2005). The antiviral drug ribavirin does not mimic the 7-methylguanosine moiety of the mRNA cap structure *in vitro*. *RNA* **11**, 1505–1513.

Wilson, J. E., Powell, M. J., Hoover, S. E., and Sarnow, P. (2000). Naturally occurring dicistronic cricket paralysis virus RNA is regulated by two internal ribosome entry sites. *Mol. Cell Biol.* **20**, 4990–4999.

Woolstencroft, R. N., Beilharz, T. H., Cook, M. A., Preiss, T., Durocher, D., and Tyers, M. (2006). Ccr4 contributes to tolerance of replication stress through control of CRT1 mRNA poly(A) tail length. *J. Cell Sci.* **119**, 5178–5192.

Wu, L., Fan, J., and Belasco, J. G. (2006). MicroRNAs direct rapid deadenylation of mRNA. *Proc. Natl. Acad. Sci. USA.* **103**, 4034–4039.

METHODS FOR STUDYING SIGNAL-DEPENDENT REGULATION OF TRANSLATION FACTOR ACTIVITY

Xuemin Wang *and* Christopher G. Proud

Contents

Department of Biochemistry and Molecular Biology, University of British Columbia, Vancouver, Canada

Methods in Enzymology, Volume 431
ISSN 0076-6879, DOI: 10.1016/S0076-6879(07)31007-0

Abstract

The translational machinery of mammalian cells is regulated through the phosphorylation of a number of its components, especially translation factor proteins. These include factors involved in the initiation and elongation stages of translation, and proteins that modify their activity. Examples include eukaryotic initiation factor (eIF) 4E, eukaryotic elongation factor (eEF) 2, and eIF4E-binding protein 1 (4E-BP1). Their phosphorylation is mediated by protein kinases that, in turn, are regulated by specific intracellular signaling pathways. These pathways include those mediated via the mammalian target of rapamycin (mTOR), the ERK and p38 MAP kinase pathways, and protein kinase B (Akt). These pathways are activated by hormones (e.g., insulin), growth factors, mitogens, and other extracellular stimuli. In some cases, amino acids also modulate the pathway (e.g., mTOR).

Procedures are described for determining the states of phosphorylation and/or activity of several translation factors, and of kinases that phosphorylate them. We also outline procedures for assessing the states of activation of relevant signaling pathways. In addition, we provide guidelines on using small molecule inhibitors to assess the involvement of specific signaling pathways in controlling translation factors and protein synthesis.

1. INTRODUCTION TO SIGNALING PATHWAYS AND THE CONTROL OF TRANSLATION FACTORS

Both the overall rate of protein synthesis and the translation of certain specific mRNAs are controlled by agents such as hormones, growth factors, and other extracellular stimuli. As precursors for protein assembly, amino acids also regulate the translational machinery. Because protein synthesis consumes a high proportion of cellular metabolic energy, the energy status of the cell also modulates translation factors.

The main method through which these agents regulate the translational apparatus is via changes in the states of phosphorylation of translation factors and related proteins: phosphorylation may, for example, alter the intrinsic activity of translation factors or affect their ability to bind other components (other factors, the ribosome, or RNA).

These changes are generally mediated via signaling pathways that regulate the activities of the protein kinases acting on translation factors. These are often highly specific enzymes and may be targeted (by co-localization) to the translational machinery. The corresponding protein phosphatases may also be regulated although much less is known about them than about the kinases.

This chapter describes techniques for studying (i) the regulation of translation factors by phosphorylation; (ii) the activities of protein kinases that act upon them; and (iii) the relevant upstream signaling pathways in mammalian cells. The major relevant pathways are summarized in Fig. 7.1 as an essential prelude to describing approaches to studying their involvement in the control of translation. Briefly, these are:

I. The phosphatidylinositide 3-kinase (PI 3-kinase) pathway, in particular, effects mediated through its effector protein kinase B (PKB, also termed Akt; three isoforms);

II/III. MAP kinases, especially the "classical" MAP kinase (ERK1/2) pathway and the p38 MAP kinase α/β pathway;

IV. The mammalian target of rapamycin (mTOR) pathway.

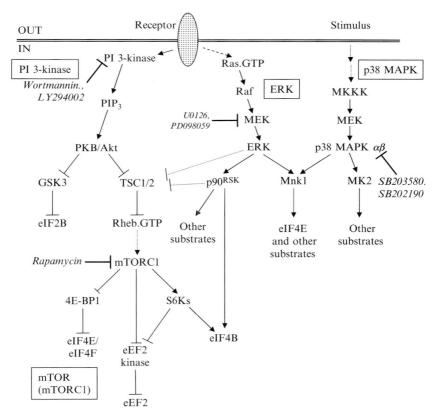

Figure 7.1 Major signaling pathways relevant to this chapter. Simplified schematic diagram of the major signaling pathways which impinge on mRNA translation: I. The phosphatidylinositide 3-kinase (PI 3-kinase) pathway; II/III. MAP kinases, especially the "classical" MAP kinase (ERK) pathway and the p38 MAP kinase pathway; IV. The mammalian target of rapamycin (mTOR) pathway. Strictly, this diagram shows the rapamycin-sensitive events linked to mTORC1. Selected inhibitors and their sites of action are shown. A number of components and cross-connections have been omitted for clarity.

In each case, a brief description of each pathway is provided, to orient the reader, together with information on the use (and limitations) of small molecule inhibitors for that pathway, and procedures that can be employed to assess pathway activation.

A number of examples are known of cross-talk between pathways, which may complicate the analysis of their roles in controlling translation, since inhibition of one may actually activate another. One example is the ability of the mTOR pathway (through its downstream component, the S6 kinases) to inhibit the activation of PI 3-kinase signaling by insulin (through the phosphorylation of insulin receptor substrate 1, IRS1 (Harrington *et al.*, 2004; Um *et al.*, 2004)). As a consequence of this "feedback" mechanism, inhibiting mTOR signaling with rapamycin may actually activate PI 3-kinase signaling (see next section).

1.1. The PI 3-kinase pathway

The family of PI 3-kinase-related kinases contains several enzymes (Abraham, 2004), which are related to mTOR. Several are protein kinases. The enzyme discussed here is PI 3-kinase itself, a lipid kinase that is activated at the plasma membrane. The principal product is phosphatidyl-inositide 1,3,4-trisphosphate (PIP_3). Phosphoinositides bind to a range of target proteins that contain pleckstrin homology (PH), PX, or FYVE domains (Lindmo and Stenmark, 2006). The most relevant of these for the present discussion is PKB (Akt). Activation of PI 3-kinase leads to the phosphorylation and activation of PKB which, in turn, phosphorylates a number of substrates, the most important ones for the present discussion being the tuberous sclerosis complex (protein) 2 (TSC2, an upstream regulator of mTOR, see later; Manning and Cantley, 2003) and glycogen synthase kinase 3. GSK3 also acts on TSC2 (Inoki *et al.*, 2006) and phosphorylates and inhibits the translation factor eIF2B (Welsh *et al.*, 1998) (Fig. 7.1). The phosphorylation of GSK3 by PKB impairs the activity of GSK3 toward certain substrates such as eIF2B (eukaryotic initiation factor 2B (Welsh *et al.*, 1998)). The inhibition of GSK3 activity against TSC2 and eIF2B exerts a positive regulatory influence on the translational machinery. eIF2B is a GDP-dissociation stimulator protein (Williams *et al.*, 2001) that serves to generate the active GTP-bound form of eIF2, which is required to bring the initiator methionyl-tRNA to the ribosome for every initiation event. TSC2 is a negative regulator of mTOR complex 1 (mTORC1), and GSK3 appears to activate TSC2.

1.1.1. Inhibitors

Two small cell-permeant molecule inhibitors of PI 3-kinase are widely used (Table 7.1). Wortmannin (Arcaro and Wymann, 1993) is a fungal product that reacts covalently with a residue in the active site of PI 3-kinase,

Table 7.1 Commonly used signaling inhibitors

Compound	Target used for	Reference	Comments
Wortmannin	PI 3-kinase	(Arcaro and Wymann, 1993)	May also inhibit mTOR (Brunn et al., 1996)
LY294002	PI 3-kinase	(Vlahos et al., 1994)	Also inhibits mTOR (Brunn et al., 1996)
Rapamycin	mTORC1	(Sabers et al., 1995)	Long-term treatment may also affect mTORC2 (Sarbassov et al., 2006)
PD098059	MEK1/2 (ERK pathway)	(Alessi et al., 1995; Dudley et al., 1995)	Also inhibits MEK5 (Kamakura et al., 1999) and activates AMPK (Dokladda et al., 2005)
PD184352	MEK1/2 (ERK pathway)	(Davies et al., 2000)	
U0126	MEK1/2 (ERK pathway)	(Favata et al., 1998)	Also inhibits MEK5 (Kamakura et al., 1999) and activates AMPK (Dokladda et al., 2005)
SB203580	p38 MAP kinase α/β	(Cuenda et al., 1995)	May also inhibit PDK1 (Lali et al., 2000)
SB202190	p38 MAP kinase α/β	See (Davies et al., 2000)	May also inhibit PDK1 (Lali et al., 2000)
CGP57380	Mnk1/2	(Tschopp et al., 2000)	Specificity not fully characterized
BI-D1870	$p90^{RSK}s$ (Rsks)	(Sapkota et al., 2007)	

A number of the signaling inhibitors described in the text are listed here. More detailed information on the specificities of a range of signaling (especially protein kinase) inhibitors is provided in Bain et al., 2003 and Davies et al., 2000.

inactivating it irreversibly. Wortmannin is highly unstable in aqueous solution and must be stored, for example, in dimethylsulfoxide (DMSO), and added to the cell culture medium immediately before initiating the experiment. Use of this compound in long-term experiments may require its re-addition (to inhibit any newly synthesized PI 3-kinase catalytic subunits).

LY294002 (Cheatham *et al.*, 1994) does not suffer from this drawback, but is not an irreversible inhibitor.

Both compounds inhibit other PIKKs *in vitro*, including mTOR (Brunn *et al.*, 1996). This imposes a major limitation on the interpretation of data obtained using LY294002 (since any inhibition of a downstream target that is caused by LY294002 may reflect the involvement of PI 3-kinase or of mTOR). In the case of wortmannin, one may be able to select concentrations that inhibit PI 3-kinase but not mTOR, but it is crucial to conduct appropriate tests to establish this selectivity (Wang and Proud, 2002a). For example, the phosphorylation state of PKB can be assessed by western blot analysis as a surrogate marker for PI 3-kinase activation (see later) and the phosphorylation of the N-terminal threonines in 4E-BP1 can be used as indicators for any inhibition of mTOR (Wang *et al.*, 2005). (These are Thr37/46 and Thr36/45 in the human and rodent 4E-BP1 polypeptides, respectively.)

1.1.2. Assays for pathway activation

PI 3-kinase activity may be measured in a radioactive assay (see, for example, Herbert *et al.*, 2000). The activation state of PKB can be assessed using phosphospecific antisera directed against sites that are associated with the activation of PKB[1], that is, Thr308 and Ser473. In the authors' experience, the commercially available anti-(P)Ser473 antisera are much more sensitive than the anti-(P)Thr308 antisera, and are accordingly used much more widely. One should note a 2006 report that suggests that phosphorylation of Ser473 may not be required for the activation of PKB against all substrates (Jacinto *et al.*, 2006): it is thus possible that studying the phosphorylation of this site is not a completely reliable indicator of PKB activation.

As mentioned, GSK3 is phosphorylated by PKB (resulting in its inactivation against certain substrates (Cohen and Frame, 2001)). Thus, it is possible to use phosphospecific antisera against the relevant site in GSK3 (Ser9 in GSK3β, generally the more abundant isoform; Ser21 in GSK3α) as a "surrogate"' readout of PKB activation. However, one needs to be aware that GSK3 isoforms can also be phosphorylated on the same sites through the ERK (Sutherland *et al.*, 1993) and mTOR (Sutherland and Cohen, 1994) pathways (by p90[RSK]s and S6 kinases; see later). Thus, for example,

---

[1] There are actually three different isoforms of PKB (Akt) in mammalian cells, but since their mechanisms of activation appear to be similar, this complexity need not concern us here.

failure of a PI 3-kinase inhibitor to block GSK3 phosphorylation could indicate the involvement of the ERK pathway.

When using signaling inhibitors, it is very important to determine that the compound being used is truly 100% effective: since signaling cascades often have the potential for amplification, even a relatively low level of residual activity in an upstream component may suffice to allow substantial activation of downstream targets of the pathway. This is likely to be especially problematic when looking at later timepoints.

1.2. Signaling linked to MAP kinases

MAP kinase signaling modules typically comprise a cassette of three protein kinases (Fig. 7.2A). For example, ERK (two isoforms in mammalian cells, ERK1/2) is phosphorylated (at Thr and Tyr residues in the "activation loop" of its catalytic domain by an upstream MAP kinase kinase (in this case, MEK) which is, in turn, phosphorylated and activated by a MAP kinase kinase kinase, in this instance, a member of the Raf group) (Fig. 7.2A).

The p38 MAP kinase pathway shows a similar organization (Figs. 7.1, 7.2A). In fact, there are four related p38 MAP kinase isoforms in mammals, termed p38 MAP kinases α–δ. Most work has focused on the α and β isoforms, which regulate downstream protein kinases involved in controlling mRNA translation and stability and are inhibited by compounds such as SB203580 and SB202190, which do not inhibit p38 MAP kinases γ or δ. p38 MAP kinases α/β are activated by MKK3 and MKK6, and probably also by MKK4 (Kang et al., 2006).

The ERKs and p38 MAP kinases α/β can directly phosphorylate proteins involved in cell regulation, but also phosphorylate and activate downstream kinases, in a fourth tier of the kinase cascade (Fig. 7.2A). Most relevant here are (1) the p90[RSK]s, (2) the Mnks, and (iii) MK2. The p90[RSK]s comprise a family of four kinases that are activated by ERK-mediated phosphorylation (Roux and Blenis, 2004). They phosphorylate several proteins that are relevant to translational control, including TSC2, eIF4B (Shahbazian et al., 2006), and eEF2 kinase (Wang et al., 2001). As noted, they can also phosphorylate GSK3. The name "RSK" derives from early work that found that they could phosphorylate ribosomal protein S6, although it is now clear that the major S6 kinases in mammals are the S6 kinases that are activated by mTOR signaling (Pende et al., 2004). However, p90[RSK]s may contribute to S6 phosphorylation under some conditions (Roux et al., 2007).

There are two genes for the Mnks in man and mice (Mnk1 and Mnk2) and, at least in human cells, each gives rise to two different transcripts by alternative splicing (O'Loghlen et al., 2004; Scheper et al., 2003). These, in turn, generate proteins that differ at their C-termini (the longer forms are termed Mnk1a and Mnk2a, the shorter ones, Mnk1b and Mnk2b). Mnk1a

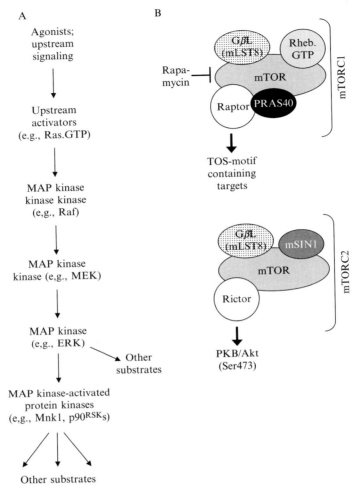

Figure 7.2 Organization of specific signaling modules. (A) The arrangement of upstream and downstream kinases in a typical MAP kinase module is shown, the named examples being from the classical MAP kinase (ERK) pathway. (B) The known composition of mTORC1 and mTORC2 is shown.

and Mnk2a each possess, in their longer C-terminal regions, sites for interaction with ERK and p38 MAP kinases α/β (although Mnk2a mainly binds ERK (Parra *et al.*, 2005; Waskiewicz *et al.*, 1997)). Mnk2a has high activity, even in serum-starved cells (Scheper *et al.*, 2001) and this is not greatly affected either by stimulation of the cells or blockade of the ERK/p38 MAP kinase α/β pathways. Mnk1a has low basal activity, which is greatly enhanced by agents that activate ERK or p38 MAP kinase α/β (Wang *et al.*, 1998; Waskiewicz *et al.*, 1999), and these effects are blocked by inhibiting these pathways. Mnk2b (Scheper *et al.*, 2003), and perhaps Mnk1b (see O'Loghlen *et al.*, 2004), have low activities and it is not clear whether or how these are controlled physiologically. The Mnks are the kinases that phosphorylate eIF4E *in vivo* (Ueda *et al.*, 2004). A small number of other Mnk substrates has been identified, including RNA-binding proteins such as HnRNP A_1 (Buxade *et al.*, 2005).

MK2 (also termed MAP kinase-activated protein kinase 2, MAPKAP-K2) is activated by p38 MAP kinase α/β (Kotlyarov *et al.*, 2002; Roux and Blenis, 2004). MK2 plays a key role in the control of the production of certain cytokines, for example, tumor necrosis factor α. MK2 does so by phosphorylating proteins that bind specifically to the regulatory regions in the 3′-untranslated regions (UTRs) of such mRNAs (Hitti *et al.*, 2006). These regions contain "AU-rich elements" (AREs) to which proteins such as HnRNP A_1 also bind.

1.2.1. Inhibitors of MAP kinase signaling

As mentioned previously, SB203580 and SB202190 are used as specific inhibitors of p38 MAP kinases α/β (Lee *et al.*, 1994) (see Table 7.1). No specific inhibitors of the ERKs are yet available, but several compounds have been described as inhibiting the upstream kinases (MEK1/2; Table 7.1). The first of these was PD098059, which appears to inhibit activation of the MEKs rather than their catalytic activities (Alessi *et al.*, 1995). However, this compound, and another widely used MEK1/2 inhibitor, U0126, also inhibit MEK5, which lies in a distinct pathway (ERK5 signaling) (Nishimoto and Nishida, 2006). U0126 also stimulates the AMP-activated protein kinase, AMPK (Dokladda *et al.*, 2005). PD184352 is both more potent and more specific than these two compounds, but is not yet commercially available.

BI-D1870 (Table 7.1) has been reported to be a specific p90[RSK] inhibitor that lacks activity against other closely related kinases (in the so-called AGC family, which includes the S6 kinases and PKB [Sapkota *et al.*, 2007]).

CGP57380 inhibits both Mnk1 and Mnk2, of mouse or human origin (Knauf *et al.*, 2001; Lee and Proud, unpublished data). It showed some activity against other members of a panel of around 30 kinases that were tested, although its IC50 against Mnk1 or Mnk2 was 10× lower (better) than against the next most sensitive enzyme.[2] Like BI-D1870, CGP57380 is cell permeant although the concentrations required to inhibit the phosphorylation of eIF4E range up to 40 μM (see Buxade *et al.*, 2005, among others), depending upon the cell type. The specificity of CGP57380 at such concentrations remains to be established (although the available data

---

[2] The IC50 values commonly reported in the literature are derived from *in vitro* studies using ATP concentrations (e.g., 100?M) that are much lower than those that exist within cells. Since many kinase inhibitors, including CGP57380, are competitive with respect to ATP, the concentrations required to block kinase activity *in vivo* will often be much higher than those that are effective *in vitro*.

indicate that it did not interfere with the activation of ERK, p38 MAP kinases α/β or PKB, at least in Jurkat T-cells; Buxadé *et al.*, 2005).

To date, no specific inhibitors of MK2 have been reported. However, since MK2 is activated by p38 MAP kinases α/β, use of SB203580 will prevent its activation (but also interferes with the control of other kinases, such as Mnk1a, that are activated by the same pathway).

"Dominant-interfering" mutants are often employed to explore the roles of specific pathway components in cellular regulation. For example, ectopic overexpression of a kinase-dead mutant of a particular kinase may block downstream signaling by binding upstream regulators and preventing them from turning on the endogenous wildtype version of that kinase within the cells. However, this approach is fraught with problems, especially when a pathway divides. For example, an overexpressed kinase-dead version of Mnk1a would be expected to bind ERK and p38 MAP kinase, and thus to block activation of the endogenous Mnk1a (Fig. 7.2A). However, since ERK and p38 MAP kinases are also required for activation of Mnk2a, as well as of kinases such as the p90^{RSK}s and MK2, such an approach could actually interfere with multiple downstream components of these pathways and the resulting data must be treated with great caution. In particular, it is crucial to establish that the expression of a specific protein does not interfere with the activation or function of other components in the same or related signaling pathways.

1.2.2. Assays for pathway activation

Phosphospecific antisera are available that recognize the T-loop (activating) phosphorylation sites in both the ERKs (ERK1/2) and p38 MAP kinases α/β, as well as in the Mnks. Since compounds such as PD098059 and PD184352 block the ERK pathway at MEK, above the level of ERK, the effectiveness of pathway blockade can be assessed by looking at the T-loop phosphorylation of ERK1/2. However, because SB203580 (etc.) actually act to inhibit the activity of p38 MAP kinases α/β, studying the phosphorylation of these kinases themselves is inappropriate as a guide to pathway blockade and one must study a downstream kinase to verify the efficacy of pathway inhibition. MK2 is a good choice because it is activated only by p38 MAP kinases α/β (unlike Mnk1a, which is also activated by ERK). Cross-talk may be observed (for one example, see Zhang *et al.*, 2001).

Direct kinase assays can be performed for all of these enzymes following their immunoprecipitation from cell lysates (see protocols later).

1.3. mTOR

mTOR is a large, multidomain protein that possesses a kinase domain related to the lipid kinase PI 3-kinase but shows protein kinase activity *in vitro*. mTOR binds several different proteins, including raptor and rictor, and forms two distinct types of complex; mTORC1 (contains raptor, Fig. 7.2B) and mTORC2 (contains rictor, Fig. 7.2B). Rapamycin inhibits at least some functions of mTORC1, while mTORC2 is not sensitive to this drug (Wullschleger *et al.*, 2006). mTORC1 regulates several proteins linked to the control of the translational machinery, including 4E-BP1, the S6 kinases, and eEF2 kinase (Wang and Proud, 2006). mTORC2 phosphorylates PKB/Akt on Ser473 (Sarbassov *et al.*, 2005). Signaling through mTORC1 is activated by a variety of agents including insulin, growth factors, and some G-protein-coupled receptor agonists. Amino acids, especially leucine, also positively regulate mTORC1 signaling. In many types of cells, agents such as insulin are unable to turn on mTORC1 signaling in the absence of amino acids. Amino acids also maintain a basal level of signaling through mTORC1 in serum-starved cells. There is currently no information on the regulation of mTORC2.

The mechanism by which mTORC1 is activated by hormones (and other compounds) is perhaps best explained by reference to a well-studied activator of mTORC1, insulin. Insulin, via PI 3-kinase, activates PKB/Akt which, in turn, phosphorylates two main sites (S939 and T1462, in the most widely used numbering system) in TSC2. TSC2, together with TSC1, acts as a GTPase-activator protein (GAP) for the small G-protein Rheb, converting it to its GDP-bound form. Even nucleotide-free Rheb can bind mTOR but only the Rheb.GTP complex can stimulate its kinase activity (at least *in vitro*; Long *et al.*, 2005). PKB-mediated phosphorylation of TSC2 is believed to inactivate the GAP function of TSC1/2 (Cai *et al.*, 2006), thereby allowing Rheb to accumulate in its GTP-bound form and activating mTORC1. A number of other mechanisms for the control of TSC1/2 appear to exist, including its activation by signaling through ERK (perhaps mediated by the p90[RSK]s) and its inhibition by GSK3 and by the AMP-activated protein kinase, a sensitive detector of changes in cellular energy status. The latter links the control of mTOR to cellular ATP levels.

1.4. Downstream targets

mTORC1 modulates the phosphorylation of a number of proteins associated with the translational machinery. These include:

1. The ribosomal protein S6 kinases and their substrate S6. mTORC1 positively regulates the S6 kinases. There are two S6K genes in mammalian cells, S6K1 and S6K2. mTORC1 can directly phosphorylate S6K1 at a key regulatory site, Thr389 (in the shorter of the two splice variants of this enzyme). The S6Ks contain a TOR-signaling (TOS) motif that appears to interact directly with raptor. The best-known S6K substrate is S6 itself. In amino acid-fed cells, the phosphorylation of S6 is greatly stimulated by agents such as insulin and is completely blocked by rapamycin. There are

actually multiple rapamycin-sensitive sites of phosphorylation in S6K1 (and, likely, in S6K2, although this isoform has been studied in much less detail).

2. 4E–BP1 also contains a TOS motif and can be phosphorylated by mTORC1 at several sites *in vitro*. When "fully" phosphorylated, 4E–BP1 no longer binds to eIF4E. This release from eIF4E involves the phosphorylation of sites near the eIF4E-binding site, that is, Thr70[3] and, probably, Ser65. Phosphorylation of these sites requires prior phosphorylation of Thr37/46. Amino acids alone often suffice to promote the phosphorylation of these two sites, a process that is largely insensitive to rapamycin, while Ser65 is generally only phosphorylated in response to agents such as insulin, which are thought to activate mTORC1, and this is sensitive to rapamycin. The control of the phosphorylation of Thr70 is complicated, and does not always behave in a manner consistent with its being controlled by mTOR.

3. eEF2 kinase is an unusual protein kinase outside the main eukaryotic protein kinase families. It is subject to a variety of control mechanisms. Its activity is normally completely dependent upon calcium ions, which stimulate eEF2 kinase via calmodulin (CaM). Other positive inputs include direct phosphorylation of eEF2 kinase by cAMP-dependent protein kinase (PKA) and by AMPK. Conversely, eEF2 kinase is inhibited by phosphorylation at any of three sites that are controlled by mTOR (Herbert and Proud, 2006): Ser78[3] (phosphorylation of which inhibits CaM-binding); Ser359 (inhibits activity even at saturating Ca^{2+}-ion levels), and Ser366 (inhibits activity at submaximal Ca^{2+} concentrations). Ser366 is phosphorylated by S6K1, providing a link to mTOR, and also by p90[RSK], creating a connection to MEK/ERK signaling. The kinases acting at the other sites are not known: mTORC1 does not phosphorylate eEF2 kinase directly (Smith and Proud, unpublished data).

1.4.1. Inhibitors

Rapamycin is widely used as an inhibitor of mTORC1 and does indeed very effectively block a number of events downstream of mTOR, for example, the activation of the S6 kinases and the phosphorylation of S6. As mentioned, signaling downstream of mTORC2 is not sensitive to rapamycin, at least in the short term (a few hours), although in the long term, it may be affected (Sarbassov *et al.*, 2006). In addition, rapamycin does not inhibit all the effects attributed to mTORC1, such as, the phosphorylation of the N-terminal sites in 4E–BP1 (Thr37/46 in human cells; see later for further discussion). As noted already, inhibitors of PI 3-kinase, especially LY294002, also inhibit mTOR (Brunn *et al.*, 1996).

---

[3] This numbering is based on the sequence of human 4E–BP1. Numbers for the rodent protein are shifted by −1.

1.4.2. Assays for pathway activation

There are a number of ways in which the control of the mTOR pathway can be studied. One can either study upstream regulators of the pathway or downstream targets. Phosphospecific antisera are available for TSC2, for example, for the PKB sites (Ser939, Thr1462). The so-called "PKB substrate" antibody has been used for this purpose (Manning et al., 2002) but this antibody also recognizes another site in TSC2 that is phosphorylated in response to ERK signaling (Rolfe et al., 2005), likely by one or more p90[RSK] isoform(s). In principle, activation of Rheb can be assessed by radiolabeling cells with ^{32}P-orthophosphate and immunoprecipitating Rheb, followed by analysis of the amounts of GDP and GTP bound to Rheb by thin-layer chromatography (TLC) (Smith et al., 2005).

A phosphospecific antibody for Ser2448 in mTOR itself is widely used as a readout of mTOR "activation." However, there is little evidence that phosphorylation of this site actually activates mTOR. In fact, it has been shown to be a target for S6K (Holz and Blenis, 2005), and is thus likely to report S6K1 activity (an output from mTORC1).

Analysis of downstream targets of mTORC1 is generally both simpler and more informative.

1. S6 kinase–phosphorylation and activity: The activation state of S6K1 may be assessed by virtue of the reduced mobility on SDS–PAGE that it displays associated with its phosphorylation. This is best observed on gels containing 10% acrylamide and allowed to run for 40% longer than usual. Phosphospecific antisera are available for several sites in S6K1. That for (P)Thr389 is likely the most widely used. (This is the residue phosphorylated by mTOR in vitro and its phosphorylation appears to closely match overall kinase activity (Avruch et al., 2001).)
The activities of S6K1 or S6K2 may be assessed by direct (radioactive) kinase assay using a 32-amino acid peptide (Moule et al., 1995), 40 S ribosomal subunits (Chen and Blenis, 1990), or S6, expressed as a fusion protein with GST (Patti et al., 1998).
2. S6: phosphospecific antisera are available for several sites in S6 (Weng et al., 1998). Phosphorylation of all these sites is generally strongly inhibited by rapamycin (see Weng et al., 1998, for further information).
3. 4E-BP1 runs anomalously on SDS–PAGE: its true molecular mass is less than 15 kDa but it runs at around 22 kDa. Furthermore, phosphorylation alters its mobility, such that the more highly phosphorylated species migrate more slowly. Optimal resolution can be achieved using gels containing 13.5% (w/v) acrylamide and 0.36% (w/v) methylene bis-acrylamide. On such gels, three species (usually termed α–γ) can usually be resolved, sometimes more. Phosphospecific antisera are available for Ser65, Thr70, and Thr37/46. Some batches of the Ser65 antibody recognize another site in human 4E-BP1 (Ser101). The sequences

around Thr37 and Thr46 are so similar that the antiserum recognizes both sites: as far as the authors are aware, it has not been possible to generate antisera that recognize (P)Thr37 but not (P)Thr46 or vice versa. Use of these antisera to analyze 4E-BP1 isoforms resolved on the gel system just described reveals that the slowest migrating (γ) species is the only one phosphorylated upon Ser65, and does not bind to eIF4E. The β and γ species contain molecules phosphorylated on Thr70, and all three are phosphorylated on Thr37 and/or Thr46. The anti-(P)Thr37/46 antibody also recognizes the corresponding sites in 4E-BP2, which runs just below 4E-BP1 on SDS-PAGE. As remarked elsewhere in this chapter, the phosphorylation of endogenous 4E-BP1 at Thr37/46 is dependent upon amino acids but is little affected by agents such as insulin or rapamycin. It serves as a good indicator of the status of amino acid-sensitive mTOR signaling. When it is overexpressed, phosphorylation of 4E-BP1 at these sites may show a greater sensitivity to insulin/rapamycin (see, e.g., Wang *et al.*, 2003, 2005). It is crucial to note that, following transfer, blots for 4E-BP1 must be "fixed" (using glutaraldehyde) to prevent the protein's leaching from the membrane.

2. Experimental Protocols

2.1. General procedures

This section deals with the basic approaches to treating cells with hormones (or other agents) and with signaling inhibitors, and for the lysis of cells after treatment.

2.1.1. Cells

To study the regulation of the signaling events that regulate the translational machinery, mammalian cells are generally starved of serum prior to treatment with the agonist of choice. The purpose of this is to ensure that signaling pathways are minimally active prior to stimulation of the cells. It is also important to ensure that cells (in the case of one growing in monolayer) are not too confluent because this may impair their response to stimulation. Typically, cells that are estimated to be around 70% confluent are starved for serum overnight by transferring them to medium lacking serum. For some types of cells, this period of serum withdrawal may create problems because they may detach from the dish and/or start to undergo apoptosis: if so, shorter periods of serum withdrawal are indicated. (Cells should be examined under the microscope.)

In any study, it is crucial to begin by examining the time courses of activation of signaling components and translation factors, because these may, for example, display a delay or transient kinetics. Thus, by studying a fixed timepoint, one may miss important events. In general, upstream regulators such as PKB/Akt or ERK are likely to be activated earlier (within 5 min or faster), and their activation may subsequently decline, although in some cells, it can be sustained for 1 h or longer. Events downstream of mTOR (S6 kinases, and S6 or 4E–BP1 phosphorylation) are often turned on substantially later (by 30 to 60 min, although this is quite variable). Thus, one may need to stimulate cells for differing times to study different downstream events.

Having established a suitable timepoint for the component of interest, the researcher may find it useful to examine the dose-response relationship for the agonist under study.

Cell lysis   When studying signaling components, it is important that the integrity and phosphorylation states of proteins of interest are not altered during cell lysis and the subsequent preparation of the samples prior to analysis. The principal precautions include the use of inhibitors of protein phosphatases and of proteases, in addition to working speedily and keeping samples cold (0 to 4°).

Useful serine/threonine protein phosphatase inhibitors include microcystin-LR (which inhibits protein phosphatases 1, 2A, and 2C, and related enzymes) and β-glycerophosphate. Sodium fluoride may also be employed. Sodium orthovanadate inhibits protein tyrosine phosphatases.

In this laboratory, we also include the metal ion chelators EDTA (ethylene diamine tetraacetic acid; binds, e.g., Mg^{2+}-ions) and EGTA (ethylene glycol-bis(2-aminoethyl)-N,N,N',N',-tetraacetic acid; binds, e.g., Ca^{2+}-ions) in our lysis buffers. These agents help prevent phosphatase action (by the metal ion-dependent phosphatase PP2C, which is not inhibited by microcystin-LR), metal (Ca^{2+}) dependent proteinases, and protein kinases, which require divalent cations such as Mg^{2+} (and, in some cases, also Ca^{2+}). We also use a mix of proteinase inhibitors that inhibit a broad range of proteolytic enzymes, including serine and cysteine proteinases.

The concentration of protein in the lysate should be determined as a guide to even loading of gels or the amount of material to be subjected, for example, to immunoprecipitation. A simple and reliable method for this is that of Bradford (Bradford, 1976).

Lysis buffer   20 mM Tris/HCl, pH 7.5, 50 mM β-glycerophosphate, 0.5 mM EGTA, 0.5 mM EDTA, 1% (v/v) Triton-X 100, 14 mM β-mercaptoethanol, 0.5 mM Na_3VO_4, and proteinase inhibitors (1 mM benzamidine, 1 μg/ml pepstatin, 1 μg/ml leupeptin, 1 μg/ml antipain, and 0.2 mM phenylmethylsulfonyl fluoride [the last added immediately before use, because this compound rapidly hydrolyzes in aqueous solution]).

3. ANALYSIS OF THE PHOSPHORYLATION STATES OF TRANSLATION FACTORS AND SIGNALING COMPONENTS—OVERVIEW

Two principal methods are available: (1) western blotting using phosphospecific antisera and (2) isoelectric focusing followed by western blotting. Protocols based on mass spectroscopic methods may also be valuable, especially where either the phosphorylation site has not been identified or the protein contains multiple sites of phosphorylation.

3.1. Use of phosphospecific antisera

Western blotting using phosphospecific antisera is widely used. A wide range of phosphospecific antisera is available for translation factors and the signaling components that control them (e.g., upstream kinases or other regulators). This approach is relatively straightforward and allows the analysis of proteins that contain multiple phosphorylation sites. It is crucial to establish that the antibody is truly "phosphospecific" because such antisera may also recognize the nonphosphorylated form of the protein, thus indicating falsely high basal levels of phosphorylation. This will also tend to blunt any changes in phosphorylation by diminishing the dynamic range of the phosphoantibody. The recombinant protein expressed in *Escherichia coli* (which lacks kinases known to phosphorylate mammalian translation factors) may prove to be a useful "negative control." A useful maneuver is to include the nonphosphorylated version of the peptide used to make the phosphospecific antibody in the solution in which the primary antibody is diluted: this should block any reaction with the dephosphorylated protein by binding to any antibodies that recognize this form of the target protein. Since milk powder contains casein, which is multiply phosphorylated and can nonspecifically "sequester" some phosphospecific antisera, dilute the primary antibody with serum albumin instead of nonfat milk.

For any experiments involving the use of phosphospecific antisera, it is crucial to develop the blot (usually a separate one) with an antibody that detects the protein of interest irrespective of its state of phosphorylation. This "loading control" allows one to assess whether all lanes contain similar amounts of that protein. Ideally, the ratios of the signal strengths for the "phospho" and "total" antisera should be calculated from densitometric analysis of the blots. Care must be taken when doing this for proteins that resolve into multiple bands (dependent on their states of phosphorylation, such as 4E–BP1, S6K1) but, if sufficient resolution is achieved, it may be possible to calculate the ratios of the signal strengths of the different species. When no suitable total antibody is available, an alternative (abundant, cytosolic) protein may be studied instead, such as S6, tubulin, or actin, although this is clearly less satisfactory.

Two limitations of this approach are that (1) it requires both that the phosphorylation site(s) have been identified and that a truly phosphospecific antibody has been generated and (2) the method gives no information on the absolute levels of phosphorylation: a ten-fold increase in signal might reflect a change from 1 to 10% phosphorylation, or from 10 to 100%. Such information is important in interpreting the likely impact of the change on the process of translation.

3.1.1. Protein immunoblot procedures

After protein concentrations have been determined, cell lystates should be normalized for total protein content. In some cases (low abundance/weakly reacting antisera), it may be necessary to immunoprecipitate the protein under study prior to analysis. See later for the procedure.

For immunoblot analysis, 20 to 40 mg of proteins are separated by SDS-PAGE. The proteins are transferred to PVDF (Immobilon®) or nitrocellulose membrane. (Note that some membranes do not work well for certain antisera.) The membranes are then blocked with 5 (w/v) nonfat dry milk for 1 h. This is followed by incubation overnight at 4° with the appropriate primary antibody diluted in PBS-Tween containing either 5% bovine serum albumin (for phosphospecific antibodies) or 5% (w/v) nonfat dry milk (for other antibodies). After rigorous washes with PBS-Tween, the membranes are incubated for 1 h with, for example, horseradish peroxidase-conjugated secondary antibody diluted in PBS-Tween containing nonfat milk. Washes are repeated as before, and the positive signals are visualized, with chemiluminescence reagent and X-ray film or by other procedures, such as LiCor Odyssey® imager. Fluorescently-labeled secondary antisera are required for that system.

For analysis of 4E–BP1, transferred proteins must be fixed to the membrane by incubation in 0.05% glutaraldehyde in PBS-Tween (10 mM Tris/HCl, pH 8.0, 150 mM NaCl, 0.02% Tween) for 15 to 20 min prior to blocking.

3.1.2. Isoelectric focusing (IEF) analysis

The second widely used approach is isoelectric focusing (IEF): this technique relies on the fact that the addition of a phosphate group to a protein introduces an additional ionizable group, which will shift the isoelectric point (pI) of the protein to a lower pH. IEF resolves proteins by virtue of differences in their pI values. In general, addition of a single phosphate group will have a larger effect on the pI of a small protein than on a large one. Forms of a protein differing in their pIs can be resolved on a gel in which a pH gradient has been established using ampholines (ampholytes). The proteins resolved on the IEF gel are transferred to a membrane and revealed by immunoblotting with an antibody that detects the protein irrespective of its state of phosphorylation. It is important to note that (1) because the proteins are resolved in the absence of SDS, which gives proteins a substantial negative charge, electrophoretic transfer from IEF gels is less efficient than for SDS gels (thus decreasing sensitivity) and that (2) since any charged reagent that binds the protein will alter its mobility in IEF, ionic detergents are generally to be avoided. Nonetheless, the zwitterionic detergent CHAPS (3-[(3-cholamidopropyl) dimethylam-monio]-1-propanesulfonate) is frequently employed when preparing lysates and samples for IEF.

This method is really only suitable for proteins with a single (regulated) site of phosphorylation, although, luckily, several key translation factors fall into this category—eIF4E, eIF2α, and eEF2, for example. Analysis of proteins with multiple regulated phosphorylation sites is complex, both because this will create several species with distinct pIs and because, at least in principle, and probably quite frequently in practice, one site may increase in phosphorylation while another decreases, thereby potentially obscuring important changes at regulatory sites.

A significant advantage of this method is that is provides quantitative information: since the IEF technique actually separates the phospho- and dephospho- forms of the protein, the ratio of these species as detected by IEF/immunoblotting is a direct readout of the protein's phosphorylation status within the cell (assuming no selective loss of one species during isolation and analysis).

3.2. Isoelectric focusing (IEF) analysis

The IEF mixture is poured between the gel plates up to the top and the comb is inserted (no stacking gel is required). Once the gel has set, the comb is removed and the wells are rinsed with distilled water. Any potential leaking points of the gel apparatus need to be sealed with 2% agarose as communication between the upper and lower chambers, must be avoided.

The IEF gel mixture: 9.0 M urea, 2% (w/v) CHAPS, 6% (w/v) acrylamide, 0.3% (w/v) bis-acrylamide, 4.5% (v/v) ampholyte (pH 4 to 6.5), 1.5% (v/v) ampholyte (pH 3.5 to 10), 0.05% (w/v) ammonium persulphate, 0.25% (v/v) TEMED, and H$_2$O to 5 ml/gel. For analysis of eEF2 phosphorylation by IEF, gels containing 5% (w/v) acrylamide and 4% (v/v) ampholyte (pH 3.5 to 10) are used.

The IEF gel is prefocused for 1 h using reverse polarity with the following buffer (kept cold): cathode (lower tank): 0.05 M histidine; anode (upper tank): 0.01 M glutamic acid. Prefocusing is for 20 min at 200 V, then 20 min at 300 V, and finally 20 min at 400 V. The current should be around 3 to 10 mA/mini gel. After prefocusing, the wells should be washed thoroughly with water using a syringe. 10 μl of IEF overlay buffer is applied to each well first, then 20 μl of samples carefully loaded beneath the overlay.

The IEF gel is run at 500 to 750 V for 3 h (reverse polarity), the voltage being increased by 50 V every half-hour, and finally run at 1000 V for 30 min (the current should be <10 mA/gel). After the run is completed, the proteins are transferred to PVDF (Immobilon®) membrane followed by western blotting.

IEF overlay: 6 M urea in 10 mM glutamic acid

Sample preparation: 27 mg urea, 20 μl sample, 7 μl 7×7 sample buffer

Sample buffer (7×): 21% ampholytes (same pH range as in gel: 3:1 volume of pH 4 to 6.5 and pH 3.5 to 10 ampholyte), 15% β-mercaptoethanol, 35% CHAPS.

1×sample buffer: 0.54 g urea (to give 9 M), 43 μl of 7×sample buffer, make up to 1 ml with water

Purification of eIF4E from cell lysates for IEF: eIF4E and associated proteins are isolated from cell extract by affinity chromatography m^{7}GTP-Sepharose (as described later) and the beads were washed twice with 1 ml of lysis buffer. 18 μl of m^{7}GTP (100 μM) is added to the beads and incubated for 15 min at 4°. After centrifugation at 7000 rpm for 30 s, the supernatant is mixed with 7× sample buffer and urea (see previously), and loaded onto prefocused IEF gel.

Preparation of eIF2 from cell lysates for IEF: 60 μl of fast-flow Sepharose S (prewashed with lysis buffer) and 500 μg of cell lysate are mixed for 2 h at 4°. After centrifugation, the supernatant is removed and the beads are washed twice with lysis buffer containing 200 mM KCl. The eIF2 should be eluted with 50 μl of lysis buffer containing 400 mM KCl. 20 μl of eluate is then mixed with 7× sample buffer and urea (see previously), and loaded onto a prefocused IEF gel.

3.3. General immunoprecipitation method

One ml of lysis buffer should be added to 10 μl/IP of protein G–Sepharose. The tube should be flicked to mix the contents and then centrifuged at 8000 rpm for 1 min. The supernatant is removed, 500 μl of lysis buffer, 1 to 2 μg/IP of the relevant antibody are added, and the tube is rotated at 4° for 1 h. The beads are collected after a quick spin and washed twice with 1 ml of lysis buffer. Total cell lysate is added to the beads and the final volume increased to 500 μl with lysis buffer. The tube is rotated at 4° for 1 to 2 h, and the beads are washed twice with 1 ml of lysis buffer containing 0.5 M NaCl before further analysis. To reduce any nonspecific binding, the total lysate can be preabsorbed with protein G Sepharose for 1 h at 4° in order to remove components that bind nonspecifically to the beads. Alternatively, the extract may be incubated with the specific antibody for 1 h at 4°, and the preformed immune complex is then absorbed to protein G Sepharose. Antibody, protein G Sepharose, and cell lysate are all mixed together in a total volume of 500 μl of lysis buffer for 2 h at 4°, followed by extensive washing with this buffer.

3.4. Immunoprecipitation for western blot analysis

The immune complexes (beads) are washed 3 times with lysis buffer, and then 20 μl of 1× SDS-PAGE sample buffer are added and the samples boiled for 5 min. The supernatant is subjected to SDS-PAGE, followed by western blotting.

3.5. Immunoprecipitation for protein kinase assays using peptide substrates

The immobilized immunoprecipitates are washed twice with lysis buffer containing 0.5 M NaCl and twice with buffer A. The beads are resuspended in 20 μl of kinase buffer also containing the appropriate concentration of the specific peptide. Reactions should also be set up without peptide as a "negative control" for nonspecific or self-incorporation of radiolabel. To start the reactions, 5 μl of ATP is added (final concentration: 0.1 mM unlabeled ATP, 1 μCi [γ -^{32}P]ATP (per assay) in kinase buffer). The assays are allowed to proceed for 15 to 30 min at 30° with constant shaking at 900 rpm, and stopped by spotting 20 μl of the sample (slurry) onto a square (1.5 × 1.5 cm) of phosphocellulose (P81) paper. The P81 papers are immediately immersed in 500 ml of 1% (v/v) orthophosphoric acid, and then washed 3 times with the same solution (to remove the excess ATP). The washes therefore contain almost all of the radiolabel and must be handled carefully and disposed of appropriately. The papers are briefly rinsed in ethanol and air-dried. The incorporation of ^{32}P-label is measured by Cerenkov counting.

4. ASSAYS FOR SPECIFIC PROTEIN KINASES

It is essential to establish that the assay procedure is linear with time and input kinase over the range under study. This can be accomplished by performing pilot experiments using a range of amounts of kinase and taking samples for analysis at several times.

4.1. Immunoprecipitation for kinase assay against a protein substrate

The procedure is as outlined previously, but the immunoprecipitates are washed and resuspended in 20 μl of kinase buffer also containing specific substrate (e.g., purified protein). To start the reactions, 5 μl of ATP is added (final concentration: 0.1 mM unlabeled ATP, 1 μCi [γ-^{32}P]ATP in kinase buffer). The reactions are allowed to proceed for 10 to 30 min at 30° with constant shaking at 900 rpm and stopped by adding 7 μl of 5× SDS–PAGE sample buffer. Samples are boiled for 5 min and the supernatants are then analyzed by SDS–PAGE. The amount of ^{32}P incorporated into the substrate may be assessed by autoradiograph or quantitated by phosphorimager.

Buffer A: 50 mM Tris/HCl, pH 7.5, 0.1 mM EGTA, 14 mM β-mercaptoethanol

Kinase buffer: 50 mM Tris/HCl, pH 7.5, 0.1 mM EGTA, 10 mM MgCl₂, 1 mM DTT, to give a final ATP concentration of 100 μM

Peptide substrates: For each kinase assay, an appropriate substrate is used together with [γ-^{32}P]ATP. For the p90^{RSK}s, "Crosstide" (GRPTR SSFEAG; Wang and Proud, 2002b), the "long S6 peptide" (KEAKEKR-QEQIAKRRRLSSLRASTSKSGGSQK; Sapkota *et al.*, 2007), or the "short S6 peptide" (Poteet-Smith *et al.*, 1999) may be used as substrate. To check for any background or self-incorporation, it is essential to run controls from which the peptide substrate is omitted, and to subtract that radioactivity observed in these samples from the incorporation of label seen in the other samples.

For MK2, one may use a peptide substrate (see, e.g., Manke *et al.*, 2005) or recombinant heat shock protein (hsp) 25 (Stokoe *et al.*, 1992). For the peptide substrate, analysis is as described for the p90^{RSK} assay. If using hsp27 as substrate, samples of the reaction products are analyzed by SDS–PAGE and autoradiography or quantitation performed by phosphorimager.

For the Mnks, no peptide substrate is yet available, because they show very low activity, for example, against peptides based on the sequence around the phosphorylation site in eIF4E. However, a simple assay can be configured using recombinant eIF4E as substrate. eIF4E may be made in *E. coli* (Stern *et al.*, 1993). The protein is mainly in inclusion bodies that are denatured using urea and then allowed to renature. Correctly folded eIF4E can be purified by affinity chromatography on 7-methyl GTP-Sepharose (immobilized cap-analog). Much of the recombinant eIF4E may not bind to this material (because it is incorrectly folded), but may be subjected to a further cycle of denaturation–renaturation to retrieve more native eIF4E. The Mnk assay is performed essentially as for the MK2 assay but using eIF4E as substrate.

MAPKAPK2 assay: Each reaction contains 0.5 μg of hsp27 (10 μl), 10 μg of total cell extract (10 μl), and 6 μl of following mix: 5 μl of 5× kinase buffer, 0.25 μl 10 mM ATP, 1 μCi [γ-^{32}P]ATP, and H₂O. After incubation at 30° for 10 to 15 min, the reaction is stopped by adding 7 μl of 5 × SDS–PAGE sample buffer. The sample is boiled for 5 min and loaded onto 15% SDS–PAGE, followed by autoradiography.

5× kinase buffer: 100 mM HEPES/NaOH, pH7.4, 250 mM KCl, 50 mM MgCl₂, 25% (v/v) glycerol. Before use, add DTT and Na₃VO₄ (to 1 ml of 5× kinase buffer, add 5 μl 1 M DTT and 5 μl 0.5 M Na₃VO₄).

Note: Reactions are also set up without hsp27 for each condition, as a negative control.

Mnk activity assay (using eIF4E as substrate): For each reaction, 10 μl of recombinant eIF4E (total 200 ng) is added to 10 μl of Mnk1/2 (e.g., GST-tagged fusion protein expressed in *E. coli* or Myc-tagged enzyme immunoprecipitated from Mnk-transfected HEK 293 cells; see method; note that Mnk2 is more active against eIF4E *in vitro* than Mnk1, so less Mnk2 could be used). To initiate the kinase reactions, add to each reaction tube 6 μl of the following mix: 5 μl of 5× kinase buffer, 0.25 μl 10 mM ATP, 1 μCi $[\gamma\text{-}^{32}P]ATP$, and H_2O. The reaction is carried out in a thermostatted shaker at 30°, shaking at 900 rpm for 15 min, and stopped by the addition of 7 μl of 5× SDS-PAGE sample buffer. The samples are immediately boiled for 5 min and the supernatants are loaded onto 12.5% SDS-PAGE, followed by autoradiography or phosphorimager analysis.

Mnk assay (using "S6 peptide" as substrate): Each reaction contains 20 μl of immunoprecipitated or purified recombinant Mnk, as described previously, and S6 peptide (KEAKEKRQEQIAKKRRLSSLRASTSKSESSQK, final concentration = 200 μM) in kinase buffer. To initiate the reaction, 6 μl of the following mix is added: 5 μl 5× kinase buffer, 0.25 μl 10 mM ATP, 1 μCi $[\gamma\text{-}^{32}P]ATP$, and H_2O. Reactions are carried out in an orbital shaker at 30° and 900 rpm, for example, for 20 min (Mnk1) or for 5 min (Mnk2) and stopped by spotting 20 μl of the reaction mixture onto P81 paper. The filters are washed in 1% orthophosphoric acid three times. The radioactivity is determined by using a liquid scintillation counter.

S6 kinase assay using GST-S6 or 40 S subunits: This is performed according to the general method described previously but using GST-S6 (Patti *et al.*, 1998) or 0.5 A_{260} units of 40 S ribosomal subunits per assay (Chen and Blenis, 1990).

eEF2 kinase assay: eEF2 kinase is immunoprecipitated as previously (see "Immunoprecipitation for kinase assay"). After washing, the beads are resuspended in 30 μl of kinase buffer containing 50 mM MOPS pH7.2, 10 mM $MgCl_2$, 1 mM DTT, 50 to 150 pmol eEF2, 0.4 mM EGTA, 2 mM HEDTA, 0.67 mM $CaCl_2$, 0.1 mM unlabeled ATP and 1 μCi $[\gamma\text{-}^{32}P]ATP$. If required, calmodulin is added to a final concentration of 2 $\mu g/ml$.

Reactions are allowed to proceed for 25 to 30 min at 30° with constant shaking at 900 rpm and then stopped by adding 8 μl of 5× SDS-PAGE sample buffer, followed by boiling for 5 min. Supernatants are applied to 10% SDS-PAGE, followed by autoradiography or phosphorimaging.

Note: to measure Ca^{2+}-independent activity, reactions are also set up for without $CaCl_2$, but EGTA (1 mM, final) should be added instead to remove any Ca^{2+} in the solutions used for the assay.

The activation states of the p90[RSK]s, MK2, and the Mnks may also be assessed by western blotting procedures. In the authors' experience, identifying really reliable "readouts" based on substrates for the p90[RSK]s can be challenging because of overlaps in specificity between these enzymes and other related kinases, such as PKB and the S6 kinases. The phosphorylation state of Ser380 in p90^{RSK}, which is phosphorylated as part of the mechanisms of activation of the p90^{RSK}s, may provide an indicator of the activation state of the p90^{RSK}s[4]. A phosphospecific antibody is available for this site. Phosphorylation of the protein kinase LKB1 at Ser431 is a useful readout of p90^{RSK} activity (see Sapkota *et al.*, 2007). For MK2, monitoring the phosphorylation of hsp27, using the commercially available phosphospecific antisera, is useful (hsp27).

5. ASSAYS FOR INITIATION FACTOR FUNCTION

5.1. eIF2B

5.1.1. Immunoprecipitation method for eIF2B assay (Immunoprecipitation of endogenous eIF2B complexes)

The endogenous eIF2B complexes are isolated using the immunoprecipitation method as previously described. The immune complexes are washed twice with lysis buffer and once with TKD buffer, and resuspended in 10 μl TKD. To start the reactions, 10 μl of prelabeled [^{3}H]GDP–eIF2 is added. Each reaction is continued for 5 to 20 mins at 25 to 30° (on a thermostatted shaker at 900 rpm) depending on the activities. The reactions are terminated by adding 1 ml of ice-cold wash buffer + 10 mg/ml of BSA and immediately all contents are loaded onto cellulose nitrate filters (using a manifold, e.g., from Millipore). After rinsing three times with wash buffer, the filters are removed from manifold and briefly dried at 110°. The samples are counted in minivials using 3 ml of scintillation fluid.

Note: Because eIF2B activity in cell lysates is sensitive to freeze–thaw, lysates should be stored as aliquots.

Two key controls are (1) to measure the amount of [^{3}H]GDP initially bound to the eIF2 and (2) to set up a control assay, lacking added eIF2B, to assess the extent of any dissociation of the eIF2.[^{3}H]GDP complexes during the assay. Up to 10% dissociation is acceptable. Any such dissociation must be taken into account when calculating the activity of the test samples. Assays should be performed at least in triplicate. It is also crucial to test the linearity of the assay. If >30% of the [^{3}H]GDP initially bound to the eIF2 is released, the assay is unlikely to be linear.

TKD buffer: 20 mM Tris/HCl, pH7.5, 50 mM KCl, 1 mM DTT

Labeling of eIF2 with [^{3}H]GDP: Purified eIF2 (0.5 μg) is incubated with 0.3 μCi of [^{3}H]GDP in buffer containing 20 mM Tris/HCl, pH7.5,

---

[4] p90RSKs actually contain two kinase domains, the C-terminal, one of which is activated following phosphorylation by ERKs, and this domain then phosphorylates other residues in the enzyme, leading to activation of the N-terminal kinase domain, which catalyzes transphosphorylation (of other substrates).

10 mg/ml BSA, 1 mM DTT, 50 mM KCl (10 μl per reaction) for 20 min at 30°. The same volume of TKD is added, mixed, and kept on ice; MgCl$_2$ and then GTP are added at the final concentration of 5 mM and 0.5 mM, respectively, vortexed, and kept on ice after each addition.

Wash buffer: 50 mM Tris/HCl, pH7.5, 50 mM KCl, 5 mM MgCl$_2$.

5.1.2. His-pull down method for eIF2B assay

This is an alternate procedure used for studying the effects of mutations in eIF2B subunits on the function of the eIF2B complex. HEK 293 cells are transfected with five subunits of eIF2B (1 μg of DNA encodes for each subunit). We routinely use myc-tagged versions of all five subunits, with eIF2Bε also having a hexahistidine tag. After 2 days, cells are lysed with His-pull down lysis buffer. After analyzing the expression of eIF2B 5 subunits by SDS-PAGE and western blotting, equal amounts of the eIF2B complexes containing the recombinant His-tagged eIF2B subunits are isolated on Ni-nitrilotriacetic acid (NTA) agarose (Qiagen).

Typically, 15 μl of Ni-NTA beads are washed twice with His-pull down buffer supplement with 20 mM imidazole. Total cell lysates are added to the beads and the contents/tube is rotated for 1 to 1.5 h at 4°. The beads are washed twice with the buffer and once with TKD; the rest of the procedure is the same as previously described.

Lysis buffer for His-pull down: 25 mM HEPES/NaOH, pH7.5, 25 mM β-glycerophosphate, 50 mM KCl, 10% (v/v) glycerol, 0.5% (v/v) Triton X-100, 14 mM β-mercaptoethanol, 0.5 mM Na$_3$VO$_4$, 1 mM benzamidine, 1 μg/ml pepstatin, 1 μg/ml leupeptin, 1 μg/ml antipain, and 0.2 mM phenylmethylsulfonyl fluoride.

5.2. Analysis of eIF4F complex formation

4E-BP1 competes with the scaffold protein eIF4G for binding to eIF4E (Mader *et al.*, 1995). Thus, in terms of the control of translation initiation, the key parameter is not the phosphorylation state of 4E-BP1/2 per se, but rather the degree of association of 4E-BP1/2 or eIF4G with eIF4E. There are two eIF4G genes in mammals (eIF4G$_I$ and eIF4G$_{II}$), the former having been studied much more frequently.

eIF4E and its direct partners, the 4E-BPs and eIF4Gs, can easily be isolated from cell lysates by affinity chromatography using m^{7}GTP immobilized on Sepharose CL-4B, which is available commercially. Although it is, in principle, possible to elute the bound proteins from the resin using a buffer containing m^{7}GTP (or m^{7}GDP) prior to analysis by SDS-PAGE/western blot, in practice such elution is not very efficient. We therefore routinely add SDS-PAGE sample buffer directly to the packed resin, denature the proteins by heating at 95° for 5 min, and then recentrifuge the samples to allow the supernatant, containing the denatured and eluted proteins, to be applied to the gel.

5.2.1. 7-methyl GTP (m⁷GTP) pull-down for analysis of eIF4F complexes

eIF4E and associated proteins are isolated from cell extracts by affinity chromatography m^{7}GTP-Sepharose. Typically, 15 μl m^{7}GTP beads and 10 μl CL-4B Sepharose beads are mixed and washed once with 1 ml of lysis buffer (spin at 7 k for 30 s to remove supernatant). Lysates (250 to 400 μg total protein) are added to the beads and the final volume is made to 500 μl with lysis buffer. The contents are rotated for 1 h at 4°. The beads are washed three times with 1 ml of lysis buffer, and 20 μl of 1 × SDS-PAGE sample buffer is added. The samples are boiled for 5 min, and supernatant is loaded onto 12.5% SDS-PAGE. After transferring the proteins to PVDF membrane, the blot is fixed with 0.05% glutaraldehyde in PBS-Tween (0.02%) for 15 to 20 min before it is blocked with 5% (w/v) nonfat dry milk. The membrane is cut at around standard marker 70 kDa, the top and the bottom parts of the blot are incubated with eIF4G and 4E-BP1 antibodies, respectively, and the bottom part is then reprobed with anti–eIF4E. The signal for eIF4E serves as a "loading control," to which the signals for eIF4G and 4E-BP1 are normalized.

Although eIF4G (apparent MW about 220 kDa) is much larger than eIF4E (about 28 kDa) and 4E-BP1 (apparent size 22 kDa), all three proteins can successfully be analyzed on the same 10% acrylamide minigel, provided it is stopped when the dye-front is 1 cm from the bottom. Blots can then be developed with antisera for the proteins of choice, including eIF4E as a key control for normalization of the other signals. It is important to bear in mind that data from western blots for, for example, eIF4G and 4E-BP1 do not give the *actual ratios* of these proteins that are bound to eIF4E since signals from different antisera cannot be directly compared: this method can only report changes in the relative levels of these proteins associated with eIF4E.

5.3. Protein synthesis assays

Cells are maintained in suitable medium and starved of serum, if required. Agonist/inhibitors are added for 30 min first, then [^{35}S]Met (5 μCi/ml medium) is added for 60 to 120 min. Cells are then lysed and cleared lysates (equal amounts of proteins) are loaded onto 3 MM papers. After the samples have been allowed to soak into the paper, the papers are boiled in 5% (w/v) TCA three times (2 min each time), rinsed with 100% ethanol, and oven dried. [^{35}S]methionine incorporation is measured in 3 ml of scintillation fluid.

In many cases, signaling inhibitors may affect the basal rate of protein synthesis. It is crucial to take this into account when evaluating their effects on the activation of protein synthesis, and appropriate controls must be carried out to assess the effects of the inhibitors on the rate of protein synthesis in unstimulated cells. Two other important points are (i) that assays must be performed at least in duplicate and (ii) that cells should be no more than 85% confluent.

REFERENCES

Abraham, R. T. (2004). PI 3-kinase related kinases: "Big" players in stress-induced signaling pathways. *DNA Repair (Amst.)* **3,** 883–887.

Alessi, D. R., Cuenda, A., Cohen, P., Dudley, D. T., and Saltiel, A. R. (1995). PD-098059 is a specific inhibitor of the activation of mitogen-activated protein kinase *in vitro* and *in vivo*. *J. Biol. Chem.* **270,** 27489–27494.

Arcaro, A., and Wymann, M. P. (1993). Wortmannin is a potent phosphatidylinositol 3-kinase inhibitor: The role of phosphatidylinositol 3,4,5-trisphosphate in neutrophil responses. *Biochem. J.* **296,** 297–301.

Avruch, J., Belham, C., Weng, Q., Hara, K., and Yonezawa, K. (2001). The p70 S6 kinase integrates nutrient and growth signals to control translational capacity. *Prog. Mol. Subcell. Biol.* **26,** 115–154.

Bain, J., McLauchlan, H., Elliott, M., and Cohen, P. (2003). The specificities of protein kinase inhibitors: An update. *Biochem. J.* **371,** 199–204.

Bradford, M. M. (1976). A rapid and sensitive method for the quantitation of microgram quantities of protein utilizing the principle of protein-dye binding. *Anal. Biochem.* **77,** 248–254.

Brunn, G. J., Williams, J., Sabers, C., Weiderrecht, G., Lawrence, J. C., and Abraham, R. T. (1996). Direct inhibition of the signaling functions of the mammalian target of rapamycin by the phosphoinositide 3-kinase inhibitors, wortmannin and LY294002. *EMBO J.* **15,** 5256–5267.

Buxade, M., Parra, J. L., Rousseau, S., Shpiro, N., Marquez, R., Morrice, N., Bain, J., Espel, E., and Proud, C. G. (2005). The Mnks are novel components in the control of TNFalpha biosynthesis and phosphorylate and regulate hnRNP A1. *Immunity* **23,** 177–189.

Cai, S. L., Tee, A. R., Short, J. D., Bergeron, J. M., Kim, J., Shen, J., Guo, R., Johnson, C. L., Kiguchi, K., and Walker, C. L. (2006). Activity of TSC2 is inhibited by AKT-mediated phosphorylation and membrane partitioning. *J. Cell Biol.* **173,** 279–289.

Cheatham, B., Vlahos, C. J., Cheatham, L., Wang, L., Blenis, J., and Kahn, C. R. (1994). Phosphatidylinositol 3-kinase activation is required for insulin stimulation of pp70 S6 kinase, DNA synthesis, and glucose transporter translocation. *Mol. Cell. Biol.* **14,** 4902–4911.

Chen, R. H., and Blenis, J. (1990). Identification of Xenopus S6 protein kinase homologs (pp90rsk) in somatic cells: Phosphorylation and activation during initiation of cell proliferation. *Mol. Cell. Biol.* **10,** 3204–3215.

Cohen, P., and Frame, S. (2001). The renaissance of GSK3. *Nat. Rev. Mol. Cell Biol.* **2,** 769–776.

Cuenda, A., Rouse, J., Doza, Y. N., Meier, R., Cohen, P., Gallagher, T. F., Young, P. R., and Lee, J. C. (1995). SB-203580 is a specific inhibitor of a MAP kinase homolog which is activated by cellular stresses and interleukin-1. *FEBS Lett.* **364,** 229–233.

Davies, S. P., Reddy, H., Caivano, M., and Cohen, P. (2000). Specificity and mechanism of action of some commonly used protein kinase inhibitors. *Biochem. J.* **351,** 95–105.

Dokladda, K., Green, K. A., Pan, D. A., and Hardie, D. G. (2005). PD98059 and U0126 activate AMP-activated protein kinase by increasing the cellular AMP:ATP ratio and not via inhibition of the MAP kinase pathway. *FEBS Lett.* **579,** 236–240.

Dudley, D. T., Pang, L., Decker, S. J., Bridges, A. J., and Saltiel, A. R. (1995). A synthetic inhibitor of the mitogen-activated protein kinase cascade. *Proc. Natl. Acad. Sci. USA* **92,** 7686–7689.

Favata, M. F., Horiuchi, K. Y., Manos, E., Daulerio, A. J., Stradley, D. A., Feeser, W. S., Van Dyk, D. E., Pitts, W. J., Earl, R. A., Hobbs, F., Copeland, R. A., Magolda, R. L., *et al.* (1998).

Identification of a novel inhibitor of mitogen-activated protein kinase kinase. *J. Biol. Chem.* **273,** 18623–18632.

Harrington, L. S., Findlay, G. M., Tolkacheva, T., Gray, A., Wigfield, S., Rebholz, H., Barnett, J., Leslie, N. R., Cheng, S., Shepherd, P. R., Gout, I., Downes, C. P., *et al.* (2004). The TSC1-2 tumor suppressor controls insulin-PI3K signaling via regulation of IRS proteins. *J. Cell Biol.* **166,** 213–223.

Herbert, T. P., Kilhams, G. R., Batty, I. H., and Proud, C. G. (2000). Distinct signaling pathways mediate insulin and phorbol ester-stimulated eIF4F assembly and protein synthesis in HEK 293 cells. *J. Biol. Chem.* **275,** 11249–11256.

Herbert, T. P., and Proud, C. G. (2006). Regulation of translation elongation and the cotranslational protein targeting pathway. *In* "Translational Control in Biology and Medicine" (M. B. Mathews, N. Sonenberg, and J. W. B. Hershey, eds.), pp. 601–624. Cold Spring Harbor Laboratory Press, Cold Spring, NY.

Hitti, E., Iakovleva, T., Brook, M., Deppenmeier, S., Gruber, A. D., Radzioch, D., Clark, A. R., Blackshear, P. J., Kotlyarov, A., and Gaestel, M. (2006). Mitogen-activated protein kinase-activated protein kinase 2 regulates tumor necrosis factor mRNA stability and translation mainly by altering tristetraprolin expression, stability, and binding to adenine/uridine-rich element. *Mol. Cell Biol.* **26,** 2399–2407.

Holz, M. K., and Blenis, J. (2005). Identification of S6 kinase 1 as a novel mammalian target of rapamycin (mTOR)-phosphorylating kinase. *J. Biol. Chem.* **280,** 26089–26093.

Inoki, K., Ouyang, H., Zhu, T., Lindvall, C., Wang, Y., Zhang, X., Yang, Q., Bennett, C., Harada, Y., Stankunas, K., Wang, C. Y., He, X., *et al.* (2006). TSC2 integrates Wnt and energy signals via a coordinated phosphorylation by AMPK and GSK3 to regulate cell growth. *Cell* **126,** 955–968.

Jacinto, E., Facchinetti, V., Liu, D., Soto, N., Wei, S., Jung, S. Y., Huang, Q., Qin, J., and Su, B. (2006). SIN1/MIP1 maintains rictor-mTOR complex integrity and regulates Akt phosphorylation and substrate specificity. *Cell* **127,** 125–137.

Kamakura, S., Moriguchi, T., and Nishida, E. (1999). Activation of the protein kinase ERK5/BMK1 by receptor tyrosine kinases. *J. Biol. Chem.* **274,** 26563–26571.

Kang, Y. J., Seit-Nebi, A., Davis, R. J., and Han, J. (2006). Multiple activation mechanisms of p38alpha mitogen-activated protein kinase. *J. Biol. Chem.* **281,** 26225–26234.

Knauf, U., Tschopp, C., and Gram, H. (2001). Negative regulation of protein translation by mitogen-activated protein kinase-interacting kinases 1 and 2. *Mol. Cell. Biol.* **21,** 5500–5511.

Kotlyarov, A., Yannoni, Y., Fritz, S., Laass, K., Telliez, J. B., Pitman, D., Lin, L. L., and Gaestel, M. (2002). Distinct cellular functions of MK2. *Mol. Cell Biol.* **22,** 4827–4835.

Lali, F. V., Hunt, A. E., Turner, S. J., and Foxwell, B. M. (2000). The pyridinyl imidazole inhibitor SB203580 blocks phosphoinositide-dependent protein kinase activity, protein kinase B phosphorylation, and retinoblastoma hyperphosphorylation in interleukin-2-stimulated T cells independently of p38 mitogen-activated protein kinase. *J. Biol. Chem.* **275,** 7395–7402.

Lee, J. C., Laydon, J. T., McDonnell, P. C., Gallagher, T. F., Kumar, S., Green, D., McNulty, D., Blumenthal, M. J., Heys, J. R., Landvatter, S. W., Strickler, J. E., McLaughlin, M. M., *et al.* (1994). A protein kinase involved in the regulation of inflammatory cytokine biosynthesis. *Nature* **372,** 739–746.

Lindmo, K., and Stenmark, H. (2006). Regulation of membrane traffic by phosphoinositide 3-kinases. *J. Cell Sci.* **119,** 605–614.

Long, X., Lin, Y., Ortiz-Vega, S., Yonezawa, K., and Avruch, J. (2005). Rheb binds and regulates the mTOR kinase. *Curr. Biol.* **15,** 702–713.

Mader, S., Lee, H., Pause, A., and Sonenberg, N. (1995). The translation initiation factor eIF-4E binds to a common motif shared by the translation factor eIF-4gamma and the translational repressors 4E-binding proteins. *Mol. Cell. Biol.* **15,** 4990–4997.

Manke, I. A., Nguyen, A., Lim, D., Stewart, M. Q., Elia, A. E., and Yaffe, M. B. (2005). MAPKAP kinase-2 is a cell cycle checkpoint kinase that regulates the G2/M transition and S phase progression in response to UV irradiation. *Mol. Cell* **17**, 37–48.

Manning, B. D., and Cantley, L. C. (2003). United at last: The tuberous sclerosis complex gene products connect the phosphoinositide 3-kinase/Akt pathway to mammalian target of rapamycin (mTOR) signaling. *Biochem. Soc. Trans.* **31**, 573–578.

Manning, B. D., Tee, A. R., Logsdon, M. N., Blenis, J., and Cantley, L. C. (2002). Identification of the tuberous sclerosis complex-2 tumor suppressor gene product tuberin as a target of the phosphoinositide 3-kinase/akt pathway. *Mol. Cell* **10**, 151–162.

Moule, S. K., Edgell, N. J., Welsh, G. I., Diggle, T. A., Foulstone, E. J., Heesom, K. J., Proud, C. G., and Denton, R. M. (1995). Multiple signaling pathways involved in the stimulation of fatty acid and glycogen synthesis by insulin in rat epididymal fat pads. *Biochem. J.* **311**, 595–601.

Nishimoto, S., and Nishida, E. (2006). MAPK signaling: ERK5 versus ERK1/2. *EMBO Rep.* **7**, 782–786.

O'Loghlen, A., Gonzalez, V. M., Pineiro, D., Perez-Morgado, M. I., Salinas, M., and Martin, M. E. (2004). Identification and molecular characterization of Mnk1b, a splice variant of human MAP kinase-interacting kinase Mnk1. *Exp. Cell Res.* **299**, 343–355.

Parra, J. L., Buxade, M., and Proud, C. G. (2005). Features of the catalytic domains and C termini of the MAPK signal-integrating kinases Mnk1 and Mnk2 determine their differing activities and regulatory properties. *J. Biol. Chem.* **280**, 37623–37633.

Patti, M.-E., Brambilla, E., Luzi, L., Landaker, E. J., and Kahn, C. R. (1998). Bidirectional modulation of insulin action by amino acids. *J. Clin. Invest.* **101**, 1519–1529.

Pende, M., Um, S. H., Mieulet, V., Sticker, M., Goss, V. L., Mestan, J., Mueller, M., Fumagalli, S., Kozma, S. C., and Thomas, G. (2004). S6K1(−/−)/S6K2(−/−) mice exhibit perinatal lethality and rapamycin-sensitive 5′-terminal oligopyrimidine mRNA translation and reveal a mitogen-activated protein kinase-dependent S6 kinase pathway. *Mol. Cell Biol.* **24**, 3112–3124.

Poteet-Smith, C. E., Smith, J. A., Lannigan, D. A., Freed, T. A., and Sturgill, T. W. (1999). Generation of constitutively active p90 ribosomal S6 kinase *in vivo*. Implications for the mitogen-activated protein kinase-activated protein kinase family. *J. Biol. Chem.* **274**, 22135–22138.

Rolfe, M., McLeod, L. E., Pratt, P. F., and Proud, C. G. (2005). Activation of protein synthesis in cardiomyocytes by the hypertrophic agent phenylephrine requires the activation of ERK and involves phosphorylation of tuberous sclerosis complex 2 (TSC2). *Biochem. J.* **388**, 973–984.

Roux, P. P., and Blenis, J. (2004). ERK and p38 MAPK-activated protein kinases: A family of protein kinases with diverse biological functions. *Microbiol. Mol. Biol. Rev.* **68**, 320–344.

Roux, P. P., Shahbazian, D., Vu, H., Holz, M. K., Cohen, M. S., Taunton, J., Sonenberg, N., and Blenis, J. (2007). RAS/ERK signaling promotes site-specific ribosomal protein S6 phosphorylation via RSK and stimulates cap-dependent translation. *J. Biol. Chem.* **282**, 14056–14064.

Sabers, C. J., Martin, M. M., Brunn, G. J., Williams, J. M., Dumont, F. T., Wiederrecht, G., and Abraham, R. T. (1995). *J. Biol. Chem.* **270**, 815–822.

Sapkota, G. P., Cummings, L., Newell, F. S., Armstrong, C., Bain, J., Frodin, M., Grauert, M., Hoffmann, M., Schnapp, G., Steegmaier, M., Cohen, P., and Alessi, D. R. (2007). BI-D1870 is a specific inhibitor of the p90 RSK (ribosomal S6 kinase) isoforms *in vitro* and *in vivo*. *Biochem. J.* **401**, 29–38.

Sarbassov, D. D., Guertin, D. A., Ali, S. M., and Sabatini, D. M. (2005). Phosphorylation and regulation of Akt/PKB by the rictor-mTOR complex. *Science* **307**, 1098–1101.

Sarbassov, D. D., Ali, S. M., Sengupta, S., Sheen, J. H., Hsu, P. P., Bagley, A. F., Markhard, A. L., and Sabatini, D. M. (2006). Prolonged rapamycin treatment inhibits mTORC2 assembly and Akt/PKB. *Mol. Cell* **22**, 159–168.

Scheper, G. C., Morrice, N. A., Kleijn, M., and Proud, C. G. (2001). The MAP kinase signal-integrating kinase Mnk2 is an eIF4E kinase with high basal activity in mammalian cells. *Mol. Cell. Biol.* **21**, 743–754.

Scheper, G. C., Parra, J.-L., Wilson, M. L., van Kollenburg, B., Vertegaal, A. C. O., Han, Z.-G., and Proud, C. G. (2003). The N and C termini of the splice variants of the human mitogen-activated protein kinase-interacting kinase Mnk2 determine activity and localization. *Mol. Cell. Biol.* **23**, 5692–5705.

Shahbazian, D., Roux, P. P., Mieulet, V., Cohen, M. S., Raught, B., Taunton, J., Hershey, J. W., Blenis, J., Pende, M., and Sonenberg, N. (2006). The mTOR/PI3K and MAPK pathways converge on eIF4B to control its phosphorylation and activity. *EMBO J.* **25**, 2781–2791.

Smith, E. M., Finn, S. G., Tee, A. R., Browne, G. J., and Proud, C. G. (2005). The tuberous sclerosis protein TSC2 is not required for the regulation of the mammalian target of rapamycin by amino acids and certain cellular stresses. *J. Biol. Chem.* **280**, 18717–18727.

Stern, B. D., Wilson, M., and Jagus, R. (1993). Use of nonreducing SDS-PAGE for monitoring renaturation of recombinant protein synthesis initiation factor, eIF-4 alpha. *Protein Expr. Purif.* **4**, 320–327.

Stokoe, D., Engel, K., Campbell, D. G., Cohen, P., and Gaestel, M. (1992). Identification of MAPKAP kinase 2 as a major enzyme responsible for the phosphorylation of the small mammalian heat shock proteins. *FEBS Lett.* **313**, 307–313.

Sutherland, C., Alterio, J., Campbell, D. G., Le Bourdelles, B., Mallet, J., Haavik, J., and Cohen, P. (1993). Phosphorylation and activation of human tyrosine hydroxylase *in vitro* by mitogen-activated protein (MAP) kinase and MAP-kinase-activated kinases 1 and 2. *Eur. J. Biochem.* **217**, 715–722.

Sutherland, C., and Cohen, P. (1994). The alpha-isoform of glycogen synthase kinase-3 from rabbit skeletal muscle is inactivated by p70 S6 kinase or MAP kinase-activated protein kinase-1 *in vitro*. *FEBS Lett.* **338**, 37–42.

Tschopp, C., Knauf, U., Brauchle, M., Zurini, M., Ramage, P., Glueck, D., New, L., Han, J., and Gram, H. (2000). Phosphorylation of eIF-4E on Ser 209 in response to mitogenic and inflammatory stimuli is faithfully detected by specific antibodies. *Mol. Cell Biol. Res. Commun.* **3**, 205–211.

Ueda, T., Watanabe-Fukunaga, R., Fukuyama, H., Nagata, S., and Fukunaga, R. (2004). Mnk2 and Mnk1 are essential for constitutive and inducible phosphorylation of eukaryotic initiation factor 4E but not for cell growth or development. *Mol. Cell Biol.* **24**, 6539–6549.

Um, S. H., Frigerio, F., Watanabe, M., Picard, F., Joaquin, M., Sticker, M., Fumagalli, S., Allegrini, P. R., Kozma, S. C., Auwerx, J., and Thomas, G. (2004). Absence of S6K1 protects against age- and diet-induced obesity while enhancing insulin sensitivity. *Nature* **431**, 200–205.

Vlahos, C. J., Matter, W. F., Hui, K. Y., and Brown, R. F. (1994). *J. Biol. Chem.* **269**, 5241–5248.

Wang, L., and Proud, C. G. (2002a). Ras/Erk signaling is essential for activation of protein synthesis by Gq protein-coupled receptor agonists in adult cardiomyocytes. *Circ. Res.* **91**, 821–829.

Wang, L., and Proud, C. G. (2002b). Regulation of the phosphorylation of elongation factor 2 by MEK-dependent signaling in adult rat cardiomyocytes. *FEBS Lett.* **531**, 285–289.

Wang, X., Beugnet, A., Murakami, M., Yamanaka, S., and Proud, C. G. (2005). Distinct signaling events downstream of mTOR cooperate to mediate the effects of amino acids and insulin on initiation factor 4E-binding proteins. *Mol. Cell Biol.* **25**, 2558–2572.

Wang, X., Flynn, A., Waskiewicz, A. J., Webb, B. L. J., Vries, R. G., Baines, I. A., Cooper, J., and Proud, C. G. (1998). The phosphorylation of eukaryotic initiation factor eIF4E in response to phorbol esters, cell stresses, and cytokines is mediated by distinct MAP kinase pathways. *J. Biol. Chem.* **273,** 9373–9377.

Wang, X., Li, W., Parra, J.-L., Beugnet, A., and Proud, C. G. (2003). The C-terminus of initiation factor 4E-binding protein 1 contains multiple regulatory features that influence its function and phosphorylation. *Mol. Cell. Biol.* **23,** 1546–1557.

Wang, X., Li, W., Williams, M., Terada, N., Alessi, D. R., and Proud, C. G. (2001). Regulation of elongation factor 2 kinase by p90^{RSK1} and p70 S6 kinase. *EMBO J.* **20,** 4370–4379.

Wang, X., and Proud, C. G. (2006). The mTOR pathway in the control of protein synthesis. *Physiology (Beth.)* **21,** 362–369.

Waskiewicz, A. J., Flynn, A., Proud, C. G., and Cooper, J. A. (1997). Mitogen-activated kinases activate the serine/threonine kinases Mnk1 and Mnk2. *EMBO J.* **16,** 1909–1920.

Waskiewicz, A. J., Johnson, J. C., Penn, B., Mahalingam, M., Kimball, S. R., and Cooper, J. A. (1999). Phosphorylation of the cap-binding protein eukaryotic translation factor 4E by protein kinase Mnk1 *in vivo. Mol. Cell. Biol.* **19,** 1871–1880.

Welsh, G. I., Miller, C. M., Loughlin, A. J., Price, N. T., and Proud, C. G. (1998). Regulation of eukaryotic initiation factor eIF2B: Glycogen synthase kinase-3 phosphorylates a conserved serine which undergoes dephosphorylation in response to insulin. *FEBS Lett.* **421,** 125–130.

Weng, Q.-P., Kozlowski, M., Belham, C., Zhang, A., Comb, M. J., and Avruch, J. (1998). Regulation of the p70 S6 kinase by phosphorylation *in vivo*: Analysis using site-specific anti-phosphopeptide antibodies. *J. Biol. Chem.* **273,** 16621–16629.

Williams, D. D., Loughlin, A. J., and Proud, C. G. (2001). Characterization of the initiation factor eIF2B complex from mammalian cells as a GDP-dissociation stimulator protein. *J. Biol. Chem.* **276,** 24697–24703.

Wullschleger, S., Loewith, R., and Hall, M. N. (2006). TOR signaling in growth and metabolism. *Cell* **124,** 471–484.

Zhang, H., Shi, X., Hampong, M., Blanis, L., and Pelech, S. (2001). Stress-induced inhibition of ERK1 and ERK2 by direct interaction with p38 MAP kinase. *J. Biol. Chem.* **276,** 6905–6908.

Analysis of mRNA Translation in Cultured Hippocampal Neurons

Yi-Shuian Huang* *and* Joel D. Richter[†]

Contents

Abstract

Synaptic plasticity, the ability of neuronal synapses to undergo morphological and biochemical changes in response to various stimuli, forms the underlying basis of long-term memory storage. Regulated mRNA translation at synapses is required for this plasticity. However, the mechanism by which translation at synapses is controlled and how the encoded proteins modulate persistent changes in synaptic morphology and functional integration in response to different input stimulations remain mostly unclear (Schuman *et al.*, 2006; Sutton and Schuman, 2006). One approach to investigating the relationship between protein synthesis and plasticity is to identify factors, such as RNA binding proteins that control translation in the neurons and then determine the identities of the mRNAs to which they are bound. Molecular and cellular techniques have been employed in cultured neurons to study sequence-specific RNA-binding proteins, for example, the Cytoplasmic Polyadenylation Element Binding protein (CPEB) (Huang *et al.*, 2002, 2003) and the Fragile-X Mental Retardation Protein (FMRP) (Vanderklish and Edelman, 2005; Zalfa *et al.*, 2006) for their functions in localizing and regulating translation of mRNAs. Although several CPE-containing neuronal RNAs that undergo activity-dependent polyadenylation (Du and Richter, 2005; Wu *et al.*, 1998) and FMRP-interacting mRNAs have been identified (Brown *et al.*, 2001;

* Division of Neuroscience, Institute of Biomedical Sciences, Academia Sinica, Taipei, Taiwan
† Program of Molecular Medicine, University of Massachusetts Medical School, Worcester, Massachusetts

Methods in Enzymology, Volume 431
ISSN 0076-6879, DOI: 10.1016/S0076-6879(07)31008-2

Miyashiro *et al.*, 2003), the validation of these targets whose translation is important for plasticity *in vivo* remains to be demonstrated (Darnell *et al.*, 2005).

In general, primary neurons in culture are difficult to manipulate. For example, they do not proliferate and their transfection efficiency is low ($\sim$1 to 10% of cells); this low efficiency is reduced even further as the cells age in culture, which hampers their practical use for biochemical analysis. When biochemical approaches are applied, they are often carried out in other more facile model systems, such as oocytes, in the case of CPEB, or in brains derived from knockout mice, for both CPEB and FMRP. However, the development of various viral delivery systems, shRNA knockdown techniques, reporter assays with high sensitivity, and neuron culture protocols have allowed investigators to analyze translational control in these cells, which may ultimately be used to investigate key mechanisms of synaptic plasticity. We have employed these procedures to investigate the function of CPEB3, a novel RNA-binding protein, in primary rat hippocampal neurons (Huang *et al.*, 2006); here, we describe the experimental details of our methods, which could be used for any RNA binding protein.

1. Materials and Reagents for Primary Neuron Culture

Cell culture reagents with catalog numbers listed here are purchased from Invitrogen.

1× MEM medium

400 ml H$_2$O
50 ml of 10× Minimum Essential Medium with Earle's salts (cat. # 11430–030)
15 ml 20% glucose
15 ml 7.5% sodium bicarbonate
2 ml 1N HCl
5 ml 200 m*M* glutamine (cat. # 25030–081)
sterilized by filtration through a 0.22 μm filter and stored at 4° for 6 months
with no negative effects on cell culture.

Add 5 ml Antibiotic-Antimycotic (100×, cat.# 15240–062) and 50 ml heat-inactivated horse serum (cat.# 16050–122) to make complete MEM (used for neuron plating and glial culture). For neuron plating, the complete MEM could be kept for a couple of months at 4°. However, for glial culture, a freshly thawed serum aliquot should be used for each culture. Obvious growth retardation in glial culture is observed when using 1-month-old complete MEM stored at 4°.

Heat-inactivated horse serum

A 500-ml bottle of frozen serum is thawed overnight at 4° and swirled to thoroughly mix the contents prior to incubation in a 56° waterbath for 30 min. The serum is then aliquoted to 50-ml conical tubes and stored at −20° for a year without any problem.

HBSS dissecting medium

445 ml H$_2$O
50 ml 10× Hanks' Balanced Salt Saline (cat.# 14185-052)
5 ml 1 *M* Hepes, pH7.3
sterilized by filtration through a 0.22-μm filter and stored at room temperature. Pre-cool the buffer at 4° and add 5 ml of Antibiotic-Antimycotic
(100X, cat.# 15240-062) prior to dissection.

2.5% Trypsin solution (10×)

The 2.5% trypsin solution (cat.# 15090-046) is thawed overnight at 4°,
aliquoted 1 ml/tube, and stored at −20°.

Neurobasal culture medium

500 ml Neurobasal medium (cat.# 21103-049)
1.3 ml 200 m*M* glutamine (cat.# 25030-081)
5 ml Antibiotic-Antimycotic (100×, cat.# 15240-062)

Generally, a half bottle of the above medium is mixed with a half bottle
of B27 supplement (50×, 10 ml, cat.# 17504-044) and the Neurobasal
culture medium is used within 2 weeks. The other half-bottle of B27
supplement will be kept frozen at −20° and thawed one more time for
use in culture. If only a small quantity of culture medium is needed, aliquot
the B27 supplement to 2 ml/tube, which is good for 100 ml of medium.

Neurobasal transfection medium

500 ml Neurobasal medium (cat.# 21103-049)
1.3 ml 200 m*M* glutamine (cat.# 25030-081)

Dissecting tools

large scissors (cat.# 14001-18)
fine scissors (cat.# 14090-09)
Vannas spring scissors (cat.# 15003-08)
straight forceps (cat.# 11000-16)
two Dumont #5 tweezers (cat.# 11251-20)
small curved forceps (cat.# 11051-10)
spatula (cat.# 10094-13)

All tools are displayed in figures of their actual size in the Fine Science
Tools (FST) catalog. Please take a look before ordering because you might
already have tools similar to those listed in the lab.

1.1. Hippocampal cell culture

The hippocampal neurons are isolated and cultured according to published
protocols (Banker and Goslin, 1988; Brewer *et al.*, 1993), with modifications. Depending on the experiments, culture dishes with diameters of 35,

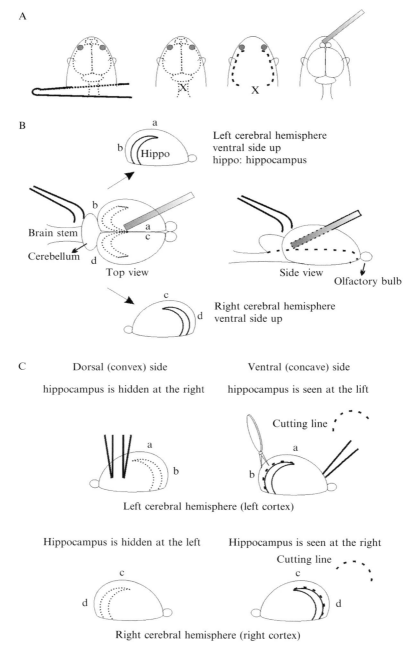

Figure 8.1 Diagram of hippocampus dissection procedure. (A) Removal of the brain from the embryo. From left to right: (1) The embryo is grasped with a pair of large straight forceps; the brain under the skull is visible. (2) Make the first cut at the brain

60, or 100 mm are incubated with 0.8, 2, or 4 ml of 0.5 mg/ml poly-L-lysine solution, respectively, at room temperature for 1 h and then washed twice with sterilized water for 15 min each. After aspiration of the second wash, 8 ml of complete MEM medium is added to the plates that are then kept at 37° in a CO_2 (5% CO_2 in air) incubator. The poly-L-lysine-coated dishes could be prepared 1 to 2 days before or on the same day as the dissection. One gram of poly-L-lysine (p-2636, Sigma Aldrich) is dissolved in 333 ml water and sterilized by filtration through a 0.22 μm filter. The 3 mg/ml poly-L-lysine stock is aliquoted to 50-ml conical tubes and stored at $-20°$ for a year without any problem. To prepare the working solution, the 3 mg/ml stock, 0.15 M sodium borate buffer, pH 8.5, and sterilized water are mixed at the ratio of 1:2:3 before use. Alternatively, the coating solution could be prepared freshly by dissolving poly-L-lysine into 50 mM sodium borate buffer and sterilized with a 0.22 μm filter. Because freezing causes sodium borate to precipitate, it is recommended to make only the required amount for each coating.

The vendor from whom we obtained timed-pregnant rats considers the appearance day of the vaginal plug as embryonic day 1 (E1); we use E19 fetuses for hippocampus dissection. The pregnant rat is sacrificed with CO_2 suffocation and the abdomen is sterilized with 70% ethanol. Once the abdomen is opened with ethanol-sterilized large scissors, the two horns of the uterus are removed to a 100-mm dish. The subsequent dissection steps are performed in a laminar flow hood with ethanol-sterilized tools (listed in "Materials and Reagents"). The fetuses removed from the embryonic sacs are placed in a 100-mm dish with ice-cold HBSS buffer. To remove the brain, the embryo is held firmly but gently with straight forceps right above the front limbs (Fig. 8.1A). Using the fine scissors, the first cut is made between the brain and spinal cord junction to kill the embryo (do not decapitate) and then two incisions are made along both sides of brain right above the ears and all the way to the eyes (Fig. 8.1A). Because the scalp and skull are soft and relatively translucent at this stage of embryogenesis, the brain is obvious and the scalp and skin could easily be cut together. Use the curved forceps to lift up the scalp and skull and gently slide the spatula caudally under the brain from the olfactory bulb side; then dislodge the brain and place it in a 100-mm dish containing ice-cold HBSS. Once the brain has been removed from the

---

and spinal cord junction indicated with an "X." (3) Make the other two incisions along areas indicated by the dotted lines. (4) From the olfactory bulb, dislodge the brain into a 100-mm dish containing ice-cold HBSS. (B) Separate the two cerebral hemispheres (cortices) by holding the brain stem with the curved forceps, slide and push the cortex laterally with a spatula. The two sides of the left and right cortices are indicated by (a), (b), and (c), (d), respectively, when they are attached to or separated from the brain stem. (C) Hippocampus dissection. The meninges are peeled away from the dorsal (convex) side of the cortex; cutting along the convex edge of the hippocampus (dotted lines) will release the hippocampus.

embryo, keep it immersed in ice-cold HBSS to slow the metabolism of the tissue. After collecting 5 to 7 brains from a half litter of embryos, the 100-mm dish is kept on ice. Repeat the procedure to finish dissecting the other embryos; keep the collected brains on ice and then place the first plate under the dissecting microscope. With the dorsal aspect of the brain facing up, hold the brain stem firmly but gently with the curved forceps, use the spatula to slide between the two cerebral hemispheres right across the top of the diencephalon and brain stem; push the spatula laterally to reveal the diencephalon and brain stem and then cut along the bottom boundary to separate the left cerebral hemisphere (i.e., cortex) (Fig. 8.1B). Repeat this procedure to remove the right cortex. We prefer to perform this step in the center of the plate so the separated left and right cortices are kept at the upper and lower sides of the dish and away from the center. Such an arrangement could help beginners to locate the hippocampus even if the brain is partially distorted or lacks a distinct olfactory bulb to indicate the front side of the cortex during the initial isolation. The hippocampus is connected at its left and right margins to the cortex (Fig. 8.1B). Knowing where the hippocampus is helps for the positioning of the fine tweezers for subsequent meninges removal without accidentally damaging the tissue. We prefer to peel the meninges from the dorsal (convex) side of the cortex where the hippocampus is invisible (Fig. 8.1C). Two sets of Dumont #5 tweezers are used to peel off the meninges from the convex side of the cortex. You may cut some cortex tissue but try not to destroy the hippocampus. Alternatively, use one set of tweezers to immobilize the cortex; hold at the opposite side of hippocampus (olfactory bulb end) and peel away the meninges with the other set of tweezers. At this stage of embryogenesis, the meninges are easily peeled away in a sheet; trace amounts of the meninges remaining attached to the cortex is possible but unlikely. After taking away the meninges from all cortices and flipping them over (concave side up), switch one set of tweezers for the spring scissors (Fig. 8.1C). Use the tweezers to immobilize the cerebral hemisphere while using the spring scissors to cut around the boundary between the hippocampus and the adjoining cortex (the convex margin of the hippocampus). The concave side of the hippocampus is not attached to the cortex, so it is released to the solution after the convex margin is cut. The dissected 6 to 7 pairs of hippocampi are scooped up using the curved forceps and placed in a 60-mm dish with ice-cold HBSS that is subsequently kept on ice. Repeat the procedures to finish the other plate of brains. After collecting all hippocampi in the 60-mm dish and ensuring that no meninges are contaminating them, pour all hippocampi with HBSS to a 50-ml conical tube and adjust the volume to 10 ml with HBSS using the scale on the tube. Add 1 ml of 2.5% trypsin to the tube and incubate in a 37° water bath for 15 min.

At the tissue culture hood, make sure that the hippocampi have settled to the bottom of the tube and aspirate off HBSS by pipetting instead of vacuum suction. Because some cell lysis releases undetectable DNA strands to the solution during trypsin digestion, it is likely that the entire tissue may be suctioned in by vacuum if contaminated by the DNA strands. Although 1 mg/ml of DNase solution could be added to the digestion mixture to mitigate the DNA problem, more economically, manual aspiration gives better control in aspirating the trypsin solution without accidentally losing the tissue. Wash the hippocampi twice with HBSS to remove the trypsin solution. Use a Pasteur pipette to separate the tissue in a crude fashion, and then use a smoothly fire-polished Pasteur pipette to triturate the tissues with 2 to 3 ml of HBSS or complete MEM. This should take no more than 20 strokes of each pipette. The opening of the fire-polished Pasteur pipette should be no smaller than half the size of the original to avoid rupturing cells with a strong shearing force. After trituration, add complete MEM medium to 20 ml. Let the visible undissociated tissue pieces, if any, settle to the bottom and take the cell suspension for counting and plating; typically, around 1 to 2 million hippocampal neurons are obtained per embryo. For the first couple of times, you will need to count cells before plating to establish the approximate number of cells you can harvest. The hippocampal neurons are plated on the dishes of your choice with complete MEM at a cell density between 30,000 and $60,000/cm^2$, depending on the experiment.

After an hour of incubation, the medium is replaced with Neurobasal culture medium. Add 2, 4, and 10 ml of Neurobasal culture medium to the 35, 60, and 100-mm plates, respectively. After 5 to 7 days in culture, the medium must be half-changed, that is, about half of the old medium is replaced with the same volume of fresh Neurobasal culture medium. Half-change the medium every 5 to 7 days until the neurons are ready to use. For neurons younger than DIV (days *in vitro*) 7, if necessary, the medium can be completely replaced without affecting cell viability. In contrast to Brewer's protocol (Brewer *et al.*, 1993), we do not include 25 μM glutamate during the first four days of culture and find no adverse effect on neurons.

2. RNA TRANSFECTION AND REPORTER ASSAYS

To investigate whether CPEB3 is involved in translational control, we employ a tethered function assay (see the chapter by Coller and Wickens, 2007) in hippocampal neurons that are transfected with several RNAs, including the dimeric MS2 coat protein (MS2CP), fused to CPEB3 or, as a control, dimeric MS2CP fused to GFP. These MS2CP RNAs are mixed with RNA encoding firefly luciferase that contained or lacked the stem-loops recognized by MS2CP (Fig. 8.2A). The mix also contains RNA encoding *Renilla* luciferase, which serves as an internal control. Because MS2CP needs to homodimerize before binding to the stem-loop sequence, we construct CPEB3 fused with two copies of MS2CPs *in cis* for two reasons.

A   Tethered function assay

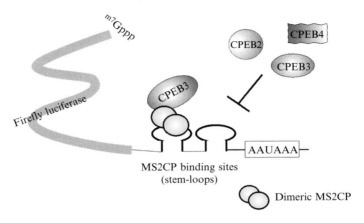

B   Translation efficiency of the reporter RNA *in vitro*

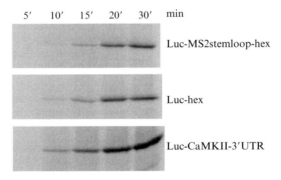

Figure 8.2   Tethered functional reporter assay in neurons. (A) Tethered functional assay using a reporter RNA containing luciferase coding region, the MS2CP binding stem-loop sequences, and a hexanucleotide AAUAAA in the 3′UTR. CPEB3 fused with the MS2 coat protein (MS2CP) is recruited to the reporter RNA through an (MS2CP)-MS2 stem-loop RNA interaction. The effect of tethering CPEB3 to the RNA is assessed by luciferase assays. The endogenous CPEB 2–4 could not interfere with the assay because they are not able to bind to the MS2 stem-loop. (B) The translation efficiency of various firefly luciferase reporter RNAs in reticulocyte lysates. Equal molar amounts of the firefly luciferase RNAs with different 3′UTR sequences were translated *in vitro* in the presence of ^{35}S-methionine at 30° for the indicated times. Luc-hex, Luc-MS2stem-loop-hex, and Luc-CaMKII-3′UTR refer to firefly luciferase RNA appended with hexanucleotide, two MS2CP binding stem-loops and hexanucleotide, and a 180-bp partial CaMKII 3′UTR, respectively.

First, we do not want to force CPEB3 to form a dimer when binding to the reporter RNA since it may not normally do so. Second, the binding of the dimeric-MS2CP fusion to the reporter RNA is less constrained by the concentration of the fusion protein since the intermolecular dimerization is replaced by an intramolecular dimerization. Both dimeric (MS2CP)-CPEB3 and (MS2CP)-GFP constructs were transfected into the cells and subsequent cellular extracts were used for *in vitro* gel retardation assays to confirm their binding to the stem-loop sequence (Huang *et al.*, 2006).

The tethered function assay is useful for studying the function of a newly identified RNA-binding protein in neurons for two reasons. First, it does not require knowledge of the binding specificity of the RNA-binding protein of interest. Second, such as in our case, the function of a group of RNA-binding proteins with similar binding specificity, that is, CPEB2, CPEB3, and CPEB4, could be distinguished by tethering each protein separately to the reporter RNA without interference from the other proteins.

The RNAs used for transfection in neurons are synthesized with mMessage mMachine and T7 Ultra mMessage mMachine kits (Ambion) following the manufacturer's protocol. The T7 Ultra synthesis kit ensures the transcribed RNA is capped in the correct orientation and is polyadenylated to achieve maximal yield of synthesized protein once introduced to neurons. In general, T7 Ultra mRNA produces 5 to 10 times more protein than RNA produced by the traditional mMessage mMachine kit. Hence, we use T7 Ultra kit to synthesize (MS2CP)-CPEB3 and (MS2CP)-GFP RNAs to ensure the MS2CP fusion proteins are rapidly and massively produced to interact with the firefly reporter RNA through the stem-loop sequence. The firefly and *Renilla* luciferase RNAs are synthesized with the traditional kit and much less is used for the transfection; with this protocol, we expect to reduce the "background" expression of firefly luciferase mRNA before it is bound by the MS2CP fusion protein. RNA instead of DNA transfection is chosen to better control the amount of heterologous RNA in the cell and to avoid any possible problem such as promoter competition among three plasmid DNAs. These RNAs are co-transfected into cultured hippocampal neurons, and translation of the reporter RNA is assayed as is the expression of (MS2CP)-CPEB3 or (MS2CP)-GFP fusions. This assay has shown that the CPEB-related proteins differentially regulate translation in cultured neurons (Huang *et al.*, 2006).

The *in vitro* transcribed RNAs are phenol–chloroform extracted, ethanol-precipitated, and washed once with 70% ethanol. The RNA pellet is resuspended in 30 μl H_2O and loaded to DyeEx 2.0 spin columns (Qiagen) to remove free nucleotides; the RNA is then quantified by spectrophotometry. Approximately 20 to 30 μg of RNA is obtained from one reaction. For the initial preparation of transcripts, we would examine the quality and quantity of synthesized RNA by separating them on 1% denaturing agarose gels with total cell RNA preparation as a size maker (28S and 18S ribosomal RNAs are around 5.1 kb and 2.0 kb in size). RNAs are kept in $-80°$ for long-term storage.

Hippocampal neurons cultured for 8 to10 days at a cell density of 30,000 to 40,000/cm^2 are used for RNA transfection. Approximately 1.3 μg of MS2CP-CPEB3 RNA, 0.2 μg of firefly luciferase RNA appended with 2 MS2CP-binding stem-loops and an AAUAAA hexanucleotide, and 0.04 μg of *Renilla* luciferase RNA, in the total amount of $\sim$1.6 μg RNA is used to transfect a 35-mm dish of hippocampal neurons using a 12-well protocol with modifications. The TransMessenger Transfection reagent from Qiagen is used for RNA transfection. On the day of transfection, 3.2 μl Enhancer R is diluted in 91.8 μl EC-R buffer in one Eppendorf tube. The 1.3-μg Ms2CP-fusion RNA is diluted to a total of 3 μl with EC-R buffer in another tube. Sufficient amounts of the both reporter RNAs are then prepared in the EC-R buffer (0.2 μg firefly and 0.04 μg *Renilla* luciferase RNAs in 2-μl volume for one transfection) and 2 μl of the diluted reporter RNA mix is added to the 3 μl of MS2CP-fusion RNA. Any leftover diluted solution is discarded. Five-μl RNA solution is added to the 95-μl mixture of Enhancer R and EC-R, vortexed for 10 s, and incubated at room temperature for 5 min; the mixture is then centrifuged for a few seconds to collect drops from the top of the tube. Add 6 μl of TransMessenger Transfection reagents to the RNA-enhancer R mixture, vortex the tube for 10 s, and the reaction is incubated for 10 min at room temperature to form RNA transfection-complexes. While waiting, rinse the neurons once with 0.5 ml of prewarmed Neurobasal transfection medium to reduce the amount of RNase, if any, in the culture. Because neurons are cultured in a serumfree medium, any agent in the serum that may degrade RNA will not cause a problem here. We have later omitted the washing step without an obviously adverse effect on transfection. Add 500 μl prewarmed Neurobasal transfection medium to the 100-μl transfection complex, pipette the 600-μl solution to a 35-mm plate of neurons after aspirating off the original culture medium, and incubate the cells in a 37° CO$_2$ incubator for 3 h. After removing the transfection complex-containing medium, add 1 ml of Neurobasal transfection medium with or without stimulating reagents (such as 50 μM N-methyl-D-aspartate (NMDA), in our case) to the transfected culture that is then incubated for another 3 h at 37° in a CO$_2$ incubator before lysis in 100 μl of 1× Passive Lysis Buffer (PLB) from the Dual-luciferase Reporter Assay System (Promega). The neurons are rinsed once with PBS and harvested immediately following the addition of PLB by scraping with a cell scraper and the lysate is transferred to an Eppendorf tube and frozen at $-80°$. Immediately before the assay, the frozen lysate is thawed in water (room temp) and centrifuged for 2 min at 12,000g at 4° to pellet cell debris. This freeze–thaw cycle helps to complete lysis of the cells. The samples are kept on ice and 10 to 20 μl of lysate supernatant is used for the dual-luciferase assay, according to the manufacturer's protocol (Promega).

The normalized luciferase activity is calculated by dividing the relative light units (RLU) of firefly luciferase activity-background luminescence by that of *Renilla* luciferase-background luminescence. Typically, the cell lysate background is around 200 to 300 RLU (similar to that of a buffer control) and about 10^4 to 10^5 and 5 to 20×10^3 RLU for firefly and *Renilla* luciferase luminescence, respectively.

Appending the firefly luciferase reporter RNA with various 3'UTRs could affect its translation even when equal molar concentrations of the reporter RNAs are used for transfection. For example, the firefly luciferase RNA appended with a partial αCaMKII 3' UTR yields higher normalized luminescence than the one appended with two MS2CP binding sites in transfected neurons. This is likely due to changes in translational competence of the reporter RNA because a similar result is also observed (Fig. 8.2B) from the *in vitro* reticulocyte lysate translation system (Promega). Notably, the firefly luciferase reporters with or without the MS2CP binding sequences show comparable translational activity in reticulocyte lysates (Fig. 8.2B) and in transfected neurons in the absence of the expression of MS2CP fusion protein (data not shown). This suggests that the MS2CP tethered function assay is a good reporter system in neurons since no endogenous RNA-binding protein is likely to bind to the stem-loop sequence and interfere with the assay.

To quantify the amount of firefly and *Renilla* luciferase RNAs, total RNA is extracted from transfected neurons, reverse transcribed, and subjected to real-time PCR amplification using QuantiTech SYBR Green PCR mixture (Qiagen).

The primers used for quantitative PCR are:

firefly sense 5'-GAGATGTATTACGCAAAGTAC and antisense 5'-C CAGTATGACCTTTATTGAGC;

Renilla sense 5''-GTTGTGTTCAAGCAGCCTGG and antisense 5'-C CAGTGAGTAAAGGTGACAG.

3. shRNA Design and Lenti-shRNA Virus Production

To assess whether the lack of an RNA-binding protein, CPEB3, in neurons might affect the translation of its potential target mRNAs by immunoblotting instead of by immunostaining, greater than ~80% knockdown of CPEB3 protein in cultured neurons has to be achieved. Efficient knockdown of an RNA in a population of neurons depends not only on the selected siRNA sequence but also on the competence of the neurons to take up the siRNA. Because of the intrinsic low transfection efficiency in cultured neurons, a pseudotyped lentiviral system expressing a small hairpin RNA (shRNA) under the control of RNA pol III promoter U6 is employed. We use the pLL3.7-syn plasmid developed by Nakagawa *et al.* (2004) where the GFP expression driven by the cytomegalovirus (CMV), as in the pLen-tiLox 3.7 (pLL3.7) plasmid (Rubinson *et al.*, 2003), is replaced with a human synapsin promoter. The infected neurons could then be traced using GFP fluorescence. Although many criteria could be referenced for the shRNA design, finding the working sequence to knock down a target of interest is an empirical test and four to five sequences should be used at a time to check the knockdown efficiency. In general, the shRNA contains a 19- to 21-mer sequence identical to the target mRNA, followed by a constant loop region and a complementary sequence of the chosen 19 to 21 nucleotides. We employed the criteria described by T. Tuschl (http://www.rockefeller.edu/labheads/tuschl/sirna.html) and Rubinson *et al.* (2003).

1. The 19-mer insert should be unique to the target transcript and the first nucleotide must be G. A 5′ guanine is required due to the constraints of the U6 promoter in the pLL3.7 plasmid, not because of a criterion for RNAi.
2. The 19-mer sequence must be flanked at the 5′ side in the target transcript by two adenine residues.
3. The 19-mer sequence must not contain four or five consecutive T or A residues, because this would act as a premature termination of transcription signal for the poly III complex.
4. The GC content of the 19-mer sequence is around 30 to 70%.
5. The 19-mer sequence is located in the coding region of the target transcript. This is our preference and is not a requirement for RNAi.
6. Preferably, the 19-mer sequence must be flanked at the 3′ side in the target transcript by two thymidine residues.

If all these criteria cannot be fulfilled, the last criterion would be dropped first (3′ TT requirement), because this does not appear to be a strong determinant for successful knockdown of gene expression. In the case of rat CPEB3, the best knockdown sequence does not fulfill this criterion. Thus, a sequence like AAGN18TT or AAGN18 in the CPEB3 coding region is selected (N is any nucleotide and N18 should contain 5 to 12 GC nucleotides and no more than four consecutive T or A according to the criteria listed above).

Example: the most effective sequence in knocking down rat CPEB3 is

GN18: GCCGTACGTGCTGGATGAT, which targets to the sequence of AAGN18: AA GCCGTACGTGCTGGATGAT CA (not TT) in the rCPEB3 mRNA.

The two oligonucleotides are designed as follows:

Sense: 5′ T-(GN18)-TTCAAGAGA-(81NC)-TTTTTTC
Antisense: 5′ *TCGA*GAAAAAA (GN18)-TCTCTTGAA-(81NC)-A
    XhoI sticky end
Thus, N18 is 5′-CCGTACGTGCTGGATGAT
81N is 5′-ATCATCCAGCACGTACGG

The first 5′ T in the sense oligonucleotide is necessary to reconstitute the −1 nucleotide of U6 promoter in the pLL3.7 plasmid. The sense and antisense oligonucleotides are 55 and 59 mers, respectively, in length and are annealed to clone to the HpaI and XhoI-linearized pLL3.7-syn plasmid. While both sequences are long (>40 nucleotides), we have successfully made these shRNA constructs using oligonucleotides without PAGE or HPLC-purification, which is normally recommended to eliminate truncated DNA polymers during long oligonucleotide synthesis. Occasionally, 1 in 5 sets of annealed oligonucleotides could not be cloned to pLL3.7-syn, likely due to the poor quality of the oligonucleotides. If so, re-synthesis of the problematic oligonucleotides is usually sufficient to solve the problem. Nonetheless, HPLC or PAGE-purification is generally a good idea when dealing with long oligonucleotides. Two-microgram pLL3.7-syn plasmid is digested with XhoI and HpaI in NEB (New England Biolab) buffer 4 and gel-purified using QiaexII resin extraction (Qiagen), according to the manufacturer's protocol. The sense and antisense oligonucleotides are diluted in $1\times$ NEB buffer 4 at 1 μM concentration (10 μl $10\times$ NEB buffer 4, 88 μl H_2O, 1 μl each of 100 μM sense and antisense oligonucleotides), heat denatured at 95° for 5 min, 70° for 10 min, and then cooled at room temperature; 1 μl of annealed oligonucleotides is diluted with 9 μl H_2O to 100 nM concentration before setting up the ligation reaction. About 100 ng purified plasmid and 100 fmol of annealed oligonucleotides (1 μl of 100 nM) are added to a 10 μl of ligation mixture that is then incubated at room temperature for 2 h or 16° overnight; 5 μl of ligation is transformed to DH5α or Stbl competent cells and selected on an ampicillin LB plate. The isolated mini-prep DNA is digested with XbaI and NotI and separated on 2% agarose gel. The clones with or without the oligonucleotide inserts are distinguished by the fragment in the size of 509 bp versus 449 bp, respectively.

The neurons are fixed and immunostained 2 days post-transfection. Because the transfected neurons are expressing GFP, a secondary antibody conjugated with proper fluorophore like Alexa594 (Invitrogen) should be used. The immunostained signals between neurons transfected with the empty pLL3.7-syn plasmid and the shRNA construct are compared to estimate the knockdown efficiency of each shRNA sequence. Alternatively, we prefer to examine the effectiveness of those shRNA plasmids in a transient heterologous system. The rat CPEB3 cDNA is reverse transcribed and PCR amplified from the total RNA isolated from hippocampal neurons and then cloned into the pcDNA3.1 plasmid with a myc-tag at the N-terminus of CPEB3. The CPEB3 cDNA plasmid along with pLL3.7-syn (as a control) or siCPEB3 shRNA constructs are co-transfected into HEK 293T cells in a 24–well plate. After overnight incubation, one-tenth of the cell extract is separated by SDS-PAGE and immunoblotted with anti–myc antibody (Fig. 8.3A). To ensure that the reduced expression of myc-CPEB3 is not due to the lower transfection efficiency of particular constructs, the transfection could be monitored and estimated by the percentage of cells displaying GFP fluorescence

A   myc-rCPEB3 knockdown test in 293T cells

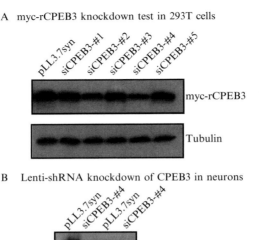

B   Lenti-shRNA knockdown of CPEB3 in neurons

Figure 8.3 CPEB3 knockdown in a heterologous expression system and in cultured neurons. (A) Comparison of the knockdown efficiency of the shRNA constructs in 293T cells. The plasmid expressing myc-rCPEB3 along with pLL3.7-syn and one of the CPEB3 shRNA constructs was transfected to 293T cells and analyzed by an immunoblot probed with anti-myc antibody. Tubulin served as a loading control. (B) Cultured hippocampal neurons infected with lentivirus containing or lacking a short hairpin sequence, such as pLL3.7-syn and siCPEB3-#4, were harvested for Western blot analysis with anti-CPEB3 antibody. [With permission from Y. S. Huang, M. C. Kan, C. L. Lin, and J. D. Richter, *EMBO J.* 25, 4865 (2006).]

when examined with an inverted fluorescent microscope before harvesting cells. If you do not have an inverted fluorescent microscope, immunoblotting with an anti–GFP antibody is an alternative. However, for cell lines such as HEK 293T, we find that the transfection efficiency between wells is quite consistent. In Fig. 8.3A, the expression level of myc-rCPEB3 is down-regulated in the presence of siCPEB3-#2 and siCPEB3-#4 shRNA constructs.

We favor the previous approach to first test the competence of our designed shRNA constructs before making lentivirus for two reasons. First, to achieve a high titer, an ultracentrifugation step is necessary to pellet the viruses, which are then resuspended in a small volume of phosphate-buffered saline (PBS) prior to infecting the neuron cultures. Because the lentivirus is pseudotyped with a vesicular stomatitis virus glycoprotein (VSVG) envelope that enables the packaged virions to infect any type of cells, a biosafety P2+ facility is recommended during production. Moreover, lentivirus production requires safety precautions and relatively tedious procedures, and thus we prefer to make the virus only for the shRNA construct that is most efficient.

Second, this is a simple way to assess efficacy of RNA knockdown if an antibody for the target protein is not available. It is fast and requires only an overnight culture of cells.

The siCPEB3-#4 construct that is the most efficacious in our test system is used for the downstream lentivirus production (Fig. 8.3A). This particular construct is used to produce lentivirus using the virapower packaging system (Invitrogen). HEK 293T/17 cells are obtained from American Type Cell Culture (ATCC, cat.# CRL-11268). The day before transfection, split a 100-mm plate of confluent 293T cells to seed three plates of the same size ($\sim 5 - 6 \times 10^6$ cells/plate). In general, the lentivirus titer is low, around 1 to 2 million virions/ml of medium or 10 million virions/100-mm plate of transfected cells. Thus, scale up the transfection according to the number of the viral particles you wish to obtain. Three μg of siCPEB3-#4 plasmid and 9 μg of virapower packaging mix are blended with 36 μl of Fugene 6 (Roche) to form a transfection complex that is subsequently added to a 100-mm plate of 293T cells, according to the manufacturer's protocol. Sixty hours after transfection, the medium is collected and centrifuged for 5 min at 600g to remove cell debris. The supernatant is filtered through a 0.45 μm filter and the resulting filtrate is added to an ultraclear centrifuge tube (cat.# 344058 or cat.# 344059 for Beckman SW28 or SW41 swinging bucket rotor), depending on the total volume of collected medium. In the culture hood, the swinging buckets are sterilized with 70% ethanol and then sprayed with 100% ethanol. This sterilization step is performed before and after virus centrifugation. There is no need to autoclave or ethanol-sterilize the centrifuge tubes. Centrifugation is carried out at 25,000 rpm for 2 h at 20°. For a 100-mm plate of transfected cells, approximately 10 to 12 ml of medium is collected and loaded into a $\sim$12-ml ultraclear centrifuge tube (cat.# 344059). Because the virus pellet is not obvious after centrifugation, carefully aspirate the supernatant without touching the bottom of the tube and then invert the tube to drain any remaining medium. We generally resuspend the virus pellet from 10 ml medium with 200 μl sterilized PBS (the titer of the concentrated stock is $\sim$5–10 $\times$ 10^7 virions/ml), aliquot to 50 μl/tube, snap frozen in liquid nitrogen, and stored at −80°. The lentivirus production and handling are conducted in conformity with institutional guidelines that are in compliance with the National Institutes of Health policies.

4. shRNA Knockdown of CPEB3 in Cultured Neurons

We have successfully used the lentivirus to infect neurons cultured from 1 to 12 days. However, infection efficiency decreases with the age of the culture. Thus, it is recommended to use neurons no more than 1 week old for the knockdown experiments. We generally infect neurons DIV 4 to 5 and wait for 3 to 4 days before harvesting the cell extract to examine the CPEB3 protein level by immunoblotting (Fig. 8.3B). To infect neurons, one-half of the culture medium is removed and collected in a dish that is then kept at 37° in a CO_2 incubator. Depending on the number of neurons plated on the dish, the amount of lentiviral particles (i.e., multiplicity of infection, or MOI) in two- to three-fold excess is added to the culture. For example, for a 60-mm dish containing 10^6 neurons, 2 ml of culture medium is removed and saved and then 2 to 3 $\times$ 10^6 viral particles (~20–40 μl of concentrated aliquot) are added to the plate. After overnight incubation in the CO_2 incubator, the virus-containing medium is aspirated to a waste bottle containing bleach. The 2 ml of the saved old medium along with 2 ml of fresh Neurobasal culture medium is added to the plate. The infected neurons are cultured for an additional 3 to 4 days before harvesting for Western blot analysis or total RNA isolation. The percentage of infected neurons could be monitored by the number of cells displaying GFP fluorescence 2 days after infection. If the infection efficiency does not reach at least 80%, consider using more virus for the next round of infection. Although polybrene is often used to enhance the retroviral and lentiviral infection efficiency, it should not be used in neuron cultures because of its severe toxicity.

The cultured hippocampal neurons infected with a control lentivirus or one expressing a short hairpin RNA against CPEB3 (siCPEB3-#4) for 3 to 4 days are lysed with 1× SDS-sample buffer (~200 μl for 10^6 neurons). The extracts are then sonicated to break the chromosomal DNA and boiled for 5 min before separation by SDS-PAGE and detection of CPEB3 (Fig. 8.3B) or other proteins, such as GluR2, αCaMKII, and synaptophysin by immunoblotting (Huang et al., 2006).

5. Synaptoneurosome Isolation and ^{35}S-met/cys Labeling

Synaptoneurosome isolation is modified from published protocols (Bagni et al., 2000; Dunkley et al., 1986). Neuron cultures or hippocampi freshly isolated from 1- to 2-month-old mice can be used as the starting material. Approximately 5 to 10 million neurons from 2- to 3-week-old cultures or one-two pairs of hippocampi are homogenized in 2 ml of homogenization buffer (0.32 M sucrose, 1 mM Hepes, pH 7.4, 1 mM MgCl$_2$, 0.1 mM EDTA, 0.25 mM DTT, 40 u/ml RNase inhibitor) with Dounce homogenizer pestle B for 10 to 12 strokes. The homogenate is centrifuged at 1000g for 5 min at 4° to pellet nuclei and unbroken cells. The supernatant is then centrifuged at 10,000g for 10 min at 4° to obtain the synaptoneurosome and mitochondrial pellet. The pellet is gently suspended in 2 ml Percoll dilution buffer (0.25 M sucrose, 0.1 mM EDTA, 2 mM Hepes, pH 7.4). To prepare a Percoll/sucrose discontinuous gradient, 40 ml of Percoll (cat. #17–0891–02, from GE Health Care Life Science Biotech) is mixed with 4.275 g sucrose, 100 μl of 1 M Hepes, pH7.4, and 10 μl of 0.5 M EDTA. Adjust the solution pH to ~7.4 with HCl and a final volume to 50 ml with H_2O to make the 80% Percoll stock that is kept at 4° and used within 2 months. The 80% stock is diluted to 3, 10, 15, and 23% with Percoll dilution buffer immediately before setting up the step gradient. The 2-ml supernatant suspension is layered over a freshly prepared 4 × 2 ml discontinuous gradient comprising 23, 15, 10, and 3% Percoll (v/v), in an ultraclear centrifuge tube (cat.# 344059, Beckman). Be careful not to disturb the interface when preparing the gradient. The gradient is then centrifuged at 32,000g (16k rpm for SW 41 rotor) for 10 min at 4°. The material containing the synaptoneurosomes between the 15 to 23% interface is collected (~0.3–0.4 ml), diluted with 3 volume of PBS, and pelleted at 12,000g for 5 min at 4°.

The synaptoneurosome pellet is suspended in 1× synaptoneurosome buffer (10 mM Tris, pH7.5, 2.2 mM $CaCl_2$, 0.5 mM Na_2HPO_4, 0.4 mM KH_2PO_4, 4 mM $NaHCO_3$, 80 mM NaCl) (Bagni $et\ al.$, 2000) in the presence or absence of stimulating agents such as 50 mM KCl or 50 μM NMDA. The treated synaptoneurosomes could then be used to isolate RNA for polyadenylation test assays or protein analysis by Western blotting (Huang $et\ al.$, 2002). Alternatively, the synaptoneurosomes could be incubated with ^{35}S–methionine to examine protein synthesis. In this case, the synaptoneurosome pellet from a pair of hippocampi isolated from wild type or CPEB knockout mice is suspended in 80 μl ice-cold 1× synaptoneurosome buffer containing 40 μCi Pro-mix L–^{35}S $in\ vitro$ cell labeling mix (GE Health Care Life Science Biotech) and 100 μg/ml chloramphenicol to inhibit protein synthesis from contaminating mitochondria. Five microliters of 1× synaptoneurosome buffer with or without the stimulating agent is added to 20 μl of the synaptoneurosome suspension and then incubated at 37° for 30 min before separating by SDS–PAGE.

6. UV-Crosslinking, Immunoprecipitation of an RNA-Binding Protein, CPEB3

By using an $in\ vitro$ UV-crosslinking approach, bacteria–purified recombinant CPEB3 has been demonstrated to interact with the 3′ UTR of GluR2 mRNA. To test whether this interaction occurs $in\ vivo$, hippocampal neuron cultures are ultraviolet (UV)-irradiated, homogenized, and immunoprecipitated with CPEB3 IgG or nonspecific IgG (Huang $et\ al.$, 2006). Three plates of 2- to 3-week-old cultured neurons (~6–8 million cells) are rinsed once with PBS. The PBS solution is then aspirated as much as possible without drying the cells, which are quickly covered with 220 μl of ice-cold immunoprecipitation buffer (20 mM Hepes, pH 7.4, 150 mM KCl, 1 mM MgCl$_2$, 0.5 mM DTT, 0.5 mM EDTA, 0.1% Triton, protease inhibitor cocktail, and 0.2 U/μl RNase inhibitor). The plates are put on ice and are subsequently irradiated with 1200 joules of UV (254 nm) light for 20 to 30 min. We generally layer several wet paper towels on ice and then place the plates on the top of the towel. Because the plates are kept on ice, the buffer layered over the neurons will not evaporate while absorbing the UV energy. After UV exposure, the neurons are scraped off the plate with a cell scraper, homogenized with Dounce homogenizer pestle B for 10 to 12 strokes, and the homogenate is centrifuged at 1000g for 5 min at 4° to remove nuclei. One-twentieth ($\sim$30 μl) of the resulting supernatant is kept on ice and saved for later RNA isolation. The RNA isolation from a portion of homogenate without immunoprecipitation is necessary because it serves as a control to test for possible RNA degradation. The remaining solution is equally divided ($\sim$280 μl each) and incubated with 30 μl of Dynabeads carrying nonspecific IgG or CPEB3 IgG. To prepare the magnetic beads for immunoprecipitation, 30 μl of Dynabeads M280 conjugated with sheep anti-rabbit IgG (cat.# 112–03D, Invitrogen) is first washed with 100 μl PBS twice, incubated with the 100 μl antibody solution (1× PBS, 30 μg bovine serum albumin, 3 μg of nonspecific or CPEB3 IgG) at room temperature for 2 h and then washed three times with PBS after removing the antibody solution. Addition of nonspecific or CPEB3-immunized serum directly to the UV-crosslinked extract followed by Dynabeads immunoprecipitation is not recommended because serum might contain RNase activity that would degrade RNA during immunoprecipitation.

After 2 h incubation of the prepared antibody beads with UV-cross-linked extract in a cold room, the beads are washed 4× with 100 μl RIPA buffer (50 mM Tris-HCl pH 7.5, 150 mM NaCl, 1% NP-40, 0.5% sodium deoxycholate, and 0.1% SDS) and 1× with genomic DNA lysis buffer (50 mM Tris, pH 7.4, 10 mM EDTA, 500 mM NaCl, 2.5 mM DTT, 0.5 mM spermidine, 1% Triton X-100). Approximately 300 μl of PK solution (1 mg/ml proteinase K in genomic DNA lysis buffer and 0.2 U/μl RNase inhibitor) is added to the total lysate previously kept on ice and the beads are then incubated at 37° for 30 min. Gently flick the tubes to resuspend the beads every 10 min during the incubation. After removal of the proteinase K solution, 300 μl of RNA extraction solution (4 M guanidine thiocyanate, 0.5% sarkosyl, and 25 mM sodium citrate, pH7) is added to the beads, incubated for 10 min and the supernatant is mixed with 30 μg yeast tRNA (as a carrier) and 30 μl of 3 M sodium acetate. The RNA solution is phenol–chloroform extracted, ethanol-precipitated, and the pellet washed once with 70% ethanol. The dry pellet is used for 1st strand cDNA synthesis, followed by PCR analysis. The removal of proteins crosslinked to the RNA is necessary for cDNA synthesis. However, there might be a short peptide still linked to the RNA after proteinase K digestion, so the PCR amplified region of the target RNA should be small, around 200 to 400 bp. In addition, random primers in place of oligo–dT may be used for the reverse transcription step.

In our case, the precipitated RNA was extracted after proteinase K digestion and subjected to RT-PCR for Arc, Map2, NF, or GluR2 mRNAs. Only the GluR2 mRNA was detected from the CPEB3 IgG immunoprecipitate (Huang *et al.*, 2006). In contrast to the immunoprecipitation of CPEB3-associated RNAs in the absence of the cross-linked step, the result obtained from this approach is much cleaner because RNAs that may be indirectly associated with CPEB3 are removed with a stringent washing buffer, such as RIPA.

REFERENCES

Bagni, C., Mannucci, L., Dotti, C. G., and Amaldi, F. (2000). Chemical stimulation of synaptosomes modulates alpha −Ca2+/calmodulin-dependent protein kinase II mRNA association to polysomes. *J. Neurosci.* **20**, RC76.

Banker, G., and Goslin, K. (1988). Developments in neuronal cell culture. *Nature* **336**, 185–186.

Brewer, G. J., Torricelli, J. R., Evege, E. K., and Price, P. J. (1993). Optimized survival of hippocampal neurons in B27-supplemented Neurobasal, a new serum-free medium combination. *J. Neurosci. Res.* **35**, 567–576.

Brown, V., Jin, P., Ceman, S., Darnell, J. C., O'Donnell, W. T., Tenenbaum, S. A., Jin, X., Feng, Y., Wilkinson, K. D., Keene, J. D., Darnell, R. B., and Warren, S. T. (2001). Microarray identification of FMRP-associated brain mRNAs and altered mRNA translational profiles in fragile X syndrome. *Cell* **107**, 477–487.

Coller, J., and Wickens, M. (2007). Tethered function assays: An adaptable approach to study RNA regulatory proteins. *Methods Enzymol.* **429**, 299–321.

Darnell, J. C., Mostovetsky, O., and Darnell, R. B. (2005). FMRP RNA targets: Identification and validation. *Genes Brain Behav.* **4**, 341–349.

Du, L., and Richter, J. D. (2005). Activity-dependent polyadenylation in neurons. *RNA* **11**, 1340–1347.

Dunkley, P. R., Jarvie, P. E., Heath, J. W., Kidd, G. J., and Rostas, J. A. (1986). A rapid method for isolation of synaptosomes on Percoll gradients. *Brain Res.* **372**, 115–129.

Huang, Y. S., Carson, J. H., Barbarese, E., and Richter, J. D. (2003). Facilitation of dendritic mRNA transport by CPEB. *Genes Dev.* **17**, 638–653.

Huang, Y. S., Jung, M. Y., Sarkissian, M., and Richter, J. D. (2002). N-methyl-D-aspartate receptor signaling results in Aurora kinase-catalyzed CPEB phosphorylation and alpha CaMKII mRNA polyadenylation at synapses. *EMBO J.* **21**, 2139–2148.

Huang, Y. S., Kan, M. C., Lin, C. L., and Richter, J. D. (2006). CPEB3 and CPEB4 in neurons: Analysis of RNA-binding specificity and translational control of AMPA receptor GluR2 mRNA. *EMBO J.* **25**, 4865–4876.

Miyashiro, K. Y., Beckel-Mitchener, A., Purk, T. P., Becker, K. G., Barret, T., Liu, L., Carbonetto, S., Weiler, I. J., Greenough, W. T., and Eberwine, J. (2003). RNA cargoes associating with FMRP reveal deficits in cellular functioning in Fmr1 null mice. *Neuron* **37**, 417–431.

Nakagawa, T., Futai, K., Lashuel, H. A., Lo, I., Okamoto, K., Walz, T., Hayashi, Y., and Sheng, M. (2004). Quaternary structure, protein dynamics, and synaptic function of SAP97 controlled by L27 domain interactions. *Neuron* **44,** 453–467.

Rubinson, D. A., Dillon, C. P., Kwiatkowski, A. V., Sievers, C., Yang, L., Kopinja, J., Rooney, D. L., Ihrig, M. M., McManus, M. T., Gertler, F. B., Scott, M. L., and Van Parijs, L. (2003). A lentivirus-based system to functionally silence genes in primary mammalian cells, stem cells, and transgenic mice by RNA interference. *Nat. Genet.* **33,** 401–406.

Schuman, E. M., Dynes, J. L., and Steward, O. (2006). Synaptic regulation of translation of dendritic mRNAs. *J. Neurosci.* **26,** 7143–7146.

Sutton, M. A., and Schuman, E. M. (2006). Dendritic protein synthesis, synaptic plasticity, and memory. *Cell* **127,** 49–58.

Vanderklish, P. W., and Edelman, G. M. (2005). Differential translation and fragile X syndrome. *Genes Brain Behav.* **4,** 360–384.

Wu, L., Wells, D., Tay, J., Mendis, D., Abbott, M. A., Barnitt, A., Quinlan, E., Heynen, A., Fallon, J. R., and Richter, J. D. (1998). CPEB-mediated cytoplasmic polyadenylation and the regulation of experience-dependent translation of alpha-CaMKII mRNA at synapses. *Neuron* **21,** 1129–1139.

Zalfa, F., Achsel, T., and Bagni, C. (2006). mRNPs, polysomes, or granules: FMRP in neuronal protein synthesis. *Curr. Opin. Neurobiol.* **16,** 265–269.

CHAPTER NINE

Detecting Ribosomal Association with the 5' Leader of mRNAs by Ribosome Density Mapping (RDM)

Naama Eldad *and* Yoav Arava

Contents

Abstract

In eukaryotes, scanning of the 5' leader by the small ribosomal subunit precedes recognition of the start codon. Thus, various sequence elements that are located within this region may affect ribosomes' progression and lead to significant effects on translation. Most notable are short ORFs located upstream of the start codon, which are known to regulate the translation of the main ORF in the transcript. The function of these elements is likely to correlate with altered ribosomal association with the 5' leader of the mRNA. Currently, the only method to determine the ribosomal association of different regions of the mRNA *in vivo* is the Ribosome Density Mapping (RDM) procedure. This method entails cleavage of the target mRNA by specific oligodeoxynucleotides and RNase H and separation of the cleavage products by velocity sedimentation in a sucrose gradient. In this chapter, we provide a detailed protocol for this procedure and discuss its feasibility.

Department of Biology, Technion—Israel Institute of Technology, Haifa, Israel

Methods in Enzymology, Volume 431

ISSN 0076-6879, DOI: 10.1016/S0076-6879(07)31009-4

1. INTRODUCTION

In the eukaryotic translation initiation process, small ribosomal sub-units scan the 5′ UTR in search of a start codon (AUG), which, once identified, serves as a start site for protein synthesis (Cigan and Donohue, 1987; Kozak, 1987). While, in most cases, this start codon is the first one encountered by the scanning subunit, an increasing number of mRNAs appear to have start codons upstream to the main ORF (Iacono *et al.*, 2005). Ribosomal subunits are likely to scan through these upstream AUGs (uAUG) and start translation at the bona fide start site (Kozak, 1987; Morris and Geballe, 2000). However, it was found for several mRNAs that the uAUG serves as a translational regulatory site at which ribosomes might initiate translation under certain environmental conditions (Morris and Geballe, 2000; Vilela and McCarthy, 2003).

Determining whether ribosomes indeed initiated translation at uAUGs is technically challenging. The common procedure involves mutating the uAUGs and inferring their functions from the levels of the encoded protein or a reporter protein. Alternatively, a toe printing assay might be used whereby the position of ribosomes is assayed using an *in vitro* transcribed mRNA (Sachs *et al.*, 2002). The Ribosome Density Mapping (RDM) procedure described later can also be applied for this purpose (Arava *et al.*, 2005). We have used it to show that ribosomes are associated with the 5′ leader of yeast GCN4 mRNA (Arava *et al.*, 2005). It allows the direct determination of ribosomal association *in vivo* and can be applied to many different mRNAs in various organisms. Moreover, it is not just restricted to the 5′ leader of the mRNA and can be applied to determine ribosomal association anywhere along the mRNA.

2. GENERAL CONCEPT

RDM is based on the well-established procedure for separating mRNAs associated with ribosomes by velocity sedimentation in a sucrose gradient (Warner *et al.*, 1963). The main difference is that while routine procedures are based on separating the entire transcript in the gradient, in RDM, the transcript, while associated with ribosomes, is cleaved at site(s) of interest by RNase H and oligodeoxynucleotide (ODN) complementary to that site, and only then separated in a gradient. This leads to sedimentation of mRNA fragments according to their ribosomal association.

Four general steps are involved in the procedure (Fig. 9.1): (1) Cell lysis and separation of polysomes (Fig. 9.1, steps 1 and 2); (2) Collection of

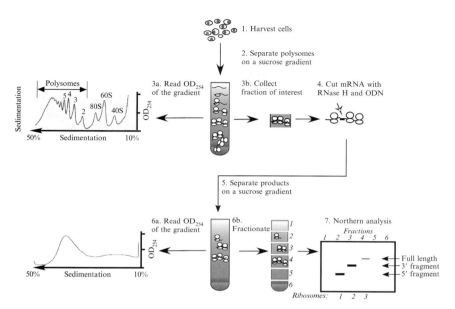

Figure 9.1 Schematic presentation of the RDM steps. See text for details.

a fraction of interest and RNase H cleavage (Fig. 9.1, steps 3 and 4); (3) Separation of the cleavage products according to their ribosomal association (Fig. 9.1, steps 5 and 6); and (4) Determination of the sedimentation position of the cleavage products by northern analysis (Fig. 9.1, step 7). Each of these steps is described later.

RDM provides the best results with mRNAs that are highly expressed, preferably from their own promoter. We observed a decrease in the quality of the results when using reporter mRNAs (e.g., GFP and luciferase) that are not native to the organism (yeast, in our case). The decrease in quality was apparent in the northern analyses, in which multiple bands or smears were observed. This is probably due to inefficient processing of these exogenous transcripts.

Each of the steps involved in RDM has been used separately in many organisms and for the study of various mRNAs. Thus, although the protocol presented herein is for the analysis of yeast cells grown at optimal conditions, it could be adapted easily to other experimental systems. It might also be applied for *in vitro* systems or extracts (e.g., reticulocytes lysate), yet we have not performed such experiments.

2.1. Reagents

Table 9.1 Solutions

Solutions	Stock	Ingredients	Storage	Remarks
Lysis buffer		20 mM Tris pH 7.4 140 mM KCl 1.5 mM MgCl$_2$ 0.5 mM DTT 1% Triton X-100 0.1 mg/ml CHX 1 mg/ml heparin	4°	Prepare one day before use
Gradient with heparin		11 ml 10–50% sucrose in a buffer: 20 mM Tris pH 7.4 140 mM KCl 5 mM MgCl$_2$ 0.5 mM DTT 0.1 mg/ml CHX 1 mg/ml heparin	4°	Prepare one day before use
Gradient without heparin		11 ml 10–50% sucrose in a buffer: 20 mM Tris pH 7.4 140 mM KCl 5 mM MgCl$_2$ 0.5 mM DTT 0.1 mg/ml CHX	4°	Prepare one day before use
RNase H buffer	5×	0.1 M Tris pH 7.4 0.5 M KCl 0.1 M MgCl$_2$ 0.5 mM DTT 2.5 mg/ml CHX	−20°	
Lysis Minus Detergent (LMD) buffer		Same as the lysis buffer but without Triton X-100	−20°	
MOPS buffer	10×	0.4 M MOPS 0.1 M Na Acetate 0.01 M EDTA Bring to pH 7.0 with acetic acid	RT	
RNA loading buffer	2×	For 15 ml stock, mix 10 ml of 100% formamide, 3 ml of 37% formaldehyde, 2 ml of 10× MOPS buffer and 25 μl of 10 mg/ml ethidium bromide	−20°	Highly toxic
Hybridization buffer	1×	0.4 M Na$_2$HPO$_4$ pH 7.2 6% SDS 1 mM EDTA	RT	Warm before using to dissolve any aggregates
Hybridization Wash 1	1×	40 mM Na$_2$HPO$_4$ pH 7.2 5% SDS 1 mM EDTA	RT	
Hybridization Wash 2	1×	40 mM Na$_2$HPO$_4$ pH 7.2 1% SDS 1 mM EDTA	RT	

Table 9.2 Enzymes and solutions

Product name	Concentration	Cat. #	Remarks
Heparin	10 mg/ml	Sigma H3393	Store at −20°. Dissolve in water.
Cyclohexamide (CHX)	10 mg/ml	Sigma C7698	Store at −20°. Dissolve in water. Vortex well. CHX is highly toxic.
RNAse Inhibitor **or** rRNAsin	40 U/μl **or** 40 U/μl	Takara 2311A **or** Promega N2515	
RNase H	5 U/μl	New England Biolabs M0297	
Antisense ODN	10 pmol/μl		~20 bases ~50%GC. Dissolve in water.
Guanidinium HCl	8 M	Sigma G4505	Stir and warm until solution becomes clear. Dissolve in water.
RNA Marker	1 mg/ml	Ambion #7145	RNA Century Plus

3. METHODS

3.1. Cell lysis and separation of polysomes

The first step in the RDM protocol is cell lysis and separation of polysomes through sucrose gradient (Fig. 9.1, steps 1 and 2), followed by the collection of a specific fraction that contains the mRNA of interest associated with ribosomes for RNase H cleavage. There are two reasons for performing RNase H cleavage on only a selected fraction and not on the entire extract: (1) Isolation of a subset of mRNAs with a specific number of bound ribosomes simplifies interpretation of the results; (2) Large amounts of heparin are used during the lysis to inhibit RNase activity. This heparin does not sediment into the gradient, and therefore the isolated fraction is clean of heparin and can be used in the following enzymatic step.

1. Grow 50 ml of yeast cells to OD_{600} 0.5 to 0.8 in YPD (1% Yeast extract, 2% Peptone, 2% Dextrose).

    The amount of cells taken for analysis is important, since too many cells will yield an overload of RNases in later steps. The growth conditions are also important since certain conditions might be enriched in RNases. Highly expressed genes (at the levels of "housekeeping genes") are likely to yield good results in RDM, but the analysis of low abundance mRNAs might be problematic. Increasing the amount of cells by increasing the volume of culture may help as long as the density remains low (OD_{600} *0.5–0.8*).

2. Add cyclohexamide (CHX) to a final concentration of 0.1 mg/ml, transfer to an ice-cold 50-ml conical tube, and immediately spin down the cells at 4000 rpm (3220g) for 4 min at 4°.

    At these concentrations, CHX inhibits the 60S subunit translocation and retains ribosomal association with mRNA.

3. Discard the supernatant, resuspend the cells' pellet in 4 ml of ice-cold lysis buffer, and spin cells again as in step 2.

    The quick spin-down and washes are important to remove leftovers of dextrose found in the rich medium. This probably helps in blocking initiation events.

4. Wash again as in step 3.

5. Resuspend the cells' pellet with 350 μl of lysis buffer, transfer to a micro tube with a screw cap, and add ice-cold glass beads (diameter 0.4–0.6 mm) to cover the cells and lysis buffer.

6. Break the cells in a bead beater by two pulses of 1.5 min at maximum level.

7. Transfer the cells' lysate into a new tube. This can be done by a quick spin-down of the glass beads and collecting the supernatant. Alternatively, it is possible to pierce the bottom of the tube with a heated needle, place it on top of a 15-ml tube, and spin at 4000 rpm for 1 min at 4° (the lysate will drip into the 15-ml tube and the glass beads will be

retained in the pierced tube). We use the cylinder of a 5-ml syringe as an adaptor between the pierced tube and the 15-ml tube.

8. Transfer the resulting lysate from the 15-ml tube to a new ice-cold micro tube. Usually, there is a small pellet that should be resuspended with the supernatant and also transferred to the micro tube.

9. Centrifuge for 5 min at 9500 rpm at 4° and transfer the supernatant to a new ice-cold micro tube.

10. Bring to a final volume of 1 ml with lysis buffer and carefully load the lysate on 10 to 50% sucrose gradient *without* heparin.

    For gradient preparation instructions, see Arava (2003).

11. Separate complexes by ultra-centrifugation using a SW41 rotor at 35,000 rpm for 2:25 h at 4°.

    The separation times may be adjusted according to the desired resolution. Many technical aspects of velocity sedimentation analysis are discussed in Rickwood (1992).

3.2. Fraction collection and RNase H cleavage

Following centrifugation, the fraction of interest is collected for cleavage by RNase H and ODN (Fig. 9.1, steps 3 and 4). This is usually the fraction that contains the majority of transcripts of the gene of interest, but could be any other fraction that contains sufficient amounts of the mRNA of interest. Prior knowledge of this fraction is therefore necessary. Such knowledge is usually achieved by performing a preliminary polysomal separation experiment in which the cells' extract is separated into fractions that are analyzed by northern blot for the mRNA of interest. In the case of yeast, a previous genome–wide analysis that characterized the ribosomal association of thousands of mRNAs could be of assistance (Arava *et al.*, 2003). For genes that are expected to be regulated through ribosomes associated with their 5′ leader, this fraction is likely to be the monosome fraction (Arava *et al.*, 2003; Kuhn *et al.*, 2001; Tzamarias *et al.*, 1989).

Collecting the correct fraction is greatly facilitated by the use of a continuous ultraviolet (UV) detector, such as the ISCO UA-6 system. The OD_{254} reading allows accurate determination of the sedimentation of various complexes (e.g., 40S, 80S, and various polysomal complexes; Fig. 9.1, step 3a). This information can be used to correct for small variations in sedimentation. Importantly, in many cases, it enables the determination of the number of ribosomes on the mRNA or on the resulting fragments, thereby allowing more accurate conclusions.

1. Collect the fraction of interest (620 μl) into an ice-cold micro tube.

    The collected polysomal fraction also contains a significant amount of RNases. An overload of RNases cannot be completed by the commonly used RNase Inhibitors (rRNasin, RNase Inhibitor, etc.) and will therefore lead to massive degradation. It is therefore recommended not to collect more than the amount indicated here.

2. Immediately add 70 μl of 0.1 M DTT (final concentration 10 mM) and 7 μl of 40 u/μl RNase inhibitor (final concentration 0.4 u/μl).

3. Transfer 600 μl of this mixture to another micro tube containing 10 μl of antisense ODN (10 pmole/μl) complementary to the cleavage site on the mRNA. The remaining 70 to 100 μl serves as a control sample ("uncut") and will be subjected to the following incubations (steps 3 to 6).

   We used many ODNs ranging in length from 18 to 25 nts and with GC content from 40 to 60%. These ODNs varied in efficiency, yet we are unable to directly link the differences in efficiency to their length or GC content. This is probably because many additional factors, such as structured cleavage site or presence of a ribosome, affect cleavage efficiency. In the case of cleavage at the translation start site region, target sequences should be selected from ~20 nts upstream to the start codon in order to avoid any inhibition by initiating ribosomes. It is recommended to examine the cleavage efficiency in a test reaction that includes polysomal RNA and the ODNs. We perform the test reactions on an isolated polysomal fraction and not on clean RNA (e.g., RNA isolated by the hot phenol or Tri-reagent methods), since many ODNs perform differently among these populations.

   Multiple ODNs complementary either to the same transcript or to different transcripts can be used in the same reaction. If several ODNs to the same transcript are used, partial products are also expected (i.e., cleavage by only one ODN; Fig. 9.2). Thus, the cleavage plan should be such that the full cleavage products and the various partial cleavage products are of distinct sizes. In cases where there is no good separation in size, it is possible to design probes that will recognize only some of the products in the northern analysis.

4. Put the sample in a beaker containing water at 37° and let it gradually cool to room temperature for 20 min.

   This supposedly allows annealing of the ODN to its target sequence.

5. Add 150 μl of 5× RNase H buffer and 10 units of RNase H.

6. Incubate the samples at 37° for 20 min.

   Longer incubation times may improve RNase H cleavage but may also result in partial degradation.

7. Stop the reaction by bringing the sample volume to 1 ml with Lysis Minus Detergent (LMD) buffer.

   The LMD contains heparin that inhibits RNase H activity. The LMD buffer will also dilute the sucrose in the sample to allow overlaying on a 10 to 50% sucrose gradient.

8. Set aside 1/10 of the reaction as a pregradient control ("cut").

3.3. Separation of cleavage product

Following cleavage, the products are separated on a sucrose gradient according to their ribosomal association (Fig. 9.1, steps 5 and 6). Since the sample includes only complexes at a fixed size (the size that was collected initially), the resulting OD profile usually has only one peak (Fig. 9.1, step 6a). It is

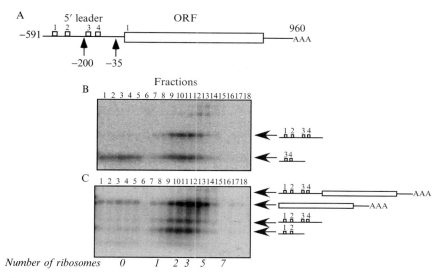

Figure 9.2 RDM analysis of GCN4 mRNA. (A) Schematic presentation of GCN4 mRNA with the four uORFs located in its 5' leader. These uORFs play an important role in regulating the translation of GCN4 ORF under different growth conditions. They exert their role by affecting the efficiency of ribosomes' scanning of the 5' leader (Hinnebusch, 2005). (B and C) To demonstrate ribosomal association with the 5' leader, yeast cells were grown under conditions that induce the GCN4 translation. Polysomal fraction containing the majority of GCN4 mRNA (with about 5–6 ribosomes) was isolated, and the GCN4 mRNA was cleaved at two positions simultaneously by adding two ODNs complementary to positions −35 and −200 upstream to the AUG of GCN4 protein (indicated by arrows in A). Cleavage products were then separated by velocity sedimentation in a sucrose gradient and 18 fractions were collected. The sedimentation position of the double-cleaved product (−200 to −35) was determined by northern analysis using a labeled oligonucleotide complementary to that region (B). The blot was then hybridized with a probe recognizing the entire transcript to determine the sedimentation position of the rest of the cleavage products (C). The identity of each cleavage product was determined according to its length and is indicated schematically to the right of each blot. The number of ribosomes associated with complexes sedimenting at each fraction is indicated below the blots. This analysis shows that under these growth conditions, the 5' leader of GCN4 sediments as associated with ribosomes. The band corresponds in size to the entire 5' leader sediments as associated with ~2 ribosomes. The 5'-most region, which contains only uORFs 1 and 2, appears to be highly associated with ribosomes since practically all of the cleavage products that correspond to it sediment in fractions 9 to 11 and almost none are in fractions 1 to 6. On the other hand, a substantial amount of the fragment that contains uORFs 3 and 4 appears to sediment in fractions 1 to 6 as free of ribosomes. These results are in agreement with the model that some of the ribosomes that scan uORFs 1 or 2 dissociate from the mRNA prior to approaching uORF3 (Abastado et al., 1991).

impossible to deduce the ribosomal association of the cleavage products from this single peak. There are two options to overcome this limitation: (1) Use of an external profile—since the sedimentation profiles are highly reproducible, it is possible to infer the number of ribosomes sedimenting at a certain position from a similar gradient that was centrifuged under similar conditions but with an entire cell extract (Fig. 9.1); (2) Spike-in extract—another option is to set aside a sample of the whole-cell extract ($\sim$1/10 of the extract) before loading on the first gradient (Fig. 9.1, step 2) and spike it into the RNase H cleavage sample before loading on the second gradient. The spiked aliquot will result in an OD_{254} profile that contains multiple peaks (as in the first profile), which can be used to determine the sedimentation position of the various complexes.

1. Carefully load the sample on a sucrose gradient containing heparin and centrifuge as in Section 3.2.
2. After centrifugation, collect multiple fractions from the gradient. The number of fractions to collect depends on the desired resolution; a reasonable number of fractions to start with is 18 (fraction vol. $\sim$0.6 ml).
3. Collect fractions into 13-ml tubes (e.g., 95 $\times$ 16.8 mm polypropylene from Sarstedt cat no. 55.518) containing 1.5 volume of 8 M GuHCl (final concentration of $\sim$5.5 M).
4. Add 2.5 sample volumes of 100% ethanol. Mix well and incubate at -20° for at least 1 h.
5. Centrifuge samples for 20 min at 12,000 rpm at 4° using a SM 24 rotor. Carefully discard the supernatant.
   The advantage of the SM 24 rotor is that it has enough room for all 18 tubes. Any other equivalent rotor will be good also.
6. Wash with 500 μl of ice-cold 80% ethanol, centrifuge as in step 5, and carefully discard the supernatant. Note that the pellet might be unstable.
7. Resuspend with 400 μl of TE, transfer to a micro tube, and precipitate again by adding 40 μl of 3 M NaAce, pH 5.3, and 1 ml of 100% ethanol. Mix well and incubate at -20° for at least 1 h.
8. Centrifuge samples for 20 min at 14,000 rpm at 4°. Discard the supernatant.
9. Wash by adding 100 μl of ice-cold 80% EtOH. Centrifuge as in step 8 and discard the supernatant.
10. Dry the pellet and resuspend in 10 μl of sterile, RNAse-free water. Keep at -20°. Take about half of the sample to northern analysis.

3.4. Determination of sedimentation position

The sedimentation of RDM cleavage products is determined by standard northern analysis (Fig. 9.3, step 7). Any northern protocol can be used for this purpose. The protocol presented here is widely used for the analysis of yeast mRNAs and is based on buffers described in Church and Gilbert (1984). It is relatively simple and provides reliable results when analyzing abundant mRNAs (>5 copies/cell [Wang *et al.*, 2002]) that yield cleavage fragments exceeding ~200 nts. For shorter fragments, it is possible to separate fragments on polyacrylamide gel and then transfer to a nylon membrane for northern blot analysis (Muhlrad and Parker, 1992).

Performing the analysis by RT-PCR in order to detect low-abundance mRNAs might be problematic. This is because in every RDM experiment there is a significant amount of primary transcript that was not cleaved and therefore sediments in the gradient with its full load of ribosomes. RT-PCR with any set of primers will not distinguish between signals derived from the cleavage products and signals from the full-length transcript.

1. Prepare 1.2 to 2.5% agarose gel (depending on the expected sizes of fragments) in 1× MOPS and formaldehyde (1.3% final concentration). Dissolve the agarose first in water, let cool to 65°, and then add the 10× MOPS and formaldehyde.
   The gel should be poured in a hood because of the use of formaldehyde.
2. Mix 5 μl of each RNA sample with 7.5 μl of RNA loading buffer and 2.5 μl of 6× loading dye.
3. Incubate at 55° for 10 min to open RNA secondary structures.
4. Load the sample on the gel. Also load an RNA size marker, the "uncut" and "cut" controls, and a sample (2–10 μg) of untreated RNA. Run the gel in 1× MOPS buffer to obtain the best resolution of the expected cleavage products.
   Running an RNA size marker is critical for the correct identification of the cleavage products. Resuspend the RNA marker in the same buffer as the samples, yet exclude any dyes that may obscure some of its bands. The uncut and cut samples are also important because they can indicate the efficiency of cleavage as well as the nonspecific bands.
   MOPS buffer has a weak buffering capacity. A long running time tends to increase pH in the upper chamber and might lead to degradation of the RNA. It is therefore advised to circulate the buffer between chambers during the running time.
5. Blot the RNA from the gel to a nylon membrane.
6. Cross link the RNA to the membrane using an UV cross linker or a dry oven.
7. The quality of RNA and transfer can be evaluated by methylene blue staining. This is done by immersing the membrane in 5% acetic acid for 5 min and then in 0.1% methylene blue in 5% acetic acid for 5 min. The membrane is then washed in water, and distinct bands of the 25S (3400 nts), 18S (1800 nts) rRNA, and RNA marker bands should appear. Mark the positions of these with a pencil because the methylene blue staining will disappear during the hybridization.
8. Prepare a radioactive probe to hybridize with the membrane.

When identification of all cleavage products is desired, the PCR product corresponding to the entire mRNA is labeled by the "random priming" method (Sambrook and Russell, 2001). When detection of only some of the fragments is desired, or when the mRNAs have high homology with unwanted sequences, specific oligonucleotides (~30 nts) are labeled by T4 kinase and radioactive $\gamma^{32}P$-ATP (Sambrook and Russell, 2001). The former labeling method yields probes with higher specific activity and is more efficient because it detects all products in one experiment. The following protocol is performed with such a probe, yet it can be modified to an oligonucleotide probe simply by lowering the temperatures of hybridization and washings.

9. Hybridize the probe with the membrane at 57° for at least 6 h in the hybridization buffer.
10. Following hybridization, wash the membrane 2× 15 min in Hybridization Wash 1 and 2× 15 min in Hybridization Wash 2 at 57°.
11. Expose to a phosphor–imager screen or a film. Three bands should appear: two that are similar in size to the cleavage products (5' and 3' to the RNase H cleavage site), and a longer band similar in size to the full-length transcript that represents the remainder of an uncut mRNA (Figs. 9.1 and 9.2).

The most common problem is the appearance of multiple bands and nonspecific signals. If there are other mRNAs that cross-hybridize with the gene of interest, a possible solution is to use a different probe that is more specific to the target sequence (e.g., oligonucleotide probe). Alternatively, increasing the hybridization or washing temperature may remove some of these nonspecific signals. Another unwanted result is the appearance of smears or degradation products that hinder the detection of the specific bands. Degradation usually occurs at the RNase H cleavage step because the fraction collected may contain an overload of RNases. In such a case, it is recommended to decrease the amount of cell extract that is loaded on the first gradient and add some more RNase inhibitor. Note that prior to loading on the gradient, the cells are in a lysis buffer that contains heparin, and after the RNase H reaction, an LMD buffer is added that also includes heparin. These steps are therefore less likely to lead to degradation. We also noticed increased degradation levels when overexpressing reporter genes. A possible solution to this is to express the reporter gene from a weaker promoter and to fuse to it UTRs from the organism into which it is inserted.

ACKNOWLEDGMENTS

Work in my laboratory is supported by grants from the Israel Science Foundation and the Ministry of Science and Technology.

REFERENCES

Abastado, J. P., Miller, P. F., and Hinnebusch, A. G. (1991). A quantitative model for translational control of the GCN4 gene of *Saccharomyces cerevisiae*. *New Biol.* **3,** 511–524.

Arava, Y. (2003). Isolation of polysomal RNA for microarray analysis. *Methods Mol. Biol.* **224,** 79–87.

Arava, Y., Boas, F. E., Brown, P. O., and Herschlag, D. (2005). Dissecting eukaryotic translation and its control by ribosome density mapping. *Nucleic Acids Res.* **33,** 2421–2432.

Arava, Y., Wang, Y., Storey, J. D., Liu, C. L., Brown, P. O., and Herschlag, D. (2003). Genome-wide analysis of mRNA translation profiles in *Saccharomyces cerevisiae*. *Proc. Natl. Acad. Sci. USA* **100,** 3889–3894.

Church, G. M., and Gilbert, W. (1984). Genomic sequencing. *Proc. Natl. Acad. Sci. USA* **81,** 1991–1995.

Cigan, A. M., and Donahue, T. F. (1987). Sequence and structural features associated with translational initiator regions in yeast—A review. *Gene* **59,** 1–18.

Hinnebusch, A. G. (2005). Translational regulation of GCN4 and the general amino acid control of yeast. *Annu. Rev. Microbiol.* **59,** 407–450.

Iacono, M., Mignone, F., and Pesole, G. (2005). uAUG and uORFs in human and rodent 5′untranslated mRNAs. *Gene* **349,** 97–105.

Kozak, M. (1987). An analysis of 5′-noncoding sequences from 699 vertebrate messenger RNAs. *Nucleic Acids Res.* **15,** 8125–8148.

Kuhn, K. M., DeRisi, J. L., Brown, P. O., and Sarnow, P. (2001). Global and specific translational regulation in the genomic response of *Saccharomyces cerevisiae* to a rapid transfer from a fermentable to a nonfermentable carbon source. *Mol. Cell Biol.* **21,** 916–927.

Morris, D. R., and Geballe, A. P. (2000). Upstream open reading frames as regulators of mRNA translation. *Mol. Cell Biol.* **20,** 8635–8642.

Muhlrad, D., and Parker, R. (1992). Mutations affecting stability and deadenylation of the yeast MFA2 transcript. *Genes Dev.* **6,** 2100–2111.

Rickwood, D. (1992). "Preparative Centrifugation, A Practical Approach," 2nd Ed. Oxford, New York.

Sachs, M. S., Wang, Z., Gaba, A., Fang, P., Belk, J., Ganesan, R., Amrani, N., and Jacobson, A. (2002). Toeprint analysis of the positioning of translation apparatus components at initiation and termination codons of fungal mRNAs. *Methods* **26,** 105–114.

Sambrook, J., and Russell, D. W. (2001). "Molecular Cloning—A Laboratory Manual." CSHL Press, New York.

Tzamarias, D., Roussou, I., and Thireos, G. (1989). Coupling of GCN4 mRNA translational activation with decreased rates of polypeptide chain initiation. *Cell* **57,** 947–954.

Vilela, C., and McCarthy, J. E. (2003). Regulation of fungal gene expression via short open reading frames in the mRNA 5′untranslated region. *Mol. Microbiol.* **49,** 859–867.

Wang, Y., Liu, C. L., Storey, J. D., Tibshirani, R. J., Herschlag, D., and Brown, P. O. (2002). Precision and functional specificity in mRNA decay. *Proc. Natl. Acad. Sci. USA* **99,** 5860–5865.

Warner, J. R., Knopf, P. M., and Rich, A. (1963). A multiple ribosomal structure in protein synthesis. *Proc. Natl. Acad. Sci. USA* **49,** 122–129.

CHAPTER TEN

GENOME-WIDE ANALYSIS OF mRNA POLYSOMAL PROFILES WITH SPOTTED DNA MICROARRAYS

Daniel Melamed *and* Yoav Arava

Contents

Department of Biology, Technion—Israel Institute of Technology, Haifa, Israel

Methods in Enzymology, Volume 431
ISSN 0076-6879, DOI: 10.1016/S0076-6879(07)31010-0

Abstract

The sedimentation of an mRNA in sucrose gradients is highly affected by its ribosomal association. Sedimentation analysis has therefore become routine for studying changes in ribosomal association of mRNAs of interest. DNA microarray technology has been combined with sedimentation analysis to characterize changes in ribosomal association for thousands of mRNAs in parallel. Such analyses revealed mRNAs that are translationally regulated and have provided new insights into the translation process. In this chapter, we describe possible experimental designs for analyzing genome-wide changes in ribosomal association, and discuss some of their advantages and disadvantages. We then provide a detailed protocol for analysis of polysomal fractions using spotted DNA microarrays.

1. Introduction

Velocity sedimentation in sucrose gradients has been used for more than four decades to assess the translational status of an mRNA (Warner *et al.*, 1963). mRNAs are separated in a sucrose gradient by ultracentrifugation according to the number of ribosomes with which they are associated, and the distribution pattern of a specific mRNA can be determined by northern analysis. Two parameters that are related to translatability of a gene can be easily obtained from such analysis: (1) The percentage of transcripts that are associated with ribosomes (ribosomal occupancy). This parameter reports of the efficiency in which transcripts of a particular gene are recruited by the translation machinery. (2) The number of ribosomes with which the mRNA is associated (ribosomal density) (Arava *et al.*, 2003). The number of ribosomes on an mRNA reports on the overall balance between the steps of ribosome binding, progression along the coding region, and dissociation. Differences in these parameters among different mRNAs reflect differences in their translation efficiency. For example, low ribosomal occupancy for a particular mRNA (low percentages of association with ribosomes) may suggest a regulatory mechanism in which the mRNAs are stored in a nontranslating pool, awaiting a signal to be recruited by the translation machinery. Moreover, a change in one of these parameters for a particular gene upon change in growth condition or a mutation is indicative of translational regulation. For example, the GCN4 mRNA was shown to have increased ribosomal density under amino acid–deprivation conditions, which is an outcome of a translational regulation mechanism.

In the last few years, DNA microarrays have been utilized to perform simultaneous analysis of the translational status of thousands of mRNAs, thereby enabling a comprehensive view of translation efficiency and regulation. Analyses were performed on mRNAs isolated from various organisms, under many different growth conditions or mutations, and at different resolutions (Arava *et al.*, 2003; Blais *et al.*, 2004; Branco-Price *et al.*, 2005; Johannes *et al.*, 1999; Kuhn *et al.*, 2001; MacKay *et al.*, 2004; Qin and Sarnow, 2004; Shenton *et al.*, 2006; Smirnova *et al.*, 2005). Such genome-wide analyses provided new insights to the process of translation in general and to the regulation of particular genes.

The data obtained has been derived from different experimental designs (namely, the number of collected fractions and the hybridization scheme), different microarray platforms (spotted versus Affymetrix arrays), and different data analysis schemes. In this chapter, we discuss various experimental designs for microarray analyses, provide detailed protocols for various experimental steps when using the spotted microarray platform, and discuss aspects of data analysis.

2. EXPERIMENTAL DESIGNS

All experimental designs for studying the translational status of a cell share the general steps of separation of complexes on a sucrose gradient, fractionation of the gradient, RNA purification from the fractions, cDNA labeling, and hybridization to a DNA microarray. The variables that are most critical for data interpretation are the number of fractions into which the gradient is separated and the method by which experimental variations are corrected ("normalization" method). In the following section, we discuss these issues in detail and provide simplified numerical examples to illustrate some points.

2.1. Number of fractions into which the gradient is separated

2.1.1. Two fractions

The simplest experimental design includes separation of the entire gradient into two fractions: free mRNAs (free) and polysome-associated mRNAs (poly) (for simplicity of this discussion, mRNAs sedimenting as associated with one ribosomes will be included with the free fraction, although this might not be correct for all mRNAs). The labeling and hybridization of the free and polysome-associated fractions can be directed one against the other, where the polysome fraction is labeled with red fluorophore and the free with green fluorophore (Fig. 10.1A). Alternatively, the labeling and hybridization can be indirect, where the free and poly are each labeled in red and an unrelated RNA sample is labeled in green and serves as a common reference (Fig. 10.1B). The advantages and disadvantages of these schemes are described later and are relevant also for other experimental designs.

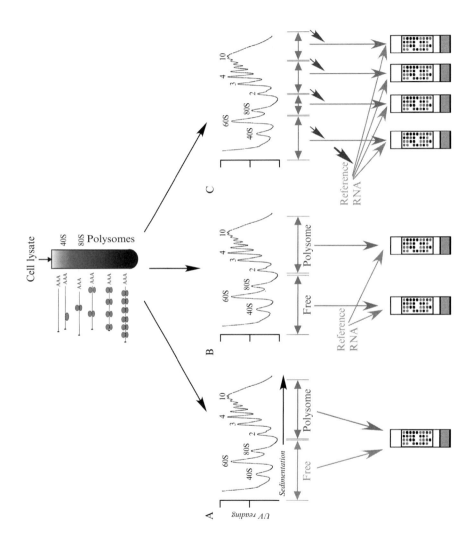

2.1.2. Direct comparison

When performing an experiment in which the polysomal fraction is labeled in red and the free fraction is labeled in green (also called a type I experiment), the resulting red-to-green ratios indicate the relative polysomal association of an mRNA. That is, genes with a high ratio are those that are highly associated with polysomes, and vice versa for low ratio. One can therefore compare the poly/free ratios of all genes at various growth conditions or treatments and determine which genes were most affected.

The main advantage of this experimental design is that both samples are compared on a single DNA microarray, and therefore errors that are due to the use of multiple microarrays, such as variation in microarray production, hybridization, and washing, are minimized. Time and money considerations are another important advantage of this design.

It should be stressed, however, that the obtained ratios from such experiments cannot be directly related to the actual distribution of an mRNA in free and polysomal fractions (i.e., a gene with a ratio of 1 does not necessarily have 50% of its molecules in free and 50% in polysomes). This is because the procedure that leads to the obtained ratios includes many steps, and each step may introduce a certain bias. Biases may arise during the steps of RNA purification, labeling, hybridization, and scanning, which results in skewed ratios. However, a reasonable assumption is that the bias is the same for all mRNAs within a sample, and therefore the ranking of the ratios is not affected by these biases. Namely, genes with higher ratios have higher polysomal association relative to other genes and therefore changes in their ranking are indicative of effects on their translation.

An inherent limitation of this design is that the information that it provides is in the form of a ratio (between the free and polysomes), and not the actual distribution in these fractions. This fact becomes a limitation when multiple treatments are compared, since the product of dividing one ratio by another does not necessarily reflect the *extent* of changes (see Example 10.1). A possible way to overcome this limitation is to extract

---

Figure 10.1 Experimental schemes for microarray analysis. All experimental schemes start with a separation step of the cell lysate by velocity sedimentation in a sucrose gradient (top scheme). Collection of the desired fractions is assisted by a continuous ultraviolet (UV) reading of the gradient (an example of such UV reading is shown in each section). This allows determination of the sedimentation position of the 40S, 60S, 80S, and polyribosomal complexes (2, 3, and more). Three general ways for fraction collection and analysis are presented (sections A, B, and C): (A) Collection of two fractions (free and polysomes) and direct comparison between them, with the free mRNA fraction labeled with green dye and the polysome fraction labeled with red dye. (B) Collection of two fractions and indirect comparison between them by utilizing an unfractionated reference RNA. (C) Collection of multiple fractions (four in this case), where each fraction is compared to an unfractionated reference sample. The blue arrows indicate the addition of spike-in RNA to each fraction and to the reference RNA. (See color insert.)

Before treatment		After treatment		
(A) Microarray results				
gene	**Rb (P/F)**	**Ra (P/F)**	**Ra/Rb**	
a	0.25	0.02	0.08	
b	1.38	0.67	0.48	
c	24.00	3.55	0.15	
d	0.02	0.02	1	
(B) mRNA distribution				
gene	**Free**	**Polysome**	**Total**	**Free**	**Polysome**	**Total**

gene	Free	Polysome	Total	Free	Polysome	Total
a	80	20	100	98	2	100
b	42	58	100	60	40	100
c	4	96	100	22	78	100
d	98	2	100	98	2	100

gene	**Pb**	**Pa**	**Pa/Pb**
a	0.2	0.02	0.1
b	0.58	0.4	0.69
c	0.96	0.78	0.81
d	0.02	0.02	1

Example 10.1 Two fractions experiment, direct comparison. This theoretical example presents possible microarray results for four genes (*a–d*) from direct comparison hybridization (A). Two fractions (free [F] and polysomes [P]) from cells before or after a certain treatment were hybridized against each other and yielded the indicated ratios (Rb (Ratio before) and Ra (Ratio after)). To recapitulate the differential effects on genes *a–d*, a trivial way would be to divide the ratio after the treatment by the ratio before the treatment (Ra/Rb). From this example, one might conclude that gene *c* was more affected than gene *b*. However, this is not necessarily correct. A possible way to draw this erroneous conclusion is demonstrated in section (B). The distributions of transcripts of genes *a–d* in the free and polysome fractions, either before or after the treatment, are indicated. Calculating the polysome/free ratio for these genes will result in ratios identical to the ones in section (A) (the microarray results). Yet, for 3 of the genes (*a*, *b*, and *c*), 18 molecules had shifted from the polysome to the free fraction, and gene *d* did not change after the treatment (for simplicity, all genes have the same number of transcripts (100), which is indicated at the Total column). Thus, in terms of percent of molecules shifting out from polysomes, gene *b* was much more affected than gene *c*. This does not agree with what we would have concluded in section A from Ra/Rb. These effects are because the extent of change for a certain gene is dependent on its polysomal distribution before the treatment. Ratio of distribution that is close to one will result in apparent weaker effects, and the farthest that the ratio is from one, the stronger the apparent effects are. One way to better recapitulate the changes is to restore the values of polysomes before (Pb) and polysomes after (Pa) (this can be done by the formula $[P = R/(1 + R)]$), and then divide Pa by Pb. This manipulation is presented in the lower four rows. The column Pa/Pb recapitulates correctly the changes of each gene in the polysome fraction.

gene	Pb			Pa	Pa/Pb
a	0.2			0.02	0.1
b	0.58			0.4	0.69
c	0.96			0.78	0.81
d	0.02			0.02	1

gene	Ra	Ra norm	Pa norm	Pa norm/Pb
a	0.02	0.02	0.02	0.08
b	0.67	0.53	0.35	0.60
c	3.55	2.84	0.74	0.77
d	0.02	0.02	0.02	0.80

Example 10.2 The effect of normalization. There are many steps in a microarray analysis, from cells' collection to obtaining the fluorescent signal, which are susceptible to experimental variation. These variations may lead to significant biases in certain samples. To allow comparison among different samples, it is critical to correct for such biases. Correction (or "normalization") is usually done by multiplying the microarray results by a constant value (k) that can be obtained by various means (see text). As a result of this multiplication, the ratios obtained from different microarrays are of similar range, and better comparisons can be made. When analyzing ratios between Polysomes and Free fractions, the step of normalization inserts some inaccuracy that should be noted. In this example, the values of Polysomes before (Pb) and after treatment (Pa) and their ratio (Pa/Pb) were taken from Example 10.1. To demonstrate the effect of normalization, the Ra values were multiplied by a (randomly chosen) constant (0.8) and presented in the "Ra norm" column (rounded to the second decimal). In reality, of course, the value will be different, yet its outcome is similar—multiplication of all ratios by the same factor. From the new Ra (Ra norm), the Pa was calculated based on the standard formula [P = R/(1 + R)] to yield Pa norm. Next, a new Pa/Pb (Pa norm/Pb) was calculated. Comparing the new Pa/Pb (Pa norm/Pb) values to the unnormalized (Pa/Pb) values reveals that although the overall trend is the same there are some differences. Specifically, some genes (a and d) seems to be more affected by the manipulation than others (gene c). The reason for this is that the normalization involves multiplication of the ratios by a constant (k). Considering the formula for deriving Pa norm [P = kR/(1 + kR)], the numerator and denominator are affected differently, depending on the initial Ratio and the normalization factor (k). Overall, since the general trend is kept, this is a good method to obtain first approximation of affected genes.

the underlying distributions in the two fractions from the experimental ratios, and from these values, determine the extent of changes (Example 10.1). This may provide more accurate results; yet, in such a case, one should be alert to the effects of normalization (see Example 10.2).

2.1.3. Indirect comparison

In an indirect comparison design, the free and polysome samples are labeled with a red fluorescent dye, and each is hybridized independently to a DNA microarray against an RNA sample that is labeled with green fluorescent dye (Fig. 10.1B). Because the green-labeled sample is the same for both samples, it serves as a common reference and the polysome-to-free ratio can be obtained simply by dividing the ratio obtained from one microarray by the other (the common denominator cancels out, in such a case). The source of this RNA sample is not important because it serves only as a reference point: it can be an unfractionated RNA from the same cells or from other cells. The most important parameters for choosing this sample is that it will yield high-quality signals on the DNA microarray and that the same sample will be used in all analyses.

This experimental design (also called type II experiment) is more complex than the direct comparison design, because it requires twice the amount of DNA microarrays and an additional reference RNA sample. Thus, variation that is due to the DNA microarray production or its data acquisition might be a source for error. On the other hand, obtaining separate values for the free and poly fractions allows additional comparisons to be made. Specifically, one can compare each of these fractions before and after treatment and determine the relative changes for each gene in each fraction (thereby overcoming some of the limitations indicated in Example 10.1). This information helps in interpreting the biological significance of the results. In many cases, when multiple conditions are tested, type II experiments will provide the most consistent results since the use of a common reference to all samples allows correction for experimental variations.

It should be noted that a change in one fraction will not necessarily be reflected by a corresponding change in another fraction, even when the total amounts of mRNAs did not change. This is because the fold of change in a certain fraction is highly dependent on the initial amounts of RNA in this fraction, and these amounts are usually not similar in the two fractions. Therefore, a 10-fold decrease in polysomes will not necessarily be reflected by a 10-fold increase in the free fraction (see Example 10.3). Thus, a reciprocal effect in two fractions is not a prerequisite for assigning genes that were affected by a certain treatment.

2.1.4. Three or more fractions

When higher resolution is desired, more than two fractions need to be collected. Gradients can be divided to 3, 4, and even 25 fractions (Arava et al., 2003; MacKay et al., 2004), usually with the aim of identifying changes in the number of ribosomes associated with each mRNA. Experiments in which more than two fractions are analyzed can be performed only as type II experiments (indirect comparisons), in which each fraction is labeled in red fluorescence and hybridized to a microarray with an unfractionated RNA sample that is labeled in green (Fig. 10.1C). This experimental design leads to a set of ratios for each gene, which describes the relative abundance of its mRNA in each fraction. Ultimately, it would be desired to create from these ratios a distribution profile for each gene in the gradient, where a high ratio suggests a high abundance in this fraction and a low ratio suggests a low abundance. Such a distribution profile would provide a wealth of information

	Before treatment				After treatment		
gene	**F**	**P**	**Total**	**gene**	**F**	**P**	**Total**
a	80	20	100	*a*	98	2	100
b	42	58	100	*b*	60	40	100
c	4	96	100	*c*	22	78	100
d	98	2	100	*d*	98	2	100
	F/T	**P/T**			**F/T**	**P/T**	
a	0.80	0.200			0.98	0.020	
b	0.42	0.580			0.60	0.400	
c	0.04	0.960			0.22	0.780	
d	0.98	0.020			0.98	0.020	
			Fa/Fb	**Pa/Pb**			
		a	1.2	0.1			
		b	1.4	0.69			
		c	5.5	0.81			
		d	1	1			

Example 10.3   Indirect comparison. In this example, each fraction (free or polysome) was hybridized together with an unfractionated sample (total). The expected ratios for genes *a–d* are presented as F/T (free over total) and P/T (polysome over total). These ratios reflect the abundance of molecules in each fraction. Analysis of the changes in each fraction (Fa/Fb and Pa/Pb) reveals the genes that were most affected in each fraction. For example, gene *c* was induced 5.5-fold in the Free fraction, and gene *a* was repressed 10-fold in the polysome fraction. As can be clearly seen, there is no correspondence between the extent of change in the polysome and the change in the free fraction. Gene *a*, that was repressed 10 times in the polysome fraction, is increased by only 1.2 in the Free fraction. Gene *c*, on the other hand, was reduced by 0.81-fold in the polysome and increased by 5.5-fold in the Free. This lack of correspondence is due to the differences in the initial amounts of mRNAs of each gene in each fraction. Genes that had low amounts of mRNA in a certain fraction will show a relatively high effect and genes with high amounts will have a relatively small effects. Thus, genes that were strongly affected in one fraction will not necessarily appear affected in another fraction.

regarding the ribosomal association of each mRNA and its ribosome occupancy. Practically, however, experimental variations in the steps of RNA purification, labeling, hybridization, or scanning might skew the ratios in a particular fraction and therefore hamper the construction of an accurate distribution profile. For example, a better RNA purification of a certain fraction will lead to a higher labeling efficiency, and therefore higher ratios for this sample. The apparent high ratios may lead to an incorrect conclusion regarding the amounts of mRNAs in this fraction.

Various "normalization" protocols have been developed to correct for such biases in microarray experiments (Quackenbush, 2002). Most of them are based on the assumption that the mRNA levels of most genes, or of a

(A) mRNAs in each fraction

	Before treatment			After treatment		
	Free	**Mono**	**Poly**	**Free**	**Mono**	**Poly**
a	78	2	20	78	20	2
b	2	2	96	2	20	78
c	6	2	92	2	22	76
d	96	2	2	96	2	2

(B) Theoretical

(i) Ratio

	Before treatment			After treatment		
	F/T	**M/T**	**P/T**	**F/T**	**M/T**	**P/T**
a	0.78	0.02	0.20	0.78	0.20	0.02
b	0.02	0.02	0.96	0.02	0.20	0.78
c	0.06	0.02	0.92	0.02	0.22	0.76
d	0.96	0.02	0.02	0.96	0.02	0.02

(ii) Percent

	Free	**Mono**	**Poly**	**Free**	**Mono**	**Poly**
a	78	2	20	78	20	2
b	2	2	96	2	20	78
c	6	2	92	2	22	76
d	96	2	2	96	2	2

(iii) Percent change

	Free	**Mono**	**Poly**
a	0	18	−18
b	0	18	−18
c	−4	20	−16
d	0	0	0

(C) Normalized

(i) Ratio (sum of ratios = 1)

	Before treatment			After treatment		
	F/T	**M/T**	**P/T**	**F/T**	**M/T**	**P/T**
a	0.43	0.25	0.10	0.44	0.31	0.01
b	0.01	0.25	0.46	0.01	0.31	0.49
c	0.03	0.25	0.44	0.01	0.34	0.48
d	0.53	0.25	0.01	0.54	0.03	0.01
sum	1.00	1.00	1.00	1.00	1.00	1.00
norm. fac	0.55	12.50	0.48	0.56	1.56	0.63

(ii) Percent

	Free	**Mono**	**Poly**	**Free**	**Mono**	**Poly**
a	55.4	32.3	12.3	57.4	40.9	1.7
b	1.5	34.8	63.7	1.4	38.2	60.4
c	4.6	34.7	60.8	1.3	41.1	57.5
d	67.0	31.8	1.2	92.5	5.4	2.2

(iii) Percent change

	Free	**Mono**	**Poly**
a	2.0	8.6	−10.6
b	−0.2	3.4	−3.3
c	−3.2	6.4	−3.2
d	25.4	−26.4	1.0

Example 10.4   Effect of normalization on multiple fractions. The example herein presents the effect of using a "standard" normalization procedure, which is based on the assumption that all samples have the same amounts of mRNA. In this example, the gradient was divided into 3 fractions (free, monosome, and polysome). (A) The numbers of molecules in each fraction are presented for 4 genes (*a–d*) either before or after a treatment. Note that for simplicity the total amounts of mRNAs for each gene did not change. (B) The theoretical ratios from hybridization of each fraction are presented in (i), and transformed into percent in (ii). The percent change upon treatment is shown in (iii). As can be easily seen, the changes for each gene in percentages exactly reflect the changes in terms of mRNA copies. Note that gene *d* did not change its distribution upon treatment. (C) (i) The data from (B) was normalized such that the sum of ratios of all genes will be 1. The assumption behind this normalization is that deviation of certain fractions from this value (e.g., the monosome fraction) is due to experimental variation, such as mRNA losses or labeling efficiency. The factor by which each value was multiplied is indicated below each column (norm. factor). (ii) The percentages in each fraction after the normalization. (iii) The changes in distribution for each gene. As can be clearly seen, there is no relation between the actual changes and the calculated changes (compare B (iii) and C (iii)). This is most easily seen for gene *d*, which did not change at all, yet the normalized percentages suggest a shift of 26.4% out of the Mono (25.4% to the free fraction and 1% to the Poly fraction).

few specific mRNAs ("housekeeping" genes), are similar among the samples. This assumption is incorrect in the case of polysomal separation, because it is well established that mRNA levels vary significantly among fractions of a sucrose gradient and therefore using such a method will lead to incorrect results (see Example 10.4). A more reliable normalization procedure is based on the inclusion of exogenous, *in vitro* transcribed mRNA (van de Peppel *et al.*, 2003; Yang, 2006) (see Section 2.2).

2.2. Inclusion of exogenous RNA (spike-in controls)

One method to overcome some of the limitations that arise when using multiple fractions is to introduce into each fraction, immediately at its collection, known amounts of *in vitro* transcribed mRNAs ("spike-in"), which have corresponding spots on the DNA microarray (Arava *et al.*, 2003; van de Peppel *et al.*, 2003; Yang, 2006). These mRNAs should be with minimal homology to the tested mRNAs and are usually taken from an unrelated organism. Since they are introduced to each fraction immediately at its collection, they will be subjected to all of the remaining experimental steps, including RNA purification, labeling, hybridization, and scanning. Therefore, any variation in their signal between fractions is due to variations in the experimental steps following collection. Thus, a normalization factor can be derived based on the variation in their signal, by which the signals for the rest of the mRNAs are corrected. The introduction of equal amounts of external RNA has the advantage that no prior assumptions regarding the levels of mRNAs in each fraction need to be done. Following normalization for experimental variation, the obtained ratios for a gene represent its distribution within the gradient. From these ratios, one can determine the fraction in which a specific mRNA is most abundant (peak fraction) or the percent of molecules that are free of ribosomes.

While the use of spike–in RNA introduces many advantages to quantitative polysomal analysis, a few points should be noted: (1) It is important to use a pool of several RNAs, each of them present at different amounts, in order to minimize biases that are due to sequence, length, or expression levels. (2) The RNAs must be of good quality and introduced at exact amounts into each fraction. Small variations in pipetting these RNA will affect the resulting normalization factor. (3) Each DNA microarray should include many spots that correspond to these RNAs that are positioned at different regions on the microarray. These spots should yield strong signals following hybridization with no cross hybridization with mRNAs from the tested sample. Any error in the signals of these spots will affect the entire microarray and thus the entire data set, and therefore may influence the conclusions from the experiment.

2.3. Analysis of changes at the transcriptome level

All experimental designs described previously are focused on determining changes in ribosome association and are refractory to changes in the steady-state mRNA levels. To derive correct conclusions regarding the effects on translation, an analysis of changes in steady-state mRNA levels needs to be performed. This is not only because it will allow better understanding of cellular processes, but because changes in steady state are expected to differentially affect polysomal fractions. It is likely that the fraction of mRNAs free of ribosomes will be more affected by changes in mRNA levels, compared to fractions of mRNAs associated with ribosomes. Because of this differential effect, a decrease in mRNA levels in the fraction free of ribosomes can be interpreted as a shift to the polysomal fractions (increase in translation) or degradation of mRNA (decrease or no change in translation). Analysis of steady state mRNA levels may help distinguish between these options.

Analysis of steady-state mRNA levels is performed by extracting total RNA from cells and labeling the RNA without a separation step on a sucrose gradient. Following labeling, samples are hybridized to a microarray and their signals are quantified. As has been indicated, most experimental steps might introduce certain errors that will be different from one sample to another. Therefore, the resulting signals do not reflect the exact amounts of mRNA in the cell, and thus do not provide the absolute changes in mRNA levels but only changes relative to other genes. For example, a two-fold increase for a gene means that it increased by two-fold relative to the change of all other transcripts. If the levels of all other transcripts were actually reduced, then the relative two-fold increase might actually be a decrease in the absolute levels of the transcripts of that gene. The fact that the changes are not absolute, but relative to other genes, should be taken into consideration when the data (from both the total and fractionated RNA analyses) are interpreted.

3. METHODS

There are numerous protocols for polysomal gradients preparations that differ mainly at the step for harvesting the cells, and the gradient composition and separation times. The protocol presented later was optimized for isolation of polysomal mRNA from the yeast *Saccharomyces cerevisiae*, yet many steps will be similar to other eukaryotes and the procedure can easily be modified for other organisms. We will use this protocol as a template on which we will indicate and highlight points that are critical for the microarray analysis. Generally, the RNA isolated by this protocol can be used for analysis by DNA microarray, Northern blot, or RT-PCR.

3.1. Gradient preparation

We typically use 11 ml of 10 to 50% linear sucrose gradients, onto which 1 ml of cells lysate is laid. The gradients can be made by a gradient maker just before use or be prepared by hand a day before using any of several methods. The protocol that is presented here is very simple and reproducible and can be used to make multiple gradients simultaneously. The gradients are poured 12 to 24 h before use and allowed to equilibrate at 4°. Alternatively, immediately after pouring, the gradients can be stored at $-80°$ indefinitely. The gradients can be thawed at 4° 12 to 24 h before use with no change in quality.

For 1 gradient of 11 ml:

1. Prepare 3-ml mixes of 10, 20, 30, 40, and 50% sucrose solutions in a gradient buffer containing 20 mM Tris-HCl (pH 7.4), 140 mM KCl, 5 mM MgCl$_2$, 0.5 mM DTT, 0.1 mg/ml Cyclohexamide (C-7698, SIGMA), and 0.5 mg/ml Heparin (H-3393, SIGMA). For more than one gradient, increase the volume of the solutions proportionally.
2. Use a long Pasteur pipette to layer 2.2 ml of the 10% sucrose solution in a Beckman polyallomer tube (14 × 89 mm, #331372). Then, underlay 2.2 ml of the 20% sucrose solution by inserting the tip of the pipette to the bottom of the tube and slowly pipetting the 20% sucrose solution under the 10% solution. Underlay the 30, 40, and 50% sucrose solutions in the same manner. Cover the tube with aluminum foil and store overnight at 4° to establish a linear gradient.

3.2. Cell lysis

The presented protocol is for yeast cells grown in rich media and collected at the logarithmic phase of growth. It is designed for a liquid culture of 50 to 100 ml ranging from an OD$_{600}$ of 0.4 to 0.8. The necessary amount of cells will vary with respect to the aims of the experiment and the number of fractions to be analyzed. Under certain growth conditions or for certain mutations, higher amounts of cells might be needed. It is therefore advised to perform a preliminary experiment for polysomal separation, and to measure the amount of RNA in each fraction. For analysis in yeast, we found that 15 to 50 μg provides a good signal on the DNA microarray and therefore the amount of cells to be grown should yield at least that amount.

1. Grow a 50 to 100 ml culture to an OD$_{600}$ 0.4 to 0.8.
2. Add cyclohexamide to a final concentration of 0.1 mg/ml (stock solution 10 mg/ml in water) in order to arrest the elongation step of translation.
3. Transfer the culture into an ice-cold 50-ml tube and immediately spin down the cells at 4000 rpm for 4 min at 4°.
   Quick pelleting of the cells and removal of the rich medium assists in halting

the translation process. Delays in this step led, in some cases, to an increased 80S peak, presumably due to recruitment of free mRNA molecules after addition of the cyclohexamide.

4. Discard the supernatant, resuspend the cell pellet in 4 ml lysis buffer (20 mM Tris-HCL (pH 7.4), 140 mM KCl, 1.5 mM MgCl$_2$, 0.5 mM DTT, 0.1 mg/ml Cyclohexamide, 1 mg/ml Heparin, 1% Triton X-100), and pellet the cells at 4000 rpm for 4 min at 4°. Repeat this step one more time to get rid of residual culture media.

   Heparin inhibits many enzymes, including all RNases, therefore, it is critical for this step. It also inhibits the labeling step; therefore it needs to be removed later (by a LiCl precipitation). We did not find any comparable alternative (in efficiency and cost) to heparin. The composition and concentrations of other ingredients can be changed depending on the cell type or experiment.

5. Discard the supernatant, resuspend the pellet in 400 μl lysis buffer, and transfer to a screw-capped microfuge tube.

6. Add 1.5 ml of chilled glass beads (0.45–0.55 mm) and vortex vigorously in a bead beater by two pulses of 90 s at maximum level to achieve complete lysis.

7. To recover the lysate, puncture the bottom of the screw-capped tube with a flame-heated needle and place it on top of a 15-ml conical tube. Use the cylinder of a 5-ml syringe as an adaptor between the screw-capped tube and the 15-ml tube.

8. Spin down the assembly composed of the lysate-containing tube, the adaptor, and the 15-ml conical tube at 4000 rpm for 1 min at 4°.

9. Transfer the beads-free lysate from the 15-ml tube (supernatant and pellet) into a new cold microfuge tube.

10. Spin at 9500 rpm (~8200g) for 5 min at 4° and transfer the supernatant into a new cold microfuge tube. This step removes most of the cell debris, leaving a cleared lysate.

11. Bring the lysate to a final volume of 0.8 ml with lysis buffer.

12. Carefully load the lysate onto a sucrose gradient and insert the gradient into a cool SW41 rotor bucket. The centrifuge tube should be filled almost to the top to avoid problems with the tube collapsing during ultracentrifugation.

13. Centrifuge at 35000 rpm for 160 min at 4°.

   A good separation of up to ~8 to 10 ribosomes was obtained with these centrifugation parameters. Although different centrifugation times might allow better resolution for larger complexes, we did not find much improvement.

3.3. Fractions collection

Fractions from a sucrose gradient can be separated either according to the complexes that they contain or to a fixed volume. When the gradient is separated to a small number of fractions (e.g., free mRNA, monosomes, and polysomes), it is feasible to separate the fractions according to the complexes they contain. This ensures that equivalent fractions from different gradients will contain mRNA molecules associated with the same complexes. Thus, even if there are variations in gradient volumes or in complexes sedimentation, the analyzed fractions are the same. Collecting in this manner obviously necessitates knowing the sedimentation position of each complex; therefore, the use of a continuous ultraviolet (UV) detector, such as ISCO UA6, is highly recommended. When many fractions are necessary, it is advisable to separate the gradient into fractions of a constant volume, and thereby minimize the errors caused by inaccuracy in complexes identification. In some cases, a combination of the two methods will work well, for example, collecting the fractions of free mRNA and monosomes according to the complexes, and then collecting the rest of the gradient at fixed volumes. The fact that these fractions have different amounts of mRNAs does not pose any conceptual problem since these differences are eliminated or corrected by the normalization procedure. It might, however, introduce a technical problem in having the optimal amount of RNA for the labeling step.

3.4. Adding spike-in controls to fractions

Spike-in controls are *in vitro* transcribed RNAs that are added to each fraction at the time of collection. They are necessary in order to correct for differences in mRNA levels in the various fractions from the gradient. These controls should have minimal cross-reactivity with the tested genome and should be added at known (and usually equal) amounts to the tested fractions. The exact amount to be added should be such that their signals will be neither too low nor too high relative to the rest of the mRNA. This might not be trivial because some polysomal fractions contain high amounts of mRNAs while others contain low amounts, and therefore their signals on the DNA microarray will differ significantly. Although it is not optimal, it is possible, in such cases, to introduce into each fraction different amounts of the *in vitro* transcribed RNAs. Since they are added at known amounts, their signals on the DNA microarray should be relative to the amounts added, and if not, a correction factor ("normalization faction") should be imputed. It is recommended to use more than one type of RNA and to spike the different RNAs at different amounts, in order to have a better coverage of sequences and expression levels.

We routinely use a mix of five mRNAs that are derived from the *lys* (ATCC no. 87482), *trp* (ATCC no. 87485), *dap* (ATCC no. 87486), *thr* (ATCC no. 87484), and *phe* (ATCC no. 87483) clones from the bacterium *Bacillus subtilis* cloned into a vector that contains a stretch of As. These RNAs are generated by *in vitro* transcription using a T3 *in vitro* transcription kit (e.g., MEGAscript from Ambion) of the linearized DNA template with the appropriate restriction enzyme.

Table 10.1   Spike-in mix stock solution

ATCC clone	Clone name	Concentration in mix
87482	lys	80 pg/μl
87485	trp	160 pg/μl
87486	dap	200 pg/μl
87484	thr	240 pg/μl
87483	phe	320 pg/μl

PCR products, corresponding to these clones, should be present at multiple sites on the DNA microarray. We routinely put at least 20 spots for each clone.

Currently, most commercially available DNA microarrays include spots that correspond to *in vitro* transcribed RNA from an unrelated organism (Yang, 2006). These RNAs can be either purchased as a ready-to-use mix or synthesized *in vitro*.

1. Prepare a mix of the five different mRNA spikes at the concentrations indicated in Table 10.1. Store at −80°.
2. Thaw the mix on ice just before use and add the appropriate volume to each fraction immediately after collection.

    The amount of mix to add depends on the experimental setting and the number of fractions collected. We typically add 70 μl of spike-in mix into an entire sucrose gradient, where each fraction receives the relative share from that amount. For example, 35 μl of the spike-in mix will be added to each fraction of a gradient that was divided into two, and 7 μl of this mix will be added to each fraction of a gradient that was divided into 10 fractions. The added amounts should consider losses during purification steps and that a minimum of 0.2 ng of each spike is needed to yield sufficient signal in the microarray hybridization.

3.5. RNA extraction from fractionated gradients

The following protocol is adapted to precipitation of RNA from a gradient that was fractionated to four fractions: free mRNA, monosome, low polysome, and high polysome. Most of the procedure is aimed at removing proteins (e.g., RNases) or reagents that will inhibit the labeling step (such as heparin). It can be used to purify RNA from sucrose gradients of any organism.

1. Collect fractions into 50-ml Oak-ridge tubes.
2. To each fraction, add 1 volume of 8 M guanidium–HCl and 2 volumes of 100% ethanol.
3. Add 17.5 μl of spike-in mix to each fraction (regardless of the fraction's volume). Mix the samples thoroughly and store for overnight at −20°.
4. Centrifuge at >10,000 rpm for 30 min at 4°.

5. Discard supernatant, add 5 ml ice–cold 75% ethanol, and centrifuge at >10,000 rpm for 20 min at 4°.
6. Resuspend pellets in 400 μl TE (pH 8.0). Let the samples stand for few minutes at room temperature to allow efficient dissolving.

   Heparin is found in high amounts in the first fractions (remnants of the lysis buffer); therefore, make sure that these fractions are dissolved well.
7. Transfer to a microfuge tube and add 0.1 volumes of 3 M sodium acetate (pH 5.2) and 2.5 volumes of 100% ethanol.
8. Incubate for at least 1 h at $-20°$ and spin at 14,000 rpm for 30 min at 4°.
9. Discard the supernatant and resuspend the pellet in 650 μl RNase-free water. Add an equal volume of water-saturated phenol:chloroform (5:1), pH 5.2. Vortex vigorously and spin at top speed for 5 min at room temperature. Take 500 μl of the aqueous phase into a new microtube tube.

   This step removes any residual proteins. It is important to avoid contamination of phenol in the extracted phase because this will affect the reverse transcriptase reaction. It is also important to avoid the interphase, which is enriched with DNA.
10. Bring to 1 ml with RNase-free water, add lithium chloride to a final concentration of 1.5 M (175 μl from 10 M stock), and incubate overnight at $-20°$. Thaw on ice and centrifuge at top speed for 20 min at 4°.

    The LiCl precipitation is necessary to remove any residual heparin, which may interfere with the labeling reaction. Some vendors (e.g., Ambion and Qiagen) sell RNA purification columns that should remove heparin. We have not tested any of these yet.
11. Carefully discard the supernatant, add 200 μl of 75% ethanol, and centrifuge at top speed for 20 min at 4°. Discard supernatant and resuspend in 150 μl RNase-free water.

    The pellet at this step is unstable and hardly visible. Carefully decant the supernatant and spin again if the pellet becomes unstable.
12. Precipitate again with sodium acetate as in steps 8 and 9.
13. Spin down at top speed for 20 min at 4°. Wash with 75% ethanol and air dry.
14. Resuspend pellet in 25 μl RNase-free water and store at $-80°$.

3.6. Preparation of a reference sample

There are two sources of RNA that can be used as a reference sample (the green-labeled sample): RNA from cells that are not related to the experiment or unfractionated RNA from the same cells under the same treatment. It is preferable to take an unrelated RNA sample that will be used in all future experiments, thereby serving as a common reference for all conditions and treatments. The most important parameter for this sample is that it yield a strong and reliable signal for as many genes as possible on the microarray. The optimal sample will yield signals for all genes on the microarray and therefore it will be possible to assign red-to-green ratio to all genes. To minimize variations, the reference sample should be extracted at large amounts so the same sample could be used in all experiments.

The following protocol is one of the versions of the "hot phenol" method and it is scaled for 100 to 150 ml of mid-log yeast culture (OD_{600}- 0.5–0.7). For different cell amounts, simply adjust the solution volumes.

1. Grow cells to mid-log phase.
2. Pellet cells at 3500g at room temperature for 3 min.
3. Discard the supernatant, wash the cells in sterile water, and pellet again as in step 2 to get rid of residual media. Following removal of supernatant, one can freeze and store the pellet at $-80°$, or continue directly to step 4.
4. Resuspend the pellet in 4 ml lysis buffer (10 mM Tris-HCl pH 7.5, 10 mM EDTA, 0.5% SDS). Place in a phenol-resistant tube (e.g., Sarstedt 13 ml, cat # 55.518).
5. Add 4 ml acid phenol (water saturated, low pH [e.g., Sigma P4682]). Vortex well.
6. Incubate at 65° for 1 h with occasional vigorous vortexing.
7. Place on ice for 10 min. Centrifuge at 4° for 10 min at top speed.
8. Transfer the upper aqueous phase into a new tube. Be careful not to take the interphase because this will lead to a contamination with DNA and/or phenol.
9. Add equal amount of acid equilibrated phenol:chloroform 5:1 (Sigma P1944), vortex well, and centrifuge again at room temperature for 5 min at top speed.
10. Transfer the upper aqueous layer into a new tube. Add equal amount of chloroform, vortex well, and centrifuge again at room temperature for 5 min at top speed.
11. Transfer the upper aqueous layer into a new tube. Precipitate the RNA by adding 0.1 volumes of 3 M sodium acetate (pH 5.2) and 2 volumes of 100% ethanol. Incubate overnight at $-20°$.
12. Centrifuge at 4° for 20 min at top speed.
13. Remove supernatant, add 1 ml of 75% ethanol, and centrifuge again at 4° for 20 min at top speed. Repeat this step and air-dry the sample.
14. The typical yield of this protocol is 1 to 2 mg RNA. Resuspend the RNA pellet to a final concentration of 5 to 10 $\mu g/\mu l$ in nuclease-free water and divide into aliquots. Store at $-80°$.

3.7. Reverse transcription and amino–allyl coupling

Labeling of the RNA sample with a fluorescent nucleotide can be done either directly or indirectly. In a direct labeling, one of the nucleotides (usually dUTP or dCTP) has a fluorescent moiety (either Cy3 or Cy5) linked to its base group. Therefore, during the reverse transcription step, the cDNA becomes labeled. In an indirect labeling, one of these nucleotides contains an amino–allyl group linked to its base group. An additional coupling step is therefore necessary, in which the fluorescent dye is coupled. We use the latter method, since it is less expensive, it is more consistent with different dyes, and it should yield longer cDNAs due to low structural hindrance by the allyl moiety.

3.7.1. cDNA synthesis

For one reaction:

For multiple samples, simply multiply everything by the necessary factor. The reference sample should be treated as one pool during the entire labeling process and split only prior to the hybridization step.

1. Prepare the following mixture:

   5 μg Oligo dT (T_{20}VN [V = any nucleotide except T])
   15 to 50 μg RNA of interest

Exact amount depends on the mRNA "richness" of the fraction. The exact amount of RNA is not critical because the spike-in RNA will correct for any variations. For example, taking twice the amount of RNA from one fraction will be associated with taking twice the amounts of its spiked-in RNA and therefore the resulting signals will be corrected accordingly.

   Nuclease-free water to 15.5 μl

   Incubate 10 min at 70° and transfer to ice for 10 min for annealing of the Oligo-dT to the mRNAs.

2. Prepare a reverse transcription reaction mix (based on Promega ImProm-II reverse transcription system).

   6.0 μl 5× reaction buffer
   4.0 μl 25 mM MgCl$_2$
   1.2 μl 25× Amino–allyl mix (12.5 mM dATP, 12.5 mM dGTP, 12.5 mM dCTP, 5 mM dTTP, 7.5 mM amino–allyl dUTP (Ambion #8439))
   0.3 μl Nuclease-free water.
   3.0 μl Reverse transcriptase

3. Incubate at 42° for 2 h.
4. Add 10 μl of 1 N NaOH, 10 μl of 0.5 M EDTA, and incubate at 65° for 15 min to degrade the RNA following the cDNA synthesis.
5. Add 25 μl of 1 M HEPES (pH 7.0) and bring the total volume to 100 μl by adding 25 μl of nuclease-free water. Precipitate the cDNA by adding 10 μl of 3 M sodium acetate (pH 5.2) and 275 μl of 100% ethanol.
6. Wash once with 70% ethanol, air dry, and resuspend in 9 μl of nuclease-free water. The samples can be stored at −80° for at least a month.

3.7.2. Fluorescent labeling

1. Add 1 μl of sodium bicarbonate 1 M (pH 9.0) to the amino–allyl labeled cDNA solution.
2. Add 1 μl of Cy3 or Cy5 dyes (Amersham cat. #336219 or 335176, respectively). Cy dyes should be suspended in 12 μl DMSO and divided into aliquots of 1 μl.

    An aliquot can be used immediately to label one cDNA sample or be completely dried out and stored at 4° in desiccator for later use. It is important to minimize the exposure of the dyes or labeled samples to light.
3. Incubate at room temperature for 1 h to allow coupling of the dye to the amino–allyl groups.
4. Purify the Cy-labeled cDNA using a DNA clean and concentrator kit (Zymo Research), according to the kit protocol. Elute the DNA with Nuclease-free water in a final volume of 5 μl.

3.8. Microarray slides preparation

The DNA microarrays are spotted on glass slides coated with amino-silane (Corning GAPS II). They should be ready for hybridization immediately when the labeled cDNA is ready. Thus, while the dyes are coupling to the cDNA (step 3 in the previous section), it is recommended to start the following process.

1. The slides are routinely stored in dark under desiccation. Handle all slides with powder-free gloves. Before use, mark the array boundaries with a diamond pen on the back of the slide since the arrays will not be visible after processing.
2. UV cross-link the slides at 300 mJ.
3. Put the slides in a metal slide rack (Shandon cat. #113) and submerge the rack in 0.1% SDS solution. Keep the tops of the slides under the level of solution and shake carefully for 30 s.
4. Dip the slides in DDW heated to 95° and incubate for 3 min.
5. Transfer the slides into a beaker with 70% ethanol and shake for 2 min.
6. Transfer the slides from the wet rack into a new, dry rack. While doing so, remove excess solution by striking gently and swiftly the bottom of each slide on a dried^{TM} Kimwipe.
7. Centrifuge the rack at 500 rpm for 5 min at room temperature for complete drying.
8. Prehybridize the slides in a coupling jar containing preheated and filtered prehybridization buffer (1% BSA (A7906 sigma), 5 × SSC, and 1% SDS) and incubate at 42° for 1 h.
9. Transfer the slides into a clean rack and soak in DDW for few seconds.
10. Transfer the slides into a dry rack and dry, as in steps 6 and 7. The slides are ready for hybridization.

3.9. Hybridization

1. Add 5 μl of each of the Cy-labeled samples into 20 μl hybridization buffer (4.5 × SSC, 75% formamide, and 0.15% SDS).
2. Add 3 μl (30 μg) poly(A)-RNA (Sigma P-9403).
3. Mix, microfuge briefly, and incubate at 95° for 3 min.
4. Microfuge briefly and add 3 μl of 10% BSA.
5. Mix and spin down at top speed for 1 min.
6. Place a microarray slide in a hybridization chamber and pipette 25 μl of the labeled probe mixture on the slide surface near one end of the microarray print area.
    Hybridization volumes will vary depending on the print size. The volumes herein are for 32 blocks, and should be adjusted for 16 or 48 blocks.
7. Grasp one end of a dust-free 22 mm × 40 mm microscope glass coverslip with forceps. Lower one end near the cDNA probe until it touches the surface outside the printed area and slowly lower the opposite end of the coverslip onto the slide. The solution will spread across the entire print area beneath the coverslip. Use a yellow tip to carefully adjust the position of the coverslip over the printed area. Large air bubbles can be moved away from the hybridization area by a gentle tapping on the coverslip with a yellow tip. Small air bubbles will be released during hybridization.
8. Add 10 μl of sterile water to the wells in both sides of the hybridization chamber in order to keep a moist environment.
9. Close the hybridization chamber and place in a water bath that has been preheated to 42°. Incubate for at least 6 h.

3.10. Washing

1. Remove one slide from the hybridization chamber at a time. The unsealing of the chamber should be done horizontally to prevent sliding of the coverslip.
2. Quickly transfer the slide (without removing the coverslip) into a rack and submerge it in Wash buffer 1 (2 × SSC and 0.05% SDS). Similarly, transfer each of the other slides into the soaked rack. Use needle to gently assist with the falling of the coverslips from the hybridization areas of the slides. Be careful not to touch any of the printed area. Shake the slides for 5 min at room temperature.
3. Transfer the rack into Wash buffer 2 (1 × SSC). Absorb excess Wash buffer 1 with a paper towel before putting the rack in Wash buffer 2. Shake the slides for 5 min at room temperature.
4. Transfer the rack into Wash buffer 3 (0.1 × SSC). Shake the slides for 5 min at room temperature.

5. Spin dry at 500 rpm for 5 min at room temperature. The slides are ready for scanning. Store the dried slides in dark box until scanning.

4. DATA ACQUISITION AND ANALYSIS

4.1. Microarray scanning

Several types of microarray scanners exist. Generally, all utilize lasers for the excitation of fluorochromes and photomultiplier tubes (PMT) to capture their fluorescence emission. Basically, increasing the laser power and/or the PMT voltage will produce higher signal intensities. For a large range of intensities, the emitted light is proportional to the number of fluorochromes in the spot. However, due to the PMT detection system limitations, extreme signals (low and high) deviate from this linearity (Lyng *et al.*, 2004). Therefore, it is recommended to perform two scans for each slide. The first scan should be at high PMT voltage to increase the intensities of the low signals to the level of linearity, thus allowing the high signals to be saturated. The second scan should be done at low PMT voltage in order to minimize the saturated spots. Subsequently, the data for the spots that were saturated in the first scan can be recovered from the second scan. It is important to ensure that the scans provide reliable signals to the spike in RNA (not too low and not saturated), since their signals will be used to normalize the signals of all other spots.

We utilize the GenePix 4000B scanner, which allows simultaneous detection of two dyes (Cy3 and Cy5). The following parameters are used for scanning:

Laser power—100% (we never change this parameter)
PMT gain—usually from 500 to 800, depending on signal intensity
Pixel size—10 μm
Lines to average—2
Focus position—10 μm

4.2. Creating a reliable dataset

Following scanning, a "gridding" step is performed in which each spot is cataloged and its signal quantified. Importantly, during this step, spots of low quality or with unreliable signals are flagged out. This step is performed automatically by gridding software, which utilizes various parameters, including differences between the spot and its background, the spot diameter and circularity, and the homogeneity of the signal in the spots in order to create a grid for all spots. Many software applications are available for this purpose (e.g., GeneSpring, Spotfire, and ImaGene), some of which are freely available (Scanalyze and Spotfinder). We use the software "GenePix" from Molecular

Devices. Although the gridding step is done automatically, it is highly recommended to go over the analysis performed by the program and correct misjudgment events. This step ensures that the subsequent statistical analysis will be performed with high-quality spots and with correct quantification.

The stringency of the filtration criteria will, of course, determine the number of spots to be analyzed and the reliability of the results. When possible, it is recommended to apply the same parameters to each slide in order to keep a common standard among fractions. Yet, the number of spots that pass the filtration criteria may differ greatly from slide to slide. This could be the outcome of low amounts of mRNA in certain fractions (e.g., free and monosome populations) or due to bad hybridizations. Because achieving signals from all fractions is crucial for determining the polysomal distribution of all mRNA, losing information for one fraction may damage the entire analysis. Therefore, it is sometimes necessary to reduce the stringency of filtration at the cost of less accurate data.

Beyond the standard features of selection criteria employed by the program, we typically add the following parameters to achieve reliable data:

1. At least 80% of a spot's pixels should have intensities more than two standard deviations above the background intensity for that spot, at each wavelength.
2. At least 80% of feature pixels at each wavelength are not saturated.
3. The diameter of the spot is greater than 55% of the feature size.
4. The coefficient of the regression line (R^2) is greater than 0.6.

4.3. Data verification

The methods for genome-wide analysis of ribosomal association are conceptually different from those for analysis of a single mRNA (e.g., northern analysis). While in assays of a single gene by northern analysis it is simple to compare the distribution of an mRNA in different fractions, in a genome-wide assay, the mRNAs are usually compared relative to others within the fraction. Because of that, it is important to analyze by northern analysis several mRNAs that appeared to be affected to different extents, and to compare the relative effects among this group between the microarrays and the northern analysis.

In experiments where spike-in controls are added, validation is simpler because one can perform a northern analysis for one of the spiked-in RNAs and use their signals to normalize any differences between fractions. This allows direct comparison between northern blot and microarray results.

ACKNOWLEDGMENTS

We thank Dr. Sunnie R. Thompson for critically reading this manuscript and for many great comments. Work in my lab is supported by grants from the Israel Science Foundation and the Israeli Ministry of Science and Technology.

REFERENCES

Arava, Y., Wang, Y., Storey, J. D., Liu, C. L., Brown, P. O., and Herschlag, D. (2003). Genome-wide analysis of mRNA translation profiles in *Saccharomyces cerevisiae*. *Proc. Natl. Acad. Sci. USA* **100,** 3889–3894.

Blais, J. D., Filipenko, V., Bi, M., Harding, H. P., Ron, D., Koumenis, C., Wouters, B. G., and Bell, J. C. (2004). Activating transcription factor 4 is translationally regulated by hypoxic stress. *Mol. Cell Biol.* **24,** 7469–7482.

Branco-Price, C., Kawaguchi, R., Ferreira, R. B., and Bailey-Serres, J. (2005). Genome-wide analysis of transcript abundance and translation in *Arabidopsis* seedlings subjected to oxygen deprivation. *Ann. Bot. (Lond.)* **96,** 647–660.

Johannes, G., Carter, M. S., Eisen, M. B., Brown, P. O., and Sarnow, P. (1999). Identification of eukaryotic mRNAs that are translated at reduced cap binding complex eIF4F concentrations using a cDNA microarray. *Proc. Natl. Acad. Sci. USA* **96,** 13118–13123.

Kuhn, K. M., DeRisi, J. L., Brown, P. O., and Sarnow, P. (2001). Global and specific translational regulation in the genomic response of *Saccharomyces cerevisiae* to a rapid transfer from a fermentable to a nonfermentable carbon source. *Mol. Cell Biol.* **21,** 916–927.

Lyng, H., Badiee, A., Svendsrud, D. H., Hovig, E., Myklebost, O., and Stokke, T. (2004). Profound influence of microarray scanner characteristics on gene expression ratios: Analysis and procedure for correction. *BMC Genomics* **5,** 10.

MacKay, V. L., Li, X., Flory, M. R., Turcott, E., and Law, G. L. (2004). Gene expression analyzed by high-resolution state array analysis and quantitative proteomics: Response of yeast to mating pheromone. *Mol. Cell Proteomics* **3,** 478–489.

Qin, X., and Sarnow, P. (2004). Preferential translation of internal ribosome entry site-containing mRNAs during the mitotic cycle in mammalian cells. *J. Biol. Chem.* **279,** 13721–13728.

Quackenbush, J. (2002). Microarray data normalization and transformation. *Nat. Genet.* **32** (Suppl.), 496–501.

Shenton, D., Smirnova, J. B., Selley, J. N., Carroll, K., Hubbard, S. J., Pavitt, G. D., Ashe, M. P., and Grant, C. M. (2006). Global translational responses to oxidative stress impact upon multiple levels of protein synthesis. *J. Biol. Chem.* **281,** 29011–29021.

Smirnova, J. B., Selley, J. N., Sanchez-Cabo, F., Carroll, K., Eddy, A. A., McCarthy, J. E., Hubbard, S. J., Pavitt, G. D., Grant, C. M., and Ashe, M. P. (2005). Global gene expression profiling reveals widespread yet distinctive translational responses to different eukaryotic translation initiation factor 2B-targeting stress pathways. *Mol. Cell Biol.* **25,** 9340–9349.

van de Peppel, J., Kemmeren, P., van Bakel, H., Radonjic, M., van Leenen, D., and Holstege, F. C. (2003). Monitoring global messenger RNA changes in externally controlled microarray experiments. *EMBO Rep.* **4,** 387–393.

Warner, J. R., Knopf, P. M., and Rich, A. (1963). A multiple ribosomal structure in protein synthesis. *Proc. Natl. Acad. Sci. USA* **49,** 122–129.

Yang, I. V. (2006). Use of external controls in microarray experiments. *Methods Enzymol.* **411,** 50–63.

CHAPTER ELEVEN

Synthesis of Anti-Reverse Cap Analogs (ARCAs) and Their Applications in mRNA Translation and Stability

Ewa Grudzien-Nogalska,*,† Janusz Stepinski,* Jacek Jemielity,*
Joanna Zuberek,* Ryszard Stolarski,* Robert E. Rhoads,† *and*
Edward Darzynkiewicz*

Contents

* Division of Biophysics, Institute of Experimental Physics, Faculty of Physics, Warsaw University, Warsaw, Poland
† Department of Biochemistry and Molecular Biology, Louisiana State University Health Sciences Center, Shreveport, Louisiana

Methods in Enzymology, Volume 431
ISSN 0076-6879, DOI: 10.1016/S0076-6879(07)31011-2

Abstract

Synthetic capped RNA transcripts produced by *in vitro* transcription in the presence of m^7Gp_3G have found a wide application in studying such processes as mRNA translation, pre-mRNA splicing, mRNA turnover, and intracellular transport of mRNA and snRNA. However, because of the presence of a 3'-OH on both m^7Guo and Guo moieties of the cap structure, one-third to one-half of the mRNAs contain a cap incorporated in the reverse orientation. The reverse cap structures bind poorly to eIF4E, the cap binding protein, and reduce overall translational efficiency. We therefore replaced the conventional m^7Gp_3G cap by "anti-reverse" cap analogs (ARCAs) in which the 3'-OH of m^7Guo moiety was substituted by 3'-deoxy or 3'-O-methyl groups, leading to $m^73'dGp_3G$ or $m_2^{7,3'-O}Gp_3G$, respectively. The class of ARCAs was extended to analogs possessing an O-methyl group or deoxy group at C2' of m^7Guo. We have also developed a series of ARCAs containing tetra- and pentaphosphates. mRNAs capped with various ARCAs were translated 1.1- to 2.6-fold more efficiently than their counterparts capped with m^7Gp_3G in both *in vitro* and *in vivo* systems. In a separate series, a methylene group was introduced between the α- and β-, or β- and γ-phosphate moieties, leading to $m_2^{7,3'-O}Gpp_{CH2}pG$ and $m_2^{7,3'-O}Gp_{CH2}ppG$. These analogs are resistant to cleavage by the decapping enzymes Dcp1/Dcp2 and DcpS, respectively. mRNA transcripts capped with $m_2^{7,3'-O}Gpp_{CH2}pG$ were more stable when introduced into cultured mammalian cells. In this chapter, we describe the synthesis of representative ARCAs and their biophysical and biochemical characterization, with emphasis on practical applications in mRNA translation.

1. INTRODUCTION

Eukaryotic mRNAs are modified at their 5' ends by addition of a 7-methylguanosine attached by a 5'-5' triphosphate bridge to the first transcribed nucleotide of the mRNA chain. This modification, known as a cap, plays a pivotal role in the function of mRNA in a variety of cellular processes including translation (Gingras *et al.*, 1999; Rhoads, 1999), splicing (Izaurralde *et al.*, 1994), intracellular transport (Izaurralde *et al.*, 1992), and turnover (Beelman *et al.*, 1998). The best-studied role of the cap is in translation, where it is specifically recognized by the translational initiation factor eIF4E (Niedzwiecka *et al.*, 2007).

Synthetic capped mRNAs are useful tools to study all of the processes mentioned previously. To create capped mRNAs, DNA templates are transcribed with either a bacterial (Contreras *et al.*, 1982) or bacteriophage (Konarska *et al.*, 1984; Yisraeli and Melton, 1989) RNA polymerase in the presence of all four ribonucleoside triphosphates and a synthetic cap dinucleotide, m^7Gp_3G. The polymerase initiates transcription with a nucleophilic attack by the 3'-OH of the Guo moiety of m^7Gp_3G on the α-phosphate of the next encoded nucleoside triphosphate, resulting in the initial product $m^7G(5')p_3(5')GpN$. Unfortunately, nucleophilic attack can also occur by the $3'$-OH of m^7Guo, producing $G(5')p_3(5')m^7GpN$ (Pasquinelli et al., 1995). This results in an in vitro product in which 30 to 50% of the RNA transcripts are capped in the reverse orientation. Because only mRNAs possessing correctly incorporated caps are properly recognized during cellular processes such as translation initiation and intracellular transport (Pasquinelli et al., 1995), the high fraction of reverse-capped transcripts can produce misleading data. We and others solved this problem by synthesizing cap dinucleotides in which the $3'$-OH of m^7Guo is substituted by an O-methyl group (Peng et al., 2002; Stepinski et al., 2001) or hydrogen atom (Stepinski et al., 2001). These compounds were incorporated into RNA transcripts exclusively in the correct orientation. We therefore named them "anti-reverse cap analogs" (ARCAs). These transcripts show two-fold higher translational efficiency in vitro compared to RNAs capped with the conventional m^7Gp_3G. We have also synthesized ARCAs consisting of tetra- and pentaphosphate-containing dinucleotides and found they produce mRNAs of even higher translational efficiency (Jemielity et al., 2003). Analogs in which the m^7Guo moiety contains $2'$-O-methyl or $2'$-deoxy modifications also act as ARCAs (Jemielity et al., 2003). mRNAs capped with various ARCAs are also translated 1.2- to 2.5-fold more efficiently when introduced into cultured mammalian cells than are mRNAs capped with m^7Gp_3G (Grudzien et al., 2006). The ability to produce proteins more efficiently has application in both research and biotechnology.

We have also produced series of ARCAs in which methylene groups are placed between the phosphate moieties in hopes of increasing the stability of mRNA transcripts against the decapping enzymes Dcp1/Dcp2 and DcpS (Grudzien et al., 2006; Kalek et al., 2006). There are two major pathways by which polyadenylated mRNA is degraded in eukaryotic cells, a $3'{\rightarrow}5'$ pathway and a $3'{\rightarrow}5'$ pathway, as well as two specialized routes for aberrant mRNA degradation (Coller and Parker, 2004). In both the $3'{\rightarrow}5'$ and $3'{\rightarrow}5'$ pathways, shortening of the poly(A) tract initiates mRNA decay. In the $5'{\rightarrow}3'$ pathway, removal of the cap structure occurs rapidly after poly(A) tract shortening (Muhlrad et al., 1995). This process is facilitated by the decapping enzyme Dcp1/Dcp2, which, in turn, exposes the transcripts to digestion by a highly possessive $5'{\rightarrow}3'$ exonuclease Xrn1 (Hsu and Stevens, 1993). Hydrolysis by either the Dcp1/Dcp2 complex or Dcp2 alone releases m^7GDP, suggesting that the site of cleavage is between the α and β phosphate moieties of the cap but not between the β and γ moieties. In the $3'{\rightarrow}5'$ pathway, deadenylated mRNA is degraded by the exosome in a $3'{\rightarrow}5'$ direction, as demonstrated both in vitro (Chen et al., 2001; Mukherjee et al., 2002; Wang and Kiledjian, 2000, 2001) and in vivo (Wang and Kiledjian, 2001). The products are capped oligonucleotides, which are then decapped by a scavenger decapping enzyme, DcpS (Liu et al., 2002). DcpS releases m^7GMP, suggesting that cleavage occurs between the β and γ

phosphate moieties. To achieve selective resistance to these two enzymes, we developed novel cap analogs in which each of two pyrophosphate oxygens was separately replaced by a methylene group, one in the α-β position ($m_2^{7,3'-O}Gp_{CH2}pG$) and one in the β-γ position ($m_2^{7,3'-O}Gp_{CH2}ppG$). The first but not the second of these analogs stabilized mRNA *in vivo*. This type of ARCAs can be used to test the relative contribution of the $5'{\rightarrow}3'$ *versus* $3'{\rightarrow}5'$ pathways because selective blockage of one of them should stabilize mRNA.

This chapter briefly describes the general methodology for making both series of ARCAs and describes in detail the synthesis of two representative compounds: P^1-guanosine-$5'$ P^3-(7,3'-O-dimethylguanosine-$5'$) triphosphate (Fig. 11.1, compound **2**) and P^1-guanosine-$5'$ P^3-(7,3'-O-dimethylguanosine-$5'$) β:γ-methylene-triphosphate (Fig. 11.1, compound **5**).

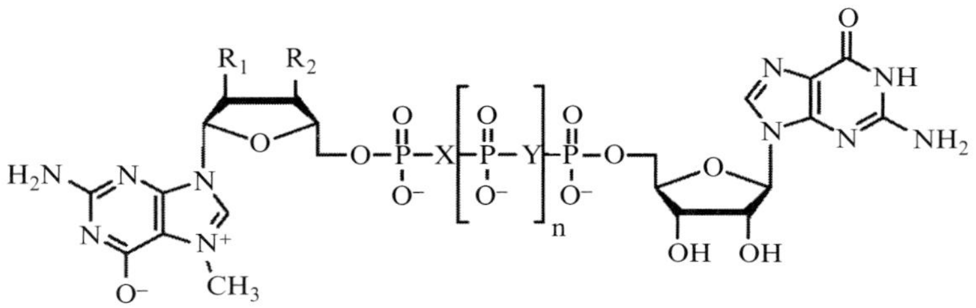

No.	Cap analog	R_1	R_2	X	Y	n
	Triphosphate series					
	m^7Gp_3G	OH	OH	O	O	1
1	$m_2^{7,2'O}Gp_3G^{(a)}$	OMe	OH	O	O	1
2	$m_2^{7,3'O}Gp_3G^{(c)}$	OH	OMe	O	O	1
3	$m^72'dGp_3G^{(a)}$	H	OH	O	O	1
4	$m^73'dGp_3G^{(c)}$	OH	H	O	O	1
5	$m_2^{7,3'O}Gp_{CH2}ppG^{(b)}$	OH	OMe	CH_2	O	1
6	$m_2^{7,3'O}Gpp_{CH2}pG^{(b)}$	OH	OMe	O	CH_2	1
	Tetraphosphate series					
	$m^7Gp_4G^{(a)}$	OH	OH	O	O	2
7	$m_2^{7,2'O}Gp_4G^{(a)}$	OMe	OH	O	O	2
8	$m_2^{7,3'O}Gp_4G^{(a)}$	OH	OMe	O	O	2
9	$m^72'dGp_4G^{(a)}$	H	OH	O	O	2
	Pentaphosphate series					
	$m^7Gp_5G^{(a)}$	OH	OH	O	O	3
10	$m_2^{7,3'O}Gp_5G^{(a)}$	OH	OMe	O	O	3

Figure 11.1   Structures of ARCAs. Denotatious to Figure 11.1:
[a]Synthesis described in Jemielity *et al.* (2003)
[b]Synthesis described in Kalek *et al.* (2006)
[c]Synthesis described in Stepinski *et al.* (2001)

The biophysical and biochemical characterization of ten ARCAs (Fig. 11.1, compounds 1–10) is presented.

2. CHEMICAL SYNTHESIS

Standard, one-dimensional ^{1}H NMR and ^{31}P NMR spectra are recorded in ^{2}H$_2$O at ambient temperature and millimolar concentration on a VarianUNITYplus 500 MHz and Varian Inova 400 MHz spectrometers, with sodium 3-trimethylsilyl-[2,2,3,3-^{2}H$_4$]propionate (TSP) as internal standard, and H$_3$PO$_4$ as external standard, respectively. Scalar, through-bond interactions between the nuclei are established by selective decoupling experiments. Mass spectra (MS) are recorded in the negative electrospray mode and are reported in mass units (m/z). Ion-exchange column chromatography is performed on DEAE-Sephadex (A-25/HCO$_3^-$ form) using a linear gradient of triethylammonium bicarbonate in water (TEAB, obtained by acidification of aqueous triethylamine to pH 7.3 using CO$_2$). Fractions are collected, and products peaks (monitored at 260 nm) are pooled and evaporated to dryness, with ethanol added repeatedly to remove the TEAB. Analytical HPLC studies are carried out using a reverse-phase column (Supelco LC-18-T, 4.6 × 250 mm, 5 μm) with a linear gradient of methanol from 0 to 25% in aqueous 0.05 M CH$_3$CO$_2$NH$_4$ pH 5.9, over 15 min (detection at 260 nm). Elutions are conducted at room temperature with 1.3 ml/min flow rate.

2.1. Synthesis of ARCAs containing conventional phosphate chains

The multistep synthesis of ARCAs 1–4 and 7–10 (Jemielity et al., 2003; Stepinski et al., 2001) is shown in Fig. 11.2. Either 2′-deoxy- and 3′-deoxyguanosine are commercial products from Sigma, whereas 2′-O-methyl- and 3′-O-methylguanosine could be prepared (Kusmierek and Shugar, 1978). The nucleosides are phosphorylated (Yoshikawa et al., 1967) and then converted to the appropriate 7-methylated 5′-diphosphates (Jemielity et al., 2003; Stepinski et al., 2001). Next, the products are coupled with guanosine 5′-mono-, di-, and triphosphate P-imidazolides to give the ARCA cap analogs with triphosphate (1–4), tetraphosphate (7–9), and pentaphosphate (10) 5′,5′-bridges, respectively. The key reactions of pyrophosphate bond formation can be achieved in anhydrous dimethylformamide solutions employing the catalytical properties of zinc chloride (Kadokura et al., 1997; Stepinski et al., 2001). A similar strategy for synthesis of m$_2$$^{7,3'-O}Gp_3$G was independently proposed (Peng et al., 2002). The key

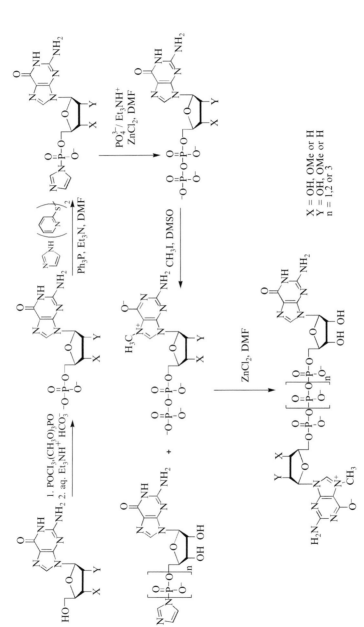

Figure 11.2 Synthesis of ARCAs.

difference between the two procedures is the final coupling reaction that Peng *et al.* (2002) carried out with GMP-morpholide as activated substrate and with 1*H*-tetrazole as catalyst.

The procedure later describes in detail the preparation of $m_2^{7,3'-O}Gp_3G$ (Fig. 11.1, compound **2**).

3′-O-methylguanosine 5′-monophosphate. 3′-O-methylguanosine (Kusmierek and Shugar, 1978) (118 mg, 0.4 mmol) is stirred overnight with trimethylphosphate (4 ml) and phosphorus oxychloride (106 μl, 1.14 mmol) at 6°. By adding 40 ml of water and neutralizing with 1 M TEAB, the reaction is quenched. DEAE-Sephadex chromatography using a linear gradient of 0 to 0.9 M TEAB affords 3′-O-methylguanosine 5′-monophosphate (TEA salt, yield: 160 mg, 69%).

3′-O-methylguanosine 5′-diphosphate. 3′-O-methylguanosine 5′-monophosphate (116 mg, TEA salt, 0.2 mmol), imidazole (68 mg, 1.0 mmol), and 2,2′-dithiodipyridine (from Aldrich, 88 mg, 0.4 mmol) are mixed in anhydrous DMF (2.4 ml) and TEA (28 μl). Triphenylphosphine (104 mg, 0.4 mmol) is added, and the mixture is stirred for 5 h at room temperature. The mixture is placed in a centrifuge tube, and sodium perchlorate (98 mg, anhydrous) dissolved in acetone (12 ml) is added. After cooling for 2 h in a refrigerator, the mixture is centrifuged and the supernatant is discarded. The precipitate is ground with a new portion of acetone, cooled, and centrifuged again. The process has to be repeated once more, and the precipitate is dried in a vacuum desiccator over P_4O_{10}. The imidazolide thus obtained is dissolved in 2.4 ml of DMF, and 400 mg of tris(triethylammonium)phosphate is added. The latter is prepared from TEA and phosphoric acid, followed by drying over P_4O_{10} in a desiccator to obtain a semicrystalline mass. Finally, 160 mg of $ZnCl_2$ is added, and the reaction mixture is stirred at room temperature for 6.5 h, poured into a beaker containing 0.5 g EDTA in 30 ml water, and neutralized with 1 M $NaHCO_3$. Chromatographic isolation on DEAE-Sephadex using a linear gradient of 0 to 1 M TEAB gives 3′-O-methylguanosine 5′-diphosphate (TEA salt, yield: 64 mg, 49%).

7,3′-O-dimethylguanosine 5′-diphosphate. 3′-O-methylguanosine 5′-diphosphate (TEA salt, 60 mg, 0.09 mmol) is mixed with 1 ml of dimethylsulfoxide, 1 ml of DMF, and 200 μl of methyl iodide at room temperature. After 3 h, the reaction mixture is treated with 30 ml of cold water and extracted three times with 10-ml portions of diethyl ether. Chromatography of the aqueous phase, after neutralization with $NaHCO_3$, on DEAE-Sephadex using a linear gradient of 0 to 0.8 M TEAB gives 7,3′-O-dimethylguanosine 5′-diphosphate (TEA salt, yield: 58 mg, 95%).

*P^1-guanosine-5′ P^3-(7,3′-O-dimethylguanosine-5′) triphosphate (**2**).* Guanosine 5′-monophosphate (purchased from Sigma, converted to the TEA salt, 69 mg, 0.15 mmol), imidazole (51 mg, 0.75 mmol), and 2,2′-dithiodipyridine (66 mg, 0.3 mmol, from Aldrich) are mixed in anhydrous DMF (3 ml) and TEA (21 μl). Triphenylphosphine (79 mg, 0.3 mmol) is added, and the

mixture is stirred for 5 h at room temperature. The mixture is placed in a centrifuge tube, and sodium perchlorate (75 mg, anhydrous) dissolved in acetone (18 ml) is added. The procedure for washing the precipitate with acetone and drying over P_4O_{10} is the same as that for 3′-O-methylguanosine 5′-monophosphate imidazolide (see previous instructions). The imidazolide of GMP thus obtained is dissolved in DMF (3 ml), and 7,3′-O-dimethyl-guanosine 5′-diphosphate (TEA salt, 34 mg, 0.055 mmol) is added. Next, $ZnCl_2$ (120 mg) is added, and the mixture is stirred at room temperature overnight, poured into a beaker containing a solution of 375 mg of EDTA in 45 ml of water, and neutralized with 1 M $NaHCO_3$. Chromatographic isolation on DEAE–Sephadex using a linear gradient of 0 to 1 M TEAB gives (2) (23 mg, 78% based on the amount of 7,3′-O-dimethylguanosine 5′-diphosphate used). The final product (2) is converted to its Na^+ salt by ion exchange using a small column of Dowex 50Wx8 (Na^+ form). Evaporation of the eluate to a small volume, precipitation with ethanol, and centrifuga-tion leads to an amorphous white powder.

The ^{1}H NMR parameters for **2** (D_2O, 500 MHz, in ppm) are: d 7.99 (1H, s, H8-G), 5.86 (1H, d, H1′-m_2G; J_{1-2} = 4.0 Hz), 5.78 (1H, d, H1′-G, $J_{1'-2'}$ = 6.3 Hz), 4.69 (1H, dd, H2′-G, $J_{1'-2'}$ = 6.3 Hz, $J_{2'-3'}$ = 5.1 Hz), 4.68 (1H, dd, H2′- m_2G; $J_{1'-2'}$ = 4.0 Hz, $J_{2'-3'}$ = 5.0 Hz), 4.47 (1H, dd, H3′-G, $J_{2'-3'}$ = 5.1 Hz, $J_{3'-4'}$ = 3.6 Hz), 4.43 (1H, m, H4′-m_2G; $J_{3'-4'}$ = 5.1 Hz, $J_{4'-5'}$ = 3.0 Hz, $J_{4'-5''}$ = 2.6 Hz, $J_{4'-P}$ ≅ 1 Hz), 4.38 (1H, m, H5′-m_2G; $J_{4'-5'}$ = 3.0 Hz, $J_{5'-5''}$ = 11.5 Hz, $J_{5'-P}$ = 4.4 Hz), 4.34 (1H, m, H4′-G; $J_{3'-4'}$ = 3.6, $J_{4'-5'}$ = 4.1 Hz, $J_{4'-5''}$ = 4.2 Hz, $J_{4'-P}$ ≅ 1 Hz), 4.28 (1H, m, H5′-G; $J_{4'-5'}$ = 4.1 Hz, $J_{5'-5''}$ = 11.8 Hz, $J_{5'-P}$ = 5.4 Hz), 4.24 (1H, m, H5″-G; $J_{4'-5''}$ = 4.2 Hz, $J_{5'-5''}$ = 11.8 Hz, $J_{5''-P}$ = 6.5 Hz), 4.22 (1H, m, H5″-m_2G; $J_{4'-5''}$ = 2.6 Hz, $J_{5'-5''}$ = 11.5 Hz, $J_{5''-P}$ = 5.9 Hz), 4.11 (1H, t, H3′- m_2G; $J_{2'-3'}$ = 5.0 Hz, $J_{3'-4'}$ = 5.1 Hz), 4.07 (3H, s, CH_3-N7), 3.48 (3H, s, CH_3-$O^{3'}$). MS: calcd for $C_{22}H_{31}N_{10}O_{18}P_3$ 816, found 815 [M-H]$^-$. HPLC Rt (retention time) 6.2 min.

2.2. Synthesis of methylene ARCAs

The route of synthesis leading to methylene ARCAs is depicted in Fig. 11.3. The crucial intermediates for synthesis of nonhydrolyzable cap analogs are 5′-bisphosphonates of nucleosides in which a bridging oxygen atom is replaced with a methylene group. We have developed simple and efficient methodol-ogy for synthesis of 5′-bisphosphonates in which nucleosides are phosphory-lated using methylenebis (phosphonic dichloride) in Yoshikawa 5′-phosphorylation conditions (Kalek *et al.*, 2005). The reaction proceeds with high 5′-regioselectivity leading to 5′-bisphosphonates with relatively high yields. Moreover, 5′-bisphosphonates of nucleosides can be easily activated as a P-imidazolides and then used in nucleophilic substitution reactions (Kalek *et al.*, 2006). 5′-bisphosphonates of nucleosides are also convenient

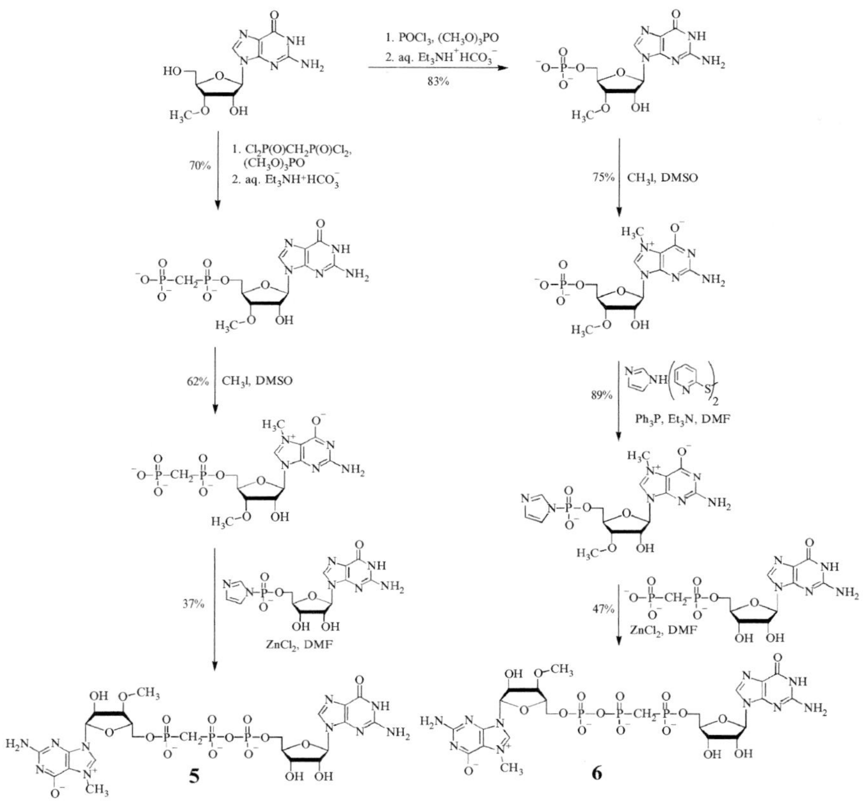

Figure 11.3   Synthesis of methylene-containing ARCAs.

nucleophilic agents in reactions with P-imidazolides of guanosine or 7,3'-O-dimethylguanosine (see Fig. 11.3, last steps in synthesis of 5 and 6). Both applications of 5'-bisphosphonates lead to 5', 5'-dinucleoside triphosphates with methylene modification, for instance, ARCAs 5 and 6. Following is presented the detailed description for preparation of $m_2^{7,3'-O}Gp_{CH2}ppG$ (Fig. 11.3, compound 5).

3'-O-methylguanosine-5'-yl (phosphono)-methylphosphonate. A solution of methylenebis(phosphonic dichloride) (30 mg, 0.12 mmol) in trimethyl phosphate (1.0 ml) cooled to 0° is added to a suspension of 3'-O-methylguanosine (36 mg, 0.12 mmol) in trimethyl phosphate (1.0 ml) at 0°. The reaction mixture is stirred at 0°. After 1 h, 0.7 M aqueous TEAB (pH 7.0) is added. Chromatographic purification on DEAE-Sephadex, using a 0 to 1 M gradient of TEAB, gives guanosine-5'-yl bisphosphonate as a white powder (TEA salt, 62 mg, 70%).

The ^1H NMR parameters for **5** (D_2O, 400 MHz, in ppm) are: d 8.17 (1H, s, H8), 5.91 (1H, d, H1′, $J_{1′-2′}$ = 6.4 Hz), 4.92 (1H, dd, H2′, $J_{1′-2′}$ = 6.4 Hz, $J_{2′-3′}$ = 5.1 Hz), 4.43 (1H, m, H4′), 4.20 (1H, dd, H3′, $J_{2′-3′}$ = 5.1 Hz, $J_{3′-4′}$ = 3.2 Hz), 4.14 (2H, m, H5′,5″), 3.53 (3H, s, CH_3O), 2.17 (2H, t, P-CH_2-P, J = 19.9 Hz). 31P NMR (163 MHz): d 19.68 (1P, α), 16.32 (1P, β). MS: calcd. for $C_{12}H_{18}N_5O_{10}P_2$ 454. Found 454 [M-H]⁻. HPLC Rt 6.3 min.

7,3′-O-dimethylguanosin-5′-yl (phosphono)-methylphosphonate. Methyl iodide (0.067 ml, 1.08 mmol) is added to a suspension of tris(triethylammo-nium) 3′-O-methylguanosin-5′-yl (phosphono)-methylphosphonate (62 mg, 0.09 mmol) in DMSO (2.5 ml) and stirred at room temperature for 5 h. The mixture is poured into water (25 ml) and extracted three times with diethyl ether (3 × 10 ml). Iodine remaining in the aqueous phase is reduced by addition of small amounts of $Na_2S_2O_5$. After neutralization to pH 7 with $NaHCO_3$, residual ether dissolved in the water is removed by evaporation under vacuum. Chromatographic isolation on DEAE-Sephadex using a 0 to 0.9 M gradient of TEAB gives 7,3′-O-dimethylguanosin-5′-yl (phosphono)-methylphosphonate as a white powder (TEA salt, 44 mg, 62%).

The ^1H NMR parameters (D_2O, 400 MHz, in ppm) are: ^1HNMR (D2O, 400 MHz): d 6.09 (1H, d, H1′, $J1′-2′$ = 3.7 Hz), 4.88 (1H, t, H2′, $J_{1′-2′}$ = $J_{2′-3′}$= 3.7 Hz), 4.51 (1H, m, H4′), 4.35–4.30 (1H, m, H5′), 4.23 (1H, m, H3′), 4.23–4.17 (1H, m, H5″), 4.16 (3H, s, CH_3N), 3.51 (3H, s, CH_3O), 2.22 (2H, t, P-CH_2-P, J = 19.7 Hz). ^{31}P NMR (163 MHz): d 18.72 (1P, α), 14.84 (1P, β). MS: calcd. for $C_{13}H_{20}N_5O_{10}P_2$ 468. Found 468 [M-H]⁻. HPLC Rt 7.9 min.

P^1-guanosine-5′ P^3-(7,3′-O-dimethylguanosine-5′) β:γ-methylene-triphosphate (5). Tris(triethylammonium) 7,3′-O-dimethylguanosin-5′-yl (phosphono)-methylphosphonate (44 mg, 0.06 mmol), sodium guanosin-5′-yl phosphor-1-imidazolidate (29 mg, 0.07 mmol), and $ZnCl_2$ (65 mg, 0.48 mmol) are stirred in anhydrous DMF (0.7 ml) at room temperature for 24 h. The mixture is poured into a cold solution of EDTA (0.179 g, 0.48 mmol) in water (20 ml) and neutralized to pH 7 by addition of $NaHCO_3$. The product is isolated by chromatography on DEAE-Sephadex using a 0 to 1.2 M gradient of TEAB. It is converted into the sodium salt using Dowex 50WX8/Na^+ form (yield: 18.5 mg, 37%).

The ^1H NMR parameters (D_2O, 400 MHz, in ppm) are: d 8.03 (1H, s, H8), 5.92 (1H, d, H1′$_{m7G}$; $J_{1′-2′}$ = 4.0 Hz), 5.82 (1H, d, H1′, $J_{1′-2′}$ = 6.0 Hz), 4.75 (1H, dd, H2′$_{m7G}$; $J_{1′-2′}$ = 4:0 Hz, $J_{2′-3′}$ = 5.0 Hz), 4.69 (1H, t, H2′, $J_{2′-1′;3′}$ = 5.7 Hz), 4.49 (1H, dd, H3′, $J_{2′-3′}$ = 5.4 Hz, $J_{3′-4′}$ = 4.0 Hz), 4.44 (1H, m, H4′$_{m7G}$), 4.33 (1H, m, H4′), 4.30–4.17 (4H, m, H5′,5′, 5′$_{m7G}$; 5′$_{m7G}$), 4.15 (1H, t, H3′$_{m7G}$; $J_{3′-2′;4′}$ =5.0 Hz), 4.08 (3H, s, CH_3N), 3.49 (3H, s, CH_3O), 2.39 (2H, t, P-CH_2-P, J = 20.4 Hz). ^{31}P NMR (163 MHz): d 28.48 (1P, β), 18.58 (1P, γ), 0.29 (1P, α). MS: calcd for $C_{23}H_{32}N_{10}O_{17}P_3$ 813. Found 813 [M-H]⁻. HPLC Rt 6.8 min.

2.3. Structure and conformation of ARCAs by NMR

High-resolution NMR spectroscopy in solution is a standard and widespread method for determining structures and conformations of organic compounds. The spectra are usually recorded for the nuclei possessing a spin quantum number $I = 1/2$, like ^1H, ^{13}C, ^{15}N, and ^{31}P. Lack of quadrupolar relaxation in the case of such nuclei gives rise to narrow, well-resolved resonance lines. Application of NMR spectroscopy to structural studies is based on magnetic interaction of nuclei with surrounding electrons as well as with other nuclei via electrons. The former type of interaction differentiates the resonance frequencies of nuclei located in different chemical environments, that is, their chemical shifts (d_i) in ppm relative to a standard. The latter type results in splitting of the resonance signals characterized by scalar coupling constants $J_{(i,j)}$, which depend on the types and positions of the interacting nuclei in the molecule.

The two NMR parameters, chemical shifts and coupling constants, allow one to identify an individual nucleus in a molecule by its resonance signal and to derive the geometric parameters of the molecule. Assignment of the resonances in the NMR spectrum to individual nuclei provides characteristic atoms and groups of atoms defining the chemical structure. In the ^1H NMR spectra of cap analogs, the following resonances are characteristic in comparison with typical nucleotides: a N7-methyl group signal at ~4.1 ppm, a sugar H1$'$ doublet at ~6 ppm with a characteristic splitting of 3 to 5 Hz, depending on the analog, and a lack of the H8 signal of the m^{7}Guo moiety due to its replacement by deuterium of the solvent. The ^{31}P NMR spectra showed one multiplet for each phosphate group, with characteristic splittings of ~19 Hz due to ^{31}P to ^{31}P scalar couplings. For the ARCA analogs, an additional signal of 2$'$-OCH$_3$ or 3$'$-OCH$_3$ is detected at ~3.5 ppm.

3. IN VITRO AND IN VIVO ASSAYS

Any newly designed cap analog must be characterized using several *in vitro* and *in vivo* assays in order to demonstrate its utility. Assays for both translational efficiency and *in vivo* stability are necessary—a cap analog that is resistant to enzymatic cleavage but poorly recognized by the translational machinery would not be particularly useful. Similarly, interpretation of results obtained from *in vivo* translation experiments requires the independent measurement of mRNA abundance and mRNA stability. Later we describe several different assays that address the recognition of novel cap analogs by RNA polymerase during the *in vitro* transcription reaction and the ability of the cap to direct the orientation of its incorporation. Methods for each of these procedures have been published previously (Cai *et al.*, 1999; Grudzien *et al.*, 2004, 2006; Jemielity *et al.*, 2003; Stepinski *et al.*, 2001).

4. BINDING AFFINITY OF ARCAs FOR eIF4E

Fluorescence titration is a commonly used technique to determine binding affinity between protein and ligand. However, this method requires a detectable change of either protein or ligand fluorescence as results of the interaction. Fortunately, eIF4E possesses eight conserved tryptophan residues, and cap binding results in quenching of intrinsic protein fluorescence. However, the overlapping of absorption and emission spectra of protein and cap analogs and their strong interaction ($K_{AS} \sim 5\text{--}500\ \mu M^{-1}$) require one to apply nonstandard filter effect corrections and to take into account the emission of free ligand. This excludes the possibility of using approximate equations to determine binding constants. The methodology of time-synchronized fluorescence titration and analysis of data to yield equilibrium association constants for complexes of eIF4E and cap analogs have been described in detail (Niedzwiecka *et al.*, 2002). See also the chapter in volume 430 (Niedzwiecka *et al.*, 2007).

Fluorescence measurements are carried out in 50 mM Hepes/KOH (pH 7.2), 100 mM KCl, 0.5 mM EDTA, and 2 mM DTT at 20°. To 0.1 or 0.05 μM mouse eIF4E(28–217) (i.e., the first 27 amino acid residues are truncated) (Niedzwiecka *et al.*, 2002) are added solutions of increasing cap analog concentrations. For this protein–ligand association, an excitation wavelength of 280 nm and an emission wavelength of 337 nm are used.

Association constants obtained for complexes of eIF4E and ARCAs show that the chemical modifications required to convert a conventional cap analog into an ARCA have only minor effects on affinity (Table 11.1). Introduction of a methyl group at C2$'$ and C3$'$ of the m^{7}Guo moiety has a slight effect on binding affinity. Extension of the phosphate chain, by contrast, causes a strong increase of association constants, while replacement of oxygen atoms by methylene groups in the polyphosphate chain results in a decrease of K_{AS}.

5. INCORPORATION OF ARCAs INTO RNA BY *IN VITRO* TRANSCRIPTION

RNAs capped with various cap analogs can be synthesized by *in vitro* transcription in the presence of a DNA template, bacterial or bacteriophage polymerase, all four nucleoside triphosphates, and different cap dinucleotides. A typical transcription reaction contains 40 mM Tris-HCl, pH 7.9, 6 mM MgCl$_2$, 2 mM spermidine, 10 mM DTT, 0.1 mg/ml BSA, 1 U/μl of RNasin (Promega), 0.5 mM ATP, 0.5 mM CTP, 0.5 mM UTP, 0.1 mM GTP, 1 mM cap analog, 15 μg/ml DNA, and 1 U/μl of SP6 or T7 polymerase (Promega). In the assays presented later, we used several

Table 11.1 Biophysical and biochemical properties of dinucleoside tri-, tetra-, and pentaphosphate cap analogs

No.	Cap analog	$K_{AS} \times 10^{-6}$ $(M^{-1})^a$	K_I $(\mu M)^b$	% Capping^c	Relative translational efficiency in vitro^d	Relative translational efficiency in vivo^e	5' degradation§ (min)^f
	Triphosphate series						
	$m^7Gp_3G^g$	12.5 ± 0.3	17.1 ± 1.0^i	69^h	1.0	1.0	156 ± 6^h
1	$m_2^{7,2'-O}Gp_3G^g$	10.8 ± 0.3	14.1 ± 0.9	73	2.1 ± 0.1	ND	ND
2	$m_2^{7,3'-O}Gp_3G^i$	10.2 ± 0.3	14.3 ± 1.9	78	1.9 ± 0.4	2.5 ± 0.1^h	282 ± 4^h
3	$m^72'dGp3G^g$	9.1 ± 0.5	27.2 ± 1.6	83	1.2 ± 0.1	ND	ND
4	$m^73'dGp3G^i$	13.1 ± 0.7	27.8 ± 7.1	ND	2.1 ± 0.5	ND	ND
5	$m_2^{7,3'-O}Gp_{CH2}ppG^h$	4.7 ± 0.1	29.3 ± 2.3	81	1.1 ± 0.3	1.2 ± 0.2	180 ± 5
6	$m_2^{7,3'-O}Gpp_{CH2}pG^h$	4.4 ± 0.2	33.5 ± 4.5	86	1.3 ± 0.3	1.3 ± 0.1	330 ± 10
	Tetraphosphate series						
	$m^7Gp_4G^g$	110.9 ± 6.0	10.8 ± 0.8	ND	1.3 ± 0.1	ND	ND
7	$m_2^{7,2'-O}Gp_4G^g$	99.8 ± 6.0	3.1 ± 0.3	69	2.6 ± 0.2	ND	ND
8	$m_2^{7,3'-O}Gp_4G^g$	84.5 ± 2.6	6.3 ± 0.9	70	2.4 ± 0.2	ND	ND
9	$m^72'dGp_4G^g$	88.0 ± 2.9	11.3 ± 0.9	72	1.8 ± 0.2	ND	ND
	Pentaphosphate series						
	$m^7Gp_5G^g$	543.9 ± 55	4.9 ± 0.1	ND	1.1 ± 0.1	ND	ND
10	$m_2^{7,3'-O}Gp_5G^g$	299.0 ± 20	3.5 ± 0.3	ND	1.4 ± 0.2	ND	ND

restriction-digested luciferase-encoding plasmids as DNA templates: pSP-*luc*+ (Promega), which is transcribed with SP6 polymerase, and two templates that are transcribed with T7 polymerase: p*luc*-A+, which contains the entire firefly luciferase mRNA sequence in pGEM4 (Promega) and a 3′-terminal 30-nt poly(A) tract, and p*luc*-A60, which is derived from p*luc*-A+ but contains a 60-nt poly(A) tract. Either p*luc*-A+ or pSP-*luc*+ is digested with NcoI for synthesis of capped oligonucleotides. Short RNAs are synthesized in the presence of 10 μCi/μl of [α-^{32}GTP] (ICN) in 50-μl reaction mixtures incubated for 45 min at 37°. Reaction mixtures are extracted with phenol and chloroform, and then RNAs are separated from unincorporated nucleotides using spin columns, according to the manufacturer's protocol (Ambion). The concentrations of mRNAs are determined by Cerenkov counting, in which the specific radioactivity of [α-^{32}P]GTP in the final transcription reaction mixture is used for conversion of cpm to pmol.

5.1. Efficiency of cap incorporation during *in vitro* transcription

The extent to which cap analogs are incorporated into RNA during transcription *in vitro* (percentage capping) along with the orientation of the cap in the RNA (percentage correct orientation) are important factors in determining the contribution of the cap to overall mRNA efficiency. The translational output results almost entirely from the capped mRNAs in a population of capped and uncapped mRNAs. The percentage capping can be determined by

---

[a] Equilibrium association constants for interaction of mouse eIF4E (28–217) with various cap analogs at 20°. The final K_{AS} value for each cap analog was calculated as a weighted average of independent titrations.
[b] Inhibitory constants for inhibition of natural globin mRNA translation in a rabbit reticulocyte lysate system. For the tri-, tetra-, and pentaphosphate series, each value for K_I was normalized by dividing with the value for K_I for the cap analog standard for the series: m^{7}Gp$_3$G, m^{7}Gp$_4$G, and m^{7}Gp$_5$G, respectively.
[c] The percentage of RNAs containing a cap, regardless of whether in the correct or reversed orientation, was calculated by comparing labeled products derived from capped mRNA (in the case of m^{7}Gp$_3$G - capped mRNA, these are m^{7}Gp$_3$Gp* + Gp$_3$m^{7}Gp*) to the total 5′-terminal products (p$_3$Gp* + m^{7}Gp$_3$Gp* + Gp$_3$m^{7}Gp*).
[d] Translational efficiency of luciferase mRNAs lacking a poly(A) tract and capped with anti reverse cap analogs in a RRL system. The relative translational efficiency was calculated as described in the text.
[e] Translational efficiency of luciferase mRNAs containing 60-nt poly(A) tract and capped with anti reverse cap analogs in MM3MG cells. Luciferase activity was normalized by the amount of luciferase RNA in the cells. Relative translational efficiency was calculated as noted previously.
[f] Degradation of 5′-terminal sequences in luciferase mRNA containing 60-nt poly(A) tract and capped with the indicated anti reverse cap analogs. The mRNAs half-lives were determined by real time PCR with primers directed against the 5′-end of luciferase mRNA as described in the text.
[g] Data from Jemielity *et al.* (2003).
[h] Data from Grudzien *et al.* (2006).
[i] Data from Cai *et al.* (1999).
[j] Data from Stepinski *et al.* (2001).

digesting capped oligonucleotides using a ribonuclease that is not base-specific, which hydrolyzes the RNA chain while leaving the cap structure intact. Cap analogs used in this study are unlabeled, so we synthesize capped oligonucleotides in the presence of $[\alpha\text{-}^{32}P]GTP$ and a DNA template in which G is the first ribonucleotide specified after the promoter in order to follow the products of the digestion reaction. The radioactivity in nucleoside monophosphates compared with cap-derived structures is kept to a minimum by use of a short DNA template. For enzymatic digestion, we use ribonuclease T2 (Invitrogen) or a cocktail of ribonucleases RiboShredder (Epicentre). Any nucleotide on the $5'$ side of a G residue acquires a ^{32}P-labeled $3'$-phosphate group after RNase T2 digestion by nearest neighbor transfer. Anion exchange chromatography on a 4.5×250 mm Partisil 10SAX/25 column (Whatman) is used to resolve the labeled $3'$-nucleoside monophosphates ($3'$-NMP*), resulting from internal positions in the RNA, from labeled $5'$-terminal products. The latter are of two types. Uncapped RNAs yield labeled guanosine $5'$-triphosphate $3'$-monophosphate (p_3Gp* in which the labeled phosphate group is indicated by *). Capped RNAs consisting of correctly capped and reverse-capped transcripts yield various cap structures, depending on the nature of the cap analog used (m^7Gp_3Gp* and Gp_3m^7Gp* when the cap analog is m^7Gp_3G). The elution program for nucleotides consists of water for the first 5 min, a linear gradient of 0 to 1.5 M KH_2PO_4 at pH 3.5 (for the RNase T2-digested triphosphate analogs) or at pH 4.5 (for the RNase T2-digested tetraphosphate analogs) for 40 min, and isocratic elution at 1.5 M KH_2PO_4 for 2 min, all at a flow rate of 1 ml/min. The amounts of uncapped and capped mRNAs can be determined by Cerenkov counting from the various digestion products obtained.

Typical data for m^7Gp_3G and $m_2^{7,3'\text{-}O}Gp_3G$ (**2**) are presented in Fig. 11.4. The RNA synthesized in the presence of the natural cap analog m^7Gp_3G yields three radioactive peaks upon RNase T2 digestion (Panels A and C). The first peak at 16 min is a mixture of all four $3'$-NMPs. The second peak at 26 min (Panel A) corresponds to a mixture of the products of correct and reverse incorporation of m^7Gp_3G (m^7Gp_3Gp* or Gp_3m^7Gp*, respectively). These two peaks elute from the anion exchange column at nearly the same time because they have almost identical charges. The third peak at 39 min corresponds to the T2-digestion product of uncapped RNA, p_3Gp*. mRNAs capped with different cap analogs yields the same $3'$-NMP* and p_3Gp* products, but the positions of the cap-derived products are different (Panels A and C).

The percentage capping for each type of analog is shown in Table 11.1 and is calculated by comparing the areas under the peaks derived from capped mRNA (for the natural cap, m^7Gp_3Gp* plus Gp_3m^7Gp*) to that derived from uncapped mRNA (p_3Gp*). The observed percentage capping varies from 69 to 86% for the triphosphate series and from 69 to 72% for tetraphosphate series, despite the fact that all cap analogs are present in the

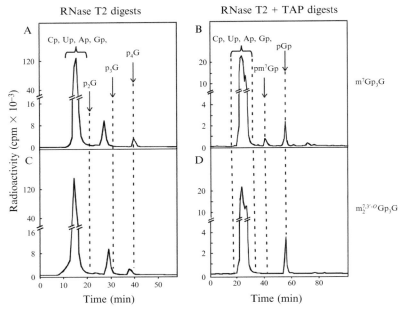

Figure 11.4   Analysis of *in vitro* synthesized RNAs. ^{32}P-Radiolabeled RNAs (48 nucleotides) capped with m[7]Gp$_3$G (A and C) or m$_2$[7,3'-O]Gp3G (B and D) were digested with either RNase T2 (A and C) or RNase T2 plus tobacco acid pyrophosphatase (TAP) (B and D) followed by anion-exchange HPLC on a Partisil 10SAX/25 column as described in the text. Fractions of 1 ml were collected, and the Cerenkov radiation was determined. The elution times of the following standard compounds, detected by ultraviolet (UV) absorption, are indicated with arrows: 3'-CMP (Cp), 3'-UMP (Up), 3'-AMP (Ap), 3'-GMP (Gp), 3',5'-m[7]GDP (pm[7]Gp), 3',5'-GDP (pGp), 5'-GDP (p$_2$G), 5'-GTP (p$_3$G), and guanosine-5'-tetraphosphate (p$_4$G).

transcription reaction at the same concentration (1 m*M*). This suggests differences in the ability of T7 polymerase to incorporate different dinucleotides of the form Np$_3$G in place of GTP. The percentage capping has not been determined for analogs in the pentaphosphate series because no chromatographic conditions have been developed to resolve the 5'-terminal products of RNase T2 digestion for this type of compound.

5.2. Analysis of cap orientation

The orientation of the cap in synthetic mRNA is the major contributor to overall translational efficiency. Cap analogs blocked at the 3'-O position of the first nucleoside moiety [m[7]3'dGp$_3$G (**4**) and m$_2$[7,3'-O]Gp$_3$G (**2**)] as well as modified in the 2'-O position (which does not participate in phosphodiester formation) [m[7]2'dGp$_3$G (**3**) and m$_2$[7,2'-O]Gp$_3$G (**1**)], are incorporated into RNA exclusively in the correct orientation (Stepinski *et al.*, 2001), which improves the efficiency of protein synthesis. To determine percentage correct orientation, labeled capped oligonucleotides must be subjected to two digestion steps. The nonspecific ribonuclease treatment, which yields labeled 3'-NMPs and 5'-terminal products, is followed by pyrophophatase digestion, which hydrolyzes pyrophosphate bonds between the first and second nucleoside moieties of the cap. For these, we used RNase T2 and tobacco acid pyrophosphatase (TAP) (Epicentre), respectively. Using the standard cap analog as an example, for those RNAs in which m^7Gp_3G is incorporated in the correct orientation, the labeled product of RNase T2 and TAP digestion is pGp*. For RNAs in which m^7Gp_3G is incorporated in the reverse orientation, the labeled product is pm^7Gp*. These nucleoside diphosphates differ in charge and can be separated by ion-exchange HPLC (Fig. 11.4B). The elution program consists of water for the first 5 min, a linear gradient of 0 to 87.5 mM of KH_2PO_4, pH 3.5, for 35 min, a linear gradient of 87.5 to 500 mM of KH_2PO_4 for 35 min, and isocratic elution at 500 mM of KH_2PO_4 for 21 min, all at a flow rate of 1 ml/min. Fractions of 1 ml are collected and the Cerenkov radiation determined. The percentage correct orientation is calculated from the ratio of pGp* to total 5'-terminal labeled products (pGp* + pm^7Gp*). Each value of pGp* is first correct for the contribution from uncapped mRNA because pGp* can arise from two sources: capped RNAs terminated with m^7Gp_3Gp* and uncapped RNAs terminated with p_3Gp*.

6. PROPERTIES OF ARCAS AND ARCA-CAPPED MRNAS IN CELL-FREE TRANSLATION SYSTEMS

ARCAs are incorporated into RNA exclusively in the correct orientation to an extent that is similar to the standard cap (see previously), which makes them potentially useful compounds in terms of increasing translational efficiency when incorporated into RNA. Similarly, they should be effective for inhibiting protein synthesis as free analogs. To test the influence of the ARCAs on protein synthesis *in vitro*, we use the microccocal nuclease treated rabbit reticulocyte lysate system (RRL system) optimized for cap-dependent translation (Cai *et al.*, 1999). Highly cap-dependent translation is achieved at 100 mM potassium acetate and 1.4 mM magnesium chloride.

p*luc*-A60 is digested with *Hpa*I for synthesis of luciferase mRNA with a 3'-terminal 60-nt poly(A) tract, *Sma*I for synthesis of luciferase mRNA lacking a poly(A) tract, and *Nco*I for synthesis of capped oligonucleotides. pSP-*luc*+ is digested with *Eco*RI to produce an SP6 RNA polymerase template that yields the RNA containing the entire firefly luciferase coding region but lacking a poly(A) tract. mRNAs are synthesized in the presence of 0.04 $\mu Ci/\mu l$ of $[\alpha-^{32}GTP]$ (ICN) in 50-μl reaction mixtures incubated for 45 min at 37°. RNAs are purified as described previously.

6.1. Inhibition of cap-dependent translation by ARCAs

ARCAs are assayed over a series of concentrations for inhibition of cap-dependent translation in the RRL system containing native β-globin mRNA. This measures competition between the cap analog and mRNA in a system containing all the proteins involved in protein synthesis. The level of β-globin synthesis is measured by incorporation of $[^{3}H]Leu$ into protein. To control for variations in the activity of the protein synthesis system, each cap analog must be tested in the same experiment over the same concentration range as that of a standard cap analog ($m^{7}Gp_{3}G$ for the triphosphate series, $m^{7}Gp_{4}G$ for the tetraphosphates series, and $m^{7}Gp_{5}G$ for pentaphosphate series). Quantitative comparisons of the cap analogs are obtained by fitting the data to a theoretical equation describing initiation of translation (Cai *et al.*, 1999). The equation describes cap-dependent translation as a function of a competitive inhibitor of mRNA binding to eIF4E. K_{I}, the value of the dissociation constant for the cap analog–eIF4E complex, is varied until the best least–squares fit is obtained. K_{I} values for all cap analogs and the fold difference compared with the K_{I} values of the cap analog standards are shown in Table 11.1. The tetraphosphate and pentaphosphate cap analogs are stronger inhibitors than are their triphosphate counterparts. However, the differences among them are not as pronounced as are the affinities for eIF4E (K_{AS}), as measured by fluorescence quenching studies. Stronger inhibition is obtained with 2'-O-methyl and 3'-O-methyl analogs than with their 2'- and 3'-deoxy counterparts.

6.2. Translational efficiency of ARCA-capped transcripts

Four parameters affect the translational efficiency of capped-mRNAs: affinity of the cap analog for eIF4E, ability of the cap analog to inhibit cell-free translation, efficiency of cap incorporation during *in vitro* transcription, and degree to which cap analog is incorporated into RNA in the correct orientation. We have measured the translational efficiency of luciferase mRNA capped with different ARCAs in the RRL system. Luciferase mRNA is obtained by *in vitro* transcription using two different templates as described previously. Protein synthesis is detected by assaying luciferase activity using beetle luciferin (Promega) as a substrate and a Monolight 2010 luminometer to detect light emission. Translation reactions are conducted under conditions in which luciferase production is linear with both time and mRNA concentration. Two types of normalization are applied to the data. First, each value for overall translation is corrected for cap-independent translation since the translation observed *in vitro* has both cap-dependent and cap-independent components. The amount of cap-independent translation is determined with a control mRNA capped with Gp$_3$G, which is not recognized by eIF4E. Second, each value for cap-dependent translational efficiency is normalized by dividing by the value for cap-dependent translational efficiency of a standard mRNA, for example, luciferase mRNA capped with m^{7}Gp$_3$G, because of day-to-day variability of the RRL system.

The translational efficiencies of mRNAs capped with each of the ARCAs are presented in Table 11.1. The results indicate that prevention of reverse incorporation by ARCAs gives a marked increase in translational efficiency. In the tri- and tetraphosphate series, 3'-O-methyl analog [m$_2^{7,3'-O}$Gp$_3$G (**2**) and m$_2^{7,3'-O}$Gp$_4$G (**8**)] and 2'-O-methyl compound [m$_2^{7,2'-O}$Gp$_3$G (**1**) and m$_2^{7,2'-O}$Gp$_4$G (**7**)] are 2-fold more active in promoting *in vitro* translation than are their non-ARCA counterparts. However, the relative translational efficiencies of m^{7}Gp$_5$G and m$_2^{7,3'-O}$Gp$_5$G (**10**) are only 1.1 and 1.4, respectively, which is less than one would have predicted from both binding affinities for eIF4E and inhibition of protein synthesis *in vitro* (Table 11.1). This may indicate that, although cap analogs with longer phosphate chains bind eIF4E more strongly, there is an upper limit beyond which high affinity for eIF4E cannot accelerate overall translation. Thus, when the rate of cap binding becomes sufficiently high, some other step in protein synthesis initiation becomes rate limiting.

7. PROPERTIES OF ARCA-CAPPED MRNAS IN MAMMALIAN CELLS

The translation experiments described previously were performed in an *in vitro* RRL translation system, but this system differs in several respects from intact cells: absence of cytoskeleton, higher level of initiation factors, very little dependence on the poly(A) tract for translational efficiency, etc. (Michel *et al.*, 2000; Rau *et al.*, 1996). A more meaningful test of translational efficiency of mRNAs capped with ARCAs is translation in living cells. As a working *in vivo* system, we chose mouse mammary epithelial cells (MM3MG) which, in contrast to many mammary gland cell lines, have normal eIF4E levels (DeFatta *et al.*, 2002). We used electroporation as a method to carry synthetic luciferase mRNA into the cells. Use of electroporation allows one to measure luciferase synthesis and luciferase mRNA level in the cells almost immediately following discharge.

The plasmid p*luc*-A60 is digested with *Hpa*I for synthesis of luciferase mRNA with a 3'-terminal 60-nt poly(A) tail. The RNAs are synthesized in 200-µl reaction mixtures incubated for 2 h at 37° using conditions described previously. Reaction mixtures are further treated with 3 U of DNase RQ1 (Promega) for 20 min at 37°, extracted with phenol and chloroform, and the RNAs purified with a RNeasy mini kit using the manufacturer's protocol (Qiagen). The concentrations of RNAs are determined spectrophotometrically.

RNA (5 µg) is introduced into 10^7 MM3MG cells by electroporation in a total volume of 400 µl serum-reduced medium Opti-MEM® I (Gibco) in a Gene Pulser cuvette (4 mm gap) with a BioRad Genepulser™ set at 0.22 kV and 960 µF. Following discharge, the cells are washed twice with PBS, centrifuged for 2 min at $300g$ at room temperature, resuspended in prewarmed complete media, divided into several Eppendorf tubes, and placed at 37°. Cells are removed at intervals following electroporation, lysed in 200 µl of Luciferase Cell Culture Lysis Reagent 1X (Promega), and luciferase activity in the supernatant is measured by luminometry as described previously. Luciferase activity is then normalized for the amount of luciferase mRNA that was actually introduced into the cells. For that purpose, the cytoplasmic RNAs are extracted from the cells using 175 µl of lysis buffer (50 mM Tris-HCl, pH 8.0, 140 mM NaCl, 1.5 mM MgCl$_2$, 0.5% (v/v) Igepal (Sigma), 1 mM DTT) and the amount of luciferase RNA is determined as described later.

Approximately 2 µg of each total RNA sample isolated from MM3MG cells is treated with DNase RQ1 according to the manufacturer's protocol. Reverse transcription is performed on 400 ng of RNA in 20-µl reaction mixtures containing 5.5 mM MgCl$_2$, 500 µM of each dNTP, 2.5 µM random hexamers, 0.2 U RNase Inhibitor, and 0.8 U MultiScribe reverse transcriptase (Applied Biosystems). The reaction mixture is incubated at 25° for 10 min, 48° for 30 min, and 95° for 5 min. Quantitative Real-time PCR is performed with specific primers designed for each gene with the Beacon Designer tool (Bio-Rad). For detecting sequences at the 5′-end of luciferase mRNA, the primers are 5′-CGTTCGGTTGGCAGAAGCTA-3′ and 5′-ACTGTTGAGCAATTCACGTTCATT-3′. These primers amplified 226 to 398 nt from the luciferase mRNA, which is 1714 nt from the cap structure to the beginning of the 3′-terminal homopolymer. Mouse GAPDH mRNA levels are measured by the same method and in the same RNA samples with the use of the primers 5′-CAATGTGTCCGTCGTGGATCT-3′ and 5′-GAAGAGTGGGGAGTTGCTGTTGA-3′. Amplification and detection are done with the iCycler IQ Real-time PCR detection system in 25-µl reaction mixtures containing 5 µl of the transcription reaction mixture (50 ng cDNA), 12.5 µl IQ SYBRgreen Supermix, and 0.3 mM primers (Bio-Rad). The incubation conditions consist of 3 min at 95° for polymerase activation, and 40 cycles of 15 s at 95° and 1 min at 60°. Luciferase mRNA levels are calculated using the absolute standard curve method described in User Bulletin No. 2 for the ABI Prism 7700 Sequence Detection System. Luciferase mRNA levels calculated from a standard curve are normalized for the amount of mouse GAPDH mRNA in each sample. Finally, luciferase mRNA at each time point after electroporation are calculated as a percent of the RNA present at zero time and the results plotted as $\log_{10}([RNA])$ versus time to determine $t_{1/2}$.

7.1. Translational efficiency of ARCA-capped mRNAs

As in *in vitro* translation experiments, we use conditions in which accumulation of luciferase is linear with mRNA concentration and time following an initial lag period of $\sim$30 min. This lag can be ascribed to assembly of the translation initiation complex, synthesis of luciferase, and release into the cytosol. Luciferase accumulation is measured for up to 75 min after electroporation. The luciferase mRNA concentration does not change significantly over this time. We determined the translational efficiency of mRNAs capped with Gp_3G, m^7Gp_3G, $m_2^{7,3'-O}Gp_3G$ (**2**), and two ARCAs in which a methylene group was substituted between the α- and β-phosphate moieties, $m_2^{7,3'-O}Gpp_{CH2}pG$ (**6**) and $m_2^{7,3'-O}Gp_{CH2}ppG$ (**5**). The latter modifications make these compounds resistant to Dcp1/Dcp2 and DcpS, respectively. mRNA capped with $m_2^{7,3'-O}Gp_3G$ (**2**) is translated 2.5-fold more efficiently than mRNA capped with m^7Gp_3G (Fig. 11.5), in good agreement with the results obtained from *in vitro* translation (Table 11.1). The increase in translational efficiency correlates with percentage correct orientation since uncapped and reverse-capped mRNAs are not recognized by eIF4E. mRNA capped with either $m_2^{7,3'-O}Gpp_{CH2}pG$ (**6**) or $m_2^{7,3'-O}Gp_{CH2}ppG$ (**5**) shows only minor increase in the translational efficiency over the mRNA

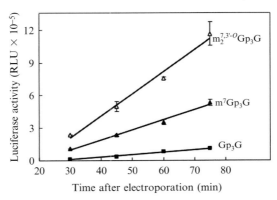

Figure 11.5 *In vivo* translational efficiency of mRNA capped with various ARCAs in MM3MG cells. Luciferase mRNAs containing a 3'-terminal 60-nt poly(A) tract and capped with Gp_3G (*filled squares*), m^7Gp_3G (*filled triangles*), or $m_2^{7,3'-O}Gp_3G$ (*open triangles*) were synthesized *in vitro*. The RNAs were electroporated into MM3MG cells, which then were lysed at the indicated time points. Equal amounts of total proteins were assayed for luciferase activity by luminometry. Relative light units (RLU) were normalized for the luciferase mRNA present in the cells 30 min after electroporation as measured by real-time PCR. The data shown represent the average of three independent experiments. Taken from Grudzien *et al.*, 2006.

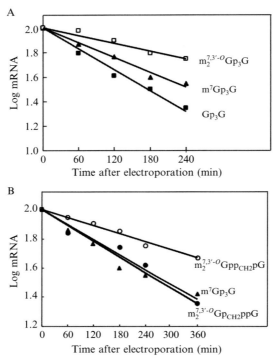

Figure 11.6 *In vivo* decay of luciferase mRNA capped with various ARCAs. Luciferase mRNAs containing a 3′-terminal 60-nt poly(A) tract and capped with Gp₃G (*filled squares*), m⁷Gp₃G (*filled triangles*), m₂^{7,3′-O}Gp₃G (*open squares*), m₂^{7,3′-O}Gpp_{CH2}pG (*open circles*), or m₂^{7,3′-O}Gp_{CH2}ppG (*filled circles*) were electroporated into MM3MG cells. RNA was isolated at the indicated times and quantitative PCR using pair of primers directed against 5′-end of luciferase mRNA was performed. The data shown represent the average of three independent experiments. Taken from Grudzien *et al.*, 2006.

capped with standard m⁷Gp₃G (Table 11.1). This is consistent with lower binding affinity for eIF4E and higher K_I value for the methylene-containing cap analogs, which offset the 2-fold increase in protein synthesis predicted from the 100% correct orientation due to the ARCA modification. Despite the lower affinity for eIF4E, the methylene-containing cap analogs are still translated 26-fold more efficiently than Gp₃G-capped mRNA.

7.2. Stability of ARCA-capped mRNAs in cultured mammalian cells

The nature of the cap structure can also influence mRNA stability. The stability of mRNA capped with different ARCAs is determined in MM3MG cells that are electroporated with luciferase mRNAs containing a 60-nt poly(A) tract and different cap analogs. The experiment is done using the same methodology as described for translation experiments with the exception that cells are harvested at various times up to 6 h after electroporation instead of 75 min. The cytoplasmic RNA is extracted, and the remaining amount of luciferase mRNA in MM3MG cells is measured by real-time PCR using pair of primers directed against the 5′-end of luciferase mRNA. The *in vivo* half-lives for various mRNAs are presented in Table 11.1. The mRNA capped with $m_2^{7,3'-O}Gp_3G$ (**2**) is more stable than the same mRNA capped either with m^7Gp_3G or Gp_3G ($t_{1/2} = 282$ versus 156 or 120 min) (Fig. 11.6A). The mRNA stability correlates with the ability of mRNA capped with different analogs to bind eIF4E, since Gp_3G is not recognized by eIF4E, m^7Gp_3G is incorporated approximately equally in the correct and reverse orientations whereas $m_2^{7,3'-O}Gp_3G$ (**2**) is incorporated entirely in the correct orientation. mRNA capped with $m_2^{7,3'-O}Gpp_{CH2}pG$ (**6**) ($t_{1/2} = 330$ min) is more stable than m^7Gp_3G or $m_2^{7,3'-O}Gp_3G$-capped mRNA ($t_{1/2} = 156$ and 282 min, respectively) (Fig. 11.6B). However, $m_2^{7,3'-O}Gp_{CH2}ppG$-capped mRNA shows no greater stability than the parent cap analog $m_2^{7,3'-O}Gp_3G$ (**2**). This is presumably because methylene substitution between the α and β phosphate moieties [$m_2^{7,3'-O}Gpp_{CH2}pG$ (**6**)] prevents decapping by Dcp2 *in vivo* whereas substitution between β and α phosphate moieties [$m_2^{7,3'-O}Gp_{CH2}ppG$ (**7**)] does not.

ACKNOWLEDGMENTS

This work was supported by Howard Hughes Medical Institute (Grant No.55005604 to E.D.), the Polish Ministry of Science and Higher Education (2 P04A 006 28 to E.D. and 3 P04A 021 25 to R.S.), and National Institutes of Health (Grant 1R03TW006446-01 to R.E.R and E.D., and 2R01GM20818 to R.E.R.)

REFERENCES

Beelman, C. A., Stevens, A., Caponigro, G., LaGrandeur, T. E., Hatfield, L., Fortner, D. M., and Parker, R. (1998). An essential component of the decapping enzyme required for normal rates of mRNA turnover. *Nature* **382,** 642–646.

Cai, A., Jankowska-Anyszka, M., Centers, A., Chlebicka, L., Stepinski, J., Stolarski, R., Darzynkiewicz, E., and Rhoads, R. E. (1999). Quantitative assessment of mRNA cap analogs as inhibitors of *in vitro* translation. *Biochemistry* **38,** 8538–8547.

Chen, C., Gherzi, R., Ong, S., Chan, E., Raijmakers, R., Pruijn, G., Stoecklin, G., Moroni, C., Mann, M., and Karin, M. (2001). AU binding proteins recruit the exosome to degrade ARE-containing mRNAs. *Cell* **107,** 451–464.

Coller, J., and Parker, R. (2004). Eukaryotic mRNA decapping. *Annu. Rev. Biochem.* **73,** 861–890.

Contreras, R., Cheroutre, H., Degrave, W., and Fiers, W. (1982). Simple, efficient *in vitro* synthesis of capped RNA useful for direct expression of cloned eukaryotic genes. *Nucl. Acids Res.* **10,** 6353–6362.

DeFatta, R., Li, Y., and De Benedetti, A. (2002). Selective killing of cancer cells based on translational control of a suicide gene. *Cancer Gene Ther.* **9,** 573–578.

Gingras, A.-C., Raught, B., and Sonenberg, N. (1999). eIF4 initiation factors: Effectors of mRNA recruitment to ribosomes and regulators of translation. *Annu. Rev. Biochem.* **68,** 913–963.

Grudzien, E., Kalek, M., Jemielity, J., Darzynkiewicz, E., and Rhoads, R. E. (2006). Differential inhibition of mRNA degradation pathways by novel cap analogs. *J. Biol. Chem.* **281,** 1857–1867.

Grudzien, E., Stepinski, J., Jankowska-Anyszka, M., Stolarski, R., Darzynkiewicz, E., and Rhoads, R. E. (2004). Novel cap analogs for *in vitro* synthesis of mRNAs with high translational efficiency. *RNA* **10,** 1479–1487.

Hsu, C., and Stevens, A. (1993). Yeast cells lacking 5′→3′ exoribonuclease 1 contain mRNA species that are poly(A) deficient and partially lack the 5′ cap structure. *Mol. Cell. Biol.* **13,** 4826–4835.

Izaurralde, E., Lewis, J., McGuigan, C., Jankowska, M., Darzynkiewicz, E., and Mattaj, I. W. (1994). A nuclear cap binding protein complex involved in pre-mRNA splicing. *Cell* **78,** 657–668.

Izaurralde, E., Stepinski, J., Darzynkiewicz, E., and Mattaj, I. W. (1992). A cap binding protein that may mediate nuclear export of RNA polymerase II-transcribed RNAs. *J. Cell Biol.* **118,** 1287–1295.

Jemielity, J., Fowler, T., Zuberek, J., Stepinski, J., Lewdorowicz, M., Niedzwiecka, A., Stolarski, R., Darzynkiewicz, E., and Rhoads, R. E. (2003). Novel "Anti-Reverse" cap analogues with superior translational properties. *RNA* **9,** 1108–1122.

Kadokura, M., Wada, T., Urashima, C., and Sekine, M. (1997). Efficient synthesis of g-methyl-capped guanosine 5′-triphosphate as a 5′-terminal unique structure of U6 RNA via a new triphosphate bond formation involving activation of methyl phosphor-imidazolidate using ZnCl$_2$ as a catalyst in DMF under anhydrous conditions. *Tetrahedron Lett.* **38,** 8359–8362.

Kalek, M., Jemielity, J., Grudzien, E., Zuberek, J., Bojarska, E., Cohen, L., Stepinski, J., Stolarski, R., Davis, R. E., Rhoads, R. E., and Darzynkiewicz, E. (2005). Synthesis and biochemical properties of novel mRNA 5′ cap analogs resistant to enzymatic hydrolysis. *Nucleosides, Nucleotides, and Nucleic Acid* **24,** 615–621.

Kalek, M., Jemielity, J., Grudzien, E., Zuberek, J., Bojarska, E., Cohen, L., Stepinski, J., Stolarski, R., Davis, R. E., Rhoads, R. E., and Darzynkiewicz, E. (2006). Enzymatically stable 5′ mRNA cap analogs: Synthesis and binding studies with human DcpS decapping enzyme. *Bioorg. Med. Chem.* **14,** 3223–3230.

Konarska, M. M., Padgett, R. A., and Sharp, P. A. (1984). Recognition of a cap structure in splicing *in vitro* of mRNA precursors. *Cell* **38,** 731–736.

Kusmierek, J., and Shugar, D. (1978). A new route to 2′(3′)-O-alkyl purine nucleosides. *Nucl. Acids Res.* s73–s77.

Liu, H., Rodgers, N. D., Jiao, X., and Kiledjian, M. (2002). The scavenger mRNA decapping enzyme DcpS is a member of the HIT family of pyrophosphatases. *EMBO J.* **21,** 4699–4708.

Michel, Y. M., Poncet, D., Piron, M., Kean, K. M., and Borman, A. M. (2000). Cap-poly (A) synergy in mammalian cell-free extracts. Investigation of the requirements for poly (A)-mediated stimulation of translation initiation. *J. Biol. Chem.* **275,** 32268–32276.

Muhlrad, D., Decker, C. J., and Parker, R. (1995). Turnover mechanisms of the stable yeast PGK1 mRNA. *Mol. Cell. Biol.* **15,** 2145–2156.

Mukherjee, D., Gao, M., O'Connor, J. P., Raijmakers, R., Pruijn, G., Lutz, C. S., and Wilusz, J. (2002). The mammalian exosome mediates the efficient degradation of mRNAs that contain AU-rich elements. *EMBO J.* **21,** 165–174.

Niedzwiecka, A., Marcotrigiano, J., Stepinski, J., Jankowska-Anyszka, M., Wyslouch-Cieszynska, A., Dadlez, M., Gingras, A.-C., Mak, P., Darzynkiewicz, E., Sonenberg, N., Burley, S. K., and Stolarski, R., et al. (2002). Biophysical studies of eIF4E cap-binding protein: Recognition of mRNA 5' cap structure and synthetic fragments of eIF4G and 4E-BP1 proteins. J. Mol. Biol. **319,** 615–635.

Niedzwiecka, A., Stepinski, J., Antosiewicz, J. M., Darzynkiewicz, E., and Stolarski, R. (2007). Biophysical approach to studying cap-eIF4E interaction by synthetic cap analogues. Methods Enzymol., **430,** 209–246.

Pasquinelli, A. E., Dahlberg, J. E., and Lund, E. (1995). Reverse 5' caps in RNAs made in vitro by phage RNA polymerases. RNA **1,** 957–967.

Peng, Z.-H., Sharma, V., Singleton, S. F., and Gershon, P. D. (2002). Synthesis and application of a chain-terminating dinucleotide mRNA cap analog. Organic Letters **4,** 161–164.

Rau, M., Ohlmann, T., Morley, S. J., and Pain, V. M. (1996). A reevaluation of the cap-binding protein, eIF4E, as a rate-limiting factor for initiation of translation in reticulocyte lysate. J. Biol. Chem. **271,** 8983–8990.

Rhoads, R. E. (1999). Minireview: Signal transduction pathways that regulate eukaryotic protein synthesis. J. Biol. Chem. **274,** 30337–30340.

Stepinski, J., Waddell, C., Stolarski, R., Darzynkiewicz, E., and Rhoads, R. E. (2001). Synthesis and properties of mRNAs containing the novel "anti-reverse" cap analogues 7-methyl(3'-O-methyl)GpppG and 7-methyl(3'-deoxy)GpppG. RNA **7,** 1486–1495.

Wang, Z., and Kiledjian, M. (2000). Identification of an erythroid-enriched endoribonuclease activity involved in specific mRNA cleavage. EMBO J. **19,** 295–305.

Wang, Z., and Kiledjian, M. (2001). Functional link between the mammalian exosome and mRNA decapping. Cell **107,** 751–762.

Yisraeli, J. K., and Melton, D. A. (1989). Synthesis of long, capped transcripts in vitro by SP6 and T7 RNA polymerases. In "Methods in Enzymology" (J. E. Dahlberg and J. N. Abelson, eds.), Vol. 180, pp. 42–50. Academic Press, San Diego.

Yoshikawa, M., Kato, T., and Takenishi, T. (1967). A novel method for phosphorylation of nucleosides to 5'-nucleotides. Tetrahedron Lett. **50,** 5065–5068.

CHAPTER TWELVE

METHODS FOR IDENTIFYING COMPOUNDS THAT SPECIFICALLY TARGET TRANSLATION

Letizia Brandi,[†] Attilio Fabbretti,[*] Pohl Milon,[*] Marcello Carotti,[*] Cynthia L. Pon,[*] and Claudio O. Gualerzi[*]

Contents

Abstract

This chapter presents methods and protocols suitable for the identification and characterization of inhibitors of the prokaryotic and/or eukaryotic translational apparatus as a whole or targeting specific, underexploited targets of the bacterial protein synthetic machinery such as translation initiation and aminoacylation. Some of the methods described have been used successfully for the high-throughput screening of libraries of natural or synthetic compounds and make use of model "universal" mRNAs that can be translated with similar efficiency by cellfree extracts of bacterial, yeast, and HeLa cells. Other methods presented here are suitable for secondary screening tests aimed at identifying a

*Department of Biology MCA, University of Camerino, Camerino, Italy
†Biotectnomics, Insubrias BioPark, Gerenzano, Italy

Methods in Enzymology, Volume 431
ISSN 0076-6879, DOI: 10.1016/S0076-6879(07)31012-4

specific target of an antibiotic within the translational pathway of prokaryotic cells.

 ## 1. INTRODUCTION

Drug-resistant, virulent bacteria and eukaryotic pathogens like protozoa and fungi continue to represent a threat to human and animal health (Barker, 2006; Bush, 2004; Walsh, 2003a,b). The simple fact that diseases caused by a single group of bacteria such as the enterobacteriaceae and a single protozoan genus such as *Plasmodium* together remain the highest causes of infant mortality in third-world countries, the estimated number being 2.8 million/year, shows that finding new anti-infective agents that might ease these problems remains a high-priority task for mankind. Thus, in spite of the relatively low economic profit that this enterprise may bring, which at least partially explains the scant interest in this problem displayed by the large, multinational pharma (Projan, 2003), the scientific community should feel the need to pursue the discovery, characterization, and development of new antibiotics.

The success achieved by Cuban physicians in the treatment of severe lesions caused by cutaneous leishmaniasis in Central American peasants using a topical formulation of a new and potent synthetic drug with combined action against sensitive and multiresistant bacteria and fungi, which was developed in the Center of Chemical Bioactives at the Central University of Santa Clara (Castañedo *et al.*, 1996, 2005), clearly indicates that discovering and developing new antibiotics can be a rewarding (not just in economical terms) enterprise, in addition to being a moral obligation that should be felt by the entire scientific community and not just by the most socially minded yet economically deprived countries.

Target-based drug discovery is generally not considered a winning strategy. Although this attitude could be partly justified by some notable failures encountered by some of the major pharmaceutical companies, the overall balance does not take into account the important successes that this approach has obtained in the development of effective inhibitors aimed at selected targets. Not considering fields other than antibacterial and antimycotic therapy (e.g., the development of anti-HIV therapeutic agents), where this strategy has been very effective, the inhibitors of peptide deformylase (Hackbarth *et al.*, 2002), FabH/FabF (Wang *et al.*, 2006; Young *et al.*, 2006), and phenylalanine tRNA synthetase (Beyer *et al.*, 2004) are just a few examples in which this strategy has been successful (for a review, see Donadio *et al.*, 2007).

Since we share the opinion that at the root of many lamented failures there might be, at least in part, "the lack of expertise, mainly in bacterial physiology and experience in drug discovery and development" (Projan, 2003), we are convinced that target-based antibiotic identification remains a potentially powerful tool that should be fully exploited. Furthermore, we regard the translation apparatus of bacteria and lower eukaryotes as an ideal target for inhibitors of potential therapeutic application. In fact, the complexity and species- and kingdom-specificity that exists beyond its well-established structural and functional conservation makes the ribosome an ideal target for antibiotics (Poehlsgaard and Douthwaite, 2005). Furthermore, the translational machinery of lower eukaryotes as a whole as well as several specific steps of prokaryotic translation (such as aminoacylation, initiation, and termination) remain underexploited targets that therefore offer opportunities for the discovery of new antibiotics directed against vital cell functions for which resistance mechanisms have not yet been developed. Finally, the species-specificity displayed by sordarins in the inhibition of fungal translation elongation factor EF2 (Shastry et al., 2001) indicates that even targets previously thought to be impracticable because of their evolutionary conservation might turn out to be ideal candidates for the successful development of inhibitors effective against pathogenic lower eukaryotes.

In the following sections, we present protocols that can be used to set up screenings specifically aimed at identifying inhibitors of these activities.

 ## 2. MATERIALS REQUIRED

2.1. Chemicals and fine chemicals

From Sigma: 3-aminoethylcarbazole (AEC); acrylamide/bis-acrylamide (30%) 37.5:1; amino acids; alumina; bentonite; benzamidine; bovine liver tRNA; bovine serum albumin (BSA); creatine phosphate (CP); diethyl pyrocarbonate (DEPC); dithiothreitol (DTT); *Escherichia coli* MRE600 tRNA; pyrophosphatase (Ppase); Ca^{++} salt of folinic acid, (5-formyl THF); HEPES; K salt of phospho-enol pyruvic acid, (PEP); creatine phospho kinase (CPK); protease inhibitor cocktail for fungal and yeast extracts; phenylmethylsulfonyl fluoride (PMSF); spermidine trihydrochloride; Tween 20.

From Oxoid: peptone; tryptone; yeast extract.

From Packard: Ultima Gold scintillation liquid.

Na salts of ribonucleotide triphosphates (Roche or Sigma); bovine serum albumin RNase-free, 20 mg/ml (Roche); RNasin ribonuclease inhibitor, 40 U/ml (Promega); both bacteriophage T7 RNA polymerase and RNA Cap structure analog m7G(5')ppp(5')G are from BioLabs; DNase-RNase-free (Roche); complete EDTA-free proteinase inhibitors cocktail (Roche); pyruvate kinase (PK) (Roche).

2.2. Biologicals

Unless otherwise specified, the methods for the preparation and purification of 70S ribosomal monomers, 30S and 50S ribosomal subunits, initiation factors IF1, IF2, and IF3, and initiator fMet-tRNA are described in detail in the chapter by Milon *et al.* (2007). Bacteriophage T7 RNA Polymerase can be purchased from New England Biolabs. The anti-mouse HRP (horseradish peroxidase)–conjugated anitbody is purchased from Amersham while the monoclonal (9F11) anti–IF2C antibody, prepared in collaboration with Areta International SpA, (Gerenzano (VA), Italy) can be purchased from this company (www.aretaint.com).

2.3. Preparation of mRNAs

The methods described here make use of several types of both natural and model mRNAs (schematically described in Fig. 12.1). Each mRNA is endowed with characteristics that make it suitable for a particular type of assay. For instance, the small size of 002mRNA (124 nt for a MW $\cong$ 41KDa) allows very high concentrations of this template to be reached, a condition that facilitates kinetics analyses in which ligand concentrations must be varied by more than one order of magnitude. The relevant characteristics of these templates are presented in Table 12.1, which also lists the figures that illustrate their experimental applications. Although these templates can be purchased from Biotectnomics, Insubrias BioPark (Gerenzano, [VA Italy]) or upon custom synthesis from other commercial sources (Curevac, Dharmacon), they can also be prepared in the laboratory by *in vitro* transcription with bacterophage T7 RNA polymerase. Here, we give the protocol routinely used for the preparation of 002mRNA, 022mRNA, and 027mRNA. For transcription, one can use either linearized plasmids derived from pTZ18R containing the 002 (pTZ-002), 022 (pTZ-022), or

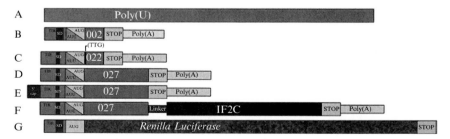

Figure 12.1 Schematic representation of the mRNA templates used in the translational tests. The relevant characteristics of the templates are indicated with different shades of gray. The lettering on the left is used to identify the template in Table 12.1.

Table 12.1 Characteristics of mRNAs used in the translational tests

mRNA	Length[a] (~nts)	S.D.	Spacer	Initiation codon	Vector	References	Example in Fig.
Poly(U)	≥200	none	none	none	none		3B
002p(A)	124	9	5	AUG/ AUU	pTZ18R	Calogero et al., 1988	5B
022p(A) (012)p(A)	119	4	9	AUG/ AUU	pTZ18R	Brandi et al., 2006c; La Teana et al., 1993	5C, D
027p(A)	197	4	9	AUG/ AUU	pTZ18R	Brandi et al., 2006c	3C, 6A
cap027p(A)	199	4	9	AUG/ AUU	pTZ18R	Brandi et al., 2007	3C
027IF2Cp(A)	913	4	9	AUG/ AUU	pTZ18R	Brandi et al., 2007	3A,B,D
Renilla Luciferase	1100	9	5	AUG	pTZ18R	Calogero et al., 1987[b], Lorenz et al., 1991[c]	4A,B,C

[a] The length includes the poly(A) tail when present.
[b] Sequence of the TIR.
[c] Coding sequence.

027 (pTZ-027) mRNA sequence or a PCR-amplicon (see later). Since the quality of the mRNA is a key element in determining the quality of the results obtained in the translation tests, special care should be taken in the preparation and purification of these mRNAs.

2.4. Preparation and linearization of the plasmid templates

E. coli JM109 cells, transformed with pTZ18R plasmid containing the sequence of 002mRNA (Calogero *et al.*, 1988), 022 mRNA (La Teana *et al.*, 1993), or 027 mRNA (Brandi *et al.*, 2006c), are grown overnight in 1 liter of LB medium supplemented with 60 μg/ml ampicillin. Plasmid purification by alkaline lysis (Sambrook and Russell, 2001) yields an average of 1 to 1.5 mg plasmid DNA, an amount sufficient for several transcription reactions. One mg of purified plasmid in 1 ml of the buffer supplied with the restriction endonuclease is digested at 37° for 2 h with 100 U of HindIII (MBI Fermentas). After the completeness of the digestion has been ascertained by analytical agarose gel electrophoresis, the DNA is subjected to phenol/chloroform extraction, ethanol precipitation, and finally resuspension in sterile H_2O and used in runoff *in vitro* transcription with T7 RNA polymerase (Calogero *et al.*, 1988).

2.5. Preparation of the amplicon templates

An alternative protocol, which avoids the time-consuming procedure for large-scale plasmid preparation, makes use of a PCR-generated template. The insert coding for 002, 022, or 027 mRNAs cloned in pTZ18R can be amplified using primers that anneal to nucleotides 143 to 160 (forward primer 5'-GCTTCCGGCTCGTATGTTGTGT G-3') and 297 to 319 (reverse primer 5'-GTAAAACGACGGCCAGT-3') of the vector. The resulting amplicons contain the phage T7 promoter sequence of the vector, preceded by a 100 nucleotides-long plasmid sequence and followed by the entire sequence of 002, 022, or 027mRNA, including the mRNA TIR (translation initiation region) and the 3'-end poly(A) tail.

The template used for the PCR amplification is plasmid DNA obtained by a simple mini-prep (Sambrook and Russell, 2001). The amplification reaction mixture consists of 50 μl of PCR buffer (provided by USB with the FideliTaq DNA polymerase) containing 1.5 mM $MgCl_2$, forward and reverse primers (0.5 μM each), the four dNTPs (0.2 mM each), 0.025 U/μl of FideliTaq DNA polymerase (USB, United States Biochemical), and 1 ng/μl of template DNA. The PCR reaction is carried out in a thermocycler (e.g., Whatman–Biometra). After an initial denaturation at 94° for 2 min, amplification is obtained with 25 cycles of denaturation (20 s at 94°), annealing (20 s at 56°), and extension (20 s at 68°), followed by a final incubation at 72° for 2 min. Each amplification reaction (50 μl) yields approximately 6 pmoles of amplicon, the quality of which is determined by appropriate restriction endonuclease digestion followed by electrophoretic analysis on agarose gel. The PCR products are purified with the GenElute PCR purification kit (Sigma) before use in transcription reactions.

2.6. *In vitro* transcription and purification of mRNAs

2.6.1. Optimization

Because the yield of transcription can vary depending upon a large number of factors (type and quality of the DNA template, T7 RNA polymerase, ribonucleotide triphosphates, etc.), it is recommended to optimize the reaction conditions on a small scale before embarking on a large-scale mRNA prep.

Thus, the optimal amounts of the various components to be used are preliminarily determined by setting up a series of small-volume (25 μl) trial reactions, each containing different amounts of the following reagents: NTPs (2–5 mM each), DNA template (5–60 nM); MgCl$_2$ (12–25 mM), and T7 RNA polymerase (1.2–1.7 U/μl). The yield of transcript can be estimated by denaturing PAGE/urea (6–8% acrylamide, 7–8 M urea) electrophoresis followed by ethidium bromide staining. No significant differences are generally found, regarding the amount of RNA produced and the level of homogeneity of the transcripts, between the plasmid and the PCR-generated template.

2.6.2. Large-scale transcription

To obtain milligram quantities of mRNA the previous *in vitro* transcription reaction is scaled up to a volume of 6 or 12 ml, depending upon the need.

A typical reaction mixture (12 ml) consists of 40 mM Tris-HCl (pH 8.1) containing 15 mM MgCl$_2$, 10 mM NaCl, 2 mM spermidine, 10 mM DTT, 0.1 mg/ml BSA, 0.05 U/μl RNase inhibitor, 0.004 U/μl Ppase, 3.75 mM each of ATP, CTP, GTP, and UTP, 0.02 μM DNA template, and 1.2 to 1.7 U/μl T7 RNA polymerase. The reaction mixtures are incubated 2 to 3 h at 37°. The cloudy Mg^{++} pyrophosphate complex present at the end of the incubation is solubilized by vigorously mixing the reaction mix before purification of the transcript by one of the methods described later.

2.6.3. Purification of the transcript

Anion exchange chromatography The reaction mixture is subjected to phenol/chloroform extraction to remove the T7 RNA polymerase using phenol equilibrated with 50 mM Na acetate (pH 4.5). After isopropanol precipitation, the pellets are resuspended in 20 mM MOPS buffer (pH 6.25) containing 350 mM NaCl. The excess unincorporated NTPs and the smaller abortive transcription products are removed by chromatography on anion exchange FPLC column (MonoQ 5/5 column, Amersham).

The sample is loaded at a flow-rate of 1 ml/min onto the FPLC column equilibrated with the same MOPS buffer used to resuspend the RNA pellets. The free nucleotides are completely removed with a 5-ml wash with 350 mM NaCl and the RNA is eluted with a 20-ml (350–750 mM NaCl) linear gradient and analyzed by PAGE/urea gel electrophoresis (see later). Up to 2 mg of RNA can be loaded onto and eluted from a 1-ml (of resin) mono Q column without loss of resolution. The homogeneity of RNA in the fractions collected, as seen by gel electrophoresis, should be >90%. The appropriate fractions are pooled and the RNA collected by ethanol precipitation. The RNA pellet is washed twice with 70% ethanol, air-dried, and finally redissolved in DEPC-treated H$_2$O. The total recovery after the entire procedure of purification is $\cong$ 90%. This protocol yields $\cong$ 800 pmoles of purified 002 mRNA/pmole template DNA.

LiCl precipitation—for large- and small-scale preparations   At the end of the transcription reaction an equal volume of 5 M LiCl is added to the reaction mixture. After standing for 30 min in an ice-bath, the transcript is collected by 10 min centrifugation at 12,000 rpm (Sorvall SA600 rotor) at room temperature. The RNA in the pellet is resuspended in DEPC-treated H$_2$O and precipitated by addition of 3 M Na acetate (pH 5.2) to a final concentration of 0.3 M and of 2.5 volumes of cold absolute EtOH. After at least 30 min at $-20°$, the sample is centrifuged (as previously) and the resulting pellet is rinsed with 70% EtOH and resuspended in DEPC-treated H$_2$O. The ethanol precipitation is repeated three times and the final pellet resuspended in DEPC-treated H$_2$O. After spectrophotometric (1 A$_{260}$ $\cong$ 40 μg/ml) determination of its concentration, the RNA is stored in small aliquots at $-80°$.

Affinity chromatography on oligo d(T) cellulose—for small-scale preparations   Oligo-d(T) cellulose (Amersham) is resuspended in DEPC-treated H$_2$O, autoclaved, packed into a small glass column (2.2 × 6.0 cm) and equilibrated with 10 column volumes of 20 mM Tris-HCl (pH 7.5) containing 0.5 M NaCl and 1 mM EDTA (pH 8). Five molar NaCl is added to the transcription mix to adjust the final NaCl concentration to 0.5 M and the resulting sample is loaded onto the column; the flow-through is collected and reloaded onto the column. The column is washed with 10 volumes of the above buffer followed by 5 volumes of 20 mM Tris-HCl (pH 7.5) containing 0.1 M NaCl and 1 mM EDTA (pH 8). The mRNA is then eluted with 3 column volumes of 20 mM Tris-HCl (pH 7.5) containing 1 mM EDTA (pH 8). Fractions of approximately 2 ml are collected and the A$_{260}$ of each fraction determined after appropriate dilution (e.g., 1:20). The fractions corresponding to the peak with highest A$_{260}$ are pooled and the RNA therein contained precipitated by addition of 3 M Na acetate (pH 5.2) to a final concentration of 0.3 M and of 2.5 volumes of cold absolute EtOH. After at least 30 min at $-20°$, the RNA is collected by 30 min centrifugation at 12,000 rpm at $4°$ (Sorvall, SA600 rotor) and the pellet washed with 70% EtOH and air-dried. The RNA is resuspended in a small volume of DEPC-treated water and, after determining its concentration (1 $A_{260} \cong 40$ $\mu g/ml$), stored in small aliquots at $-80°$.

Chemical capping of the universal mRNA   To study translation in a yeast cell-free system entailing a cap-dependent initiation step (see later), it is necessary to prepare CAP-027mRNA-p(A). Addition of a cap structure to the 5′ end of the universal mRNA is achieved during transcription, essentially as described by Dasso and Jackson (1989). For this purpose, a transcription reaction mixture is prepared, identical to that described previously except that it also contains 0.1 mM GTP and 0.5 mM m7G(5′)ppp(5′)G. After an initial incubation for 30 min at $37°$, a second addition of GTP is made (to bring the final concentration to 3.75 mM) and the reaction is allowed to continue for an additional 2 h at $37°$. At the end of the reaction, NaCl is added to the mixture to a final concentration of 0.5 M and the transcript is purified by affinity chromatography on oligo (dT)-cellulose, as described previously. The capping efficiency, which can be assessed by electrophoretic analysis under denaturing conditions, should be >65%.

Analysis of the mRNA quality by PAGE/urea   It is good practice to check the quality of the transcription product at various stages of the preparation. For this purpose, aliquots (e.g., 5 μl) of the reaction mix are taken at the beginning and end of the transcription reaction as well as after each step of the purification and mixed with an equal volume of electrophoresis sample buffer. After incubation at $65°$ for 5 min, the samples are loaded on a 6 to 8% acrylamide-7–8 M urea gel.

2.6.4. Preparation of Thr-tRNA, Ile-tRNA, and [^{14}C] Phe-tRNA

The test aimed at determining whether the target of a translational inhibitor is tRNA aminoacylation requires four precharged aa-tRNAs (fMet-tRNA, Thr-tRNA, Ile-tRNA, and [^{14}C] Phe-tRNA). Here, we present the protocol for the aminoacylation of the elongator tRNAs, while fMet-tRNA preparation is described in detail in the accompanying chapter by Milon et al., (2007).

The following components are mixed in a total volume of 2 ml in a 15-ml glass Corex tube: 10 mM Mg acetate, 30 mM Imidazole-HCl (pH 7.5), 10 mM 2-mercaptoethanol, 100 mM KCl, 5 mM ATP, 5 mM PEP, 0.025 mg/ml PK, 30 μM [^{14}C] phenylalanine ($\sim$93.8 mCi/mmol Amersham), 15 mg/ml total MRE600 tRNA, 30 μM threonine, 30 μM isoleucine, and 7 mg/ml crude "charging enzyme" protein fraction from *E. coli* MRE600 (Kaji, 1968). After 15 min incubation at $37°$, 0.8 ml of 1 M K acetate (pH 5.5), 0.7 ml of 5 M NaCl, and 7 ml of ice-cold absolute ethanol are added to the reaction mixture. After 1 h at $-20°$, the sample is centrifuged at 8000 rpm 4° for 60 min in a SA600 (Sorvall) rotor. The resulting pellet is (a) resuspended in 1.5 ml 0.2 M K acetate (pH 5.5) and 0.17 ml 5 M NaCl. The suspension obtained is (b) centrifuged at 8000 rpm at 4° for 60 min in a SA600 rotor. To the carefully collected supernatant (c) 3.3 ml cold absolute ethanol are added and the mixture kept at $-20°$ for 1 h before (d) being centrifuged at 8000 rpm at 4° for 60 min in the SA600 rotor. The procedure (a through d) is repeated once. The final pellet is air-dried and the aminoacyl-tRNA therein resuspended in 200 μl of 2 mM K acetate (pH 5.5) and stored in aliquots at $-80°$ after determining the concentration (1 $A_{260} \cong$ 43 μg or $\cong$1800 pmoles of total tRNA).

To determine the efficiency of aminoacylation of [^{14}C]Phe-tRNA, 5 μl aliquots of the aminoacylation mixture are withdrawn before and after the reaction; the samples taken from the reaction mixture at the end of the incubation are spotted onto 3-MM paper discs (Schleicher & Schuell) and processed by the "cold TCA" precipitation method, while the sample taken before the reaction is spotted on a paper disc pretreated "empty" by the same cold TCA procedure. Determination of the radioactivity present on these filters by liquid scintillation counting allows one to calculate the aminoacylation efficiency of the reaction (which, for phenylalanine, should be $\geq$2% of total tRNA). The specific activity of the [^{14}C] Phe-tRNA can be determined after one-step purification of Phe-tRNA by BD cellulose chromatography (Gillam *et al.*, 1968), followed by determination of the radioactivity and of the A_{260}.

2.7. Preparation of cellfree extracts

2.7.1. Preparation of bacterial S30

Buffer A: 10 mM Tris-HCl (pH 7.7), 10 mM Mg acetate, 60 mM NH$_4$Cl.

The *E. coli* cell-free extract (S30) fraction is prepared from *E. coli* MRE600 cells grown at 37° in LB medium to $A_{600} = 1.2$. The cells are washed three to four times with Buffer A containing 10% Glycerol and 1 mM DTT and frozen at $-80°$. The cells (50 g) are disrupted by grinding with 75 g precooled alumina (Sigma) in a prechilled mortar (diam. 20 cm) over ice for approximately 20 min. RNase-free DNAse (2.5 μg/g cells) is added during grinding. The cell slurry is resuspended in 50 ml Buffer A containing 10% glycerol, 0.5 g/l bentonite, 0.2 mM Benzamidine, and 0.2 mM PMSF (added just before use). After gentle stirring in a beaker at 4° for 10 min, the extract is centrifuged for 15 min at 4° at 12,000 rpm (SA600 Sorvall rotor) to remove alumina and cell debris. The supernatant is then centrifuged for 60 min at 4° at 12,000 rpm (SA600 rotor) to obtain an "S30 extract," which is dialyzed (3500 Da MW cutoff dialysis tubing, Spectrum Laboratories, Inc.) at 4° against 40 vol (3 changes of 2 h each) of Buffer A without bentonite. After a final centrifugation (12,000 rpm for 20 min at 4° in the SA600 rotor), the clarified S30 extract (the yield is $\cong$ 1 ml extract/g cell paste) is stored at −80° in small (50 μl, 100 μl, and 200 μl) aliquots. The optimal volume to be used in each experiment (generally, 2–6 μl/standard reaction mix) is experimentally determined for each batch of extract in a dose-response experiment in which 20 pmoles of mRNA are translated in the presence of increasing amounts of the S30 extract.

The preceding protocol can be successfully applied, essentially without modifications, to prepare active cell-free extracts from bacteria other than *Escherichia coli* (e.g., *Bacillus stearothermophilus* and clinical isolates of *Pseudomonas aeruginosa* bearing multiple antibiotic resistance).

2.7.2. Preparation of yeast S30

Buffer B: 30 mM HEPES-KOH, pH 7.4, 100 mM K acetate, 2 mM Mg acetate, 85 g/l mannitol (Difco), 20% glycerol.

Saccharomyces cerevisiae SKQ 2M cells are grown to A_{660} $\cong$1 in YPD medium, washed three to four times with Buffer B supplemented with 2 mM DTT and 0.5 mM PMSF added just before use and frozen at −80°. To maintain a low working temperature (−4 to −5°), all the following operations are performed in a cold room with incubations in an ice/NaCl mix; in addition, all operations are carried out using sterilized solutions and labware. An 8-g portion of frozen cells is resuspended in Buffer B (2.5 ml/g cells) containing 50 μl/g of cells of a protease inhibitor cocktail for fungal and yeast extracts. The cell suspension is divided into two 50-ml Sorvall centrifuge tubes and ice-cold unwashed 425 to 600 μ glass beads (Sigma) are added (7.5÷10 g/g cells). To rupture the cells, the tubes are capped and shaken by hand (or by a homemade mechanical apparatus), making 120 regular 50-cm-long vertical movements/min followed by a 1 min incubation in ice/NaCl. After repeating this shaking and resting procedure four times, the extract is centrifuged at 14,000 rpm for 20 min at −5° (SA600 rotor Sorvall) and the crude S30 phase (approximately 8–10 ml/8 g of cells) between the upper lipid layer and the pelleted glass beads is withdrawn and loaded onto a Sephadex G-25 Fine column (2 × 20 cm) pre-equilibrated with 5 vol of Buffer B without mannitol. The column is eluted with Buffer B without mannitol at a flow rate of 2.5 ml/min, collecting 1.3-ml fractions. An aliquot of each fraction is diluted 1:100 to determine the A_{260} and the fractions with highest absorbance are pooled (the final yield of S30 extract is $\cong$7–8 ml/8 grams of yeast). After determining the A_{260} of the pool, the yeast extract is divided in suitable aliquots and stored at −80°.

2.7.3. Preparation of human HeLa cells S30

Buffer C: 10 mM Hepes-KOH, pH 7.6, 10 mM K acetate, 0.5 mM Mg acetate, 5 mM DTT.

HeLa S3 (ATCC CCL-2.2) cells, a clonal derivative of the parent HeLa line (ATCC CCL-2), which are adapted to grow in suspension and therefore more suitable for large biomass production, are used for the preparation of human cell extract. The cells are maintained in suspension culture in Corning 850 cm^2 Polystyrene Roller Bottles at 37° at a concentration of 3 to 6 × 10^5 cells/ml in Eagle's Minimum Essential Medium Joklik Modification (Sigma) supplemented with 10% Fetal Bovine Serum (Invitrogen) in the presence of 5% CO_2.

To prepare the extract, 2 to 3 l of cell culture are harvested by 3 min centrifugation at 2000 rpm at 4° (GS3 rotor, Sorvall) in 250-ml sterile Falcon tubes (Corning) and resuspended in 20 ml sterile PBS. The suspension is divided into two 15-ml Falcon tubes and centrifuged for 2 min at 2000 rpm at 4° (SA600, Sorvall). This step is repeated 3 times. The final pellet is resuspended in a volume of Buffer C (containing 1× complete EDTA-free proteinase inhibitors cocktail) equal to one volume of cell pellet and kept 5 min on ice before being transferred to an ice-cold, tight-fitting Dounce homogenizer. The cells are broken with 15 to 18 strokes, pausing a few seconds between each stroke to avoid heating. The homogenate is then centrifuged at 10,500 rpm for 5 min at 4° (SA600, Sorvall). The supernatant is collected, divided into aliquots, and stored at −80°.

3. Methods and Tests

3.1. Assessment of the *in vivo* target of an inhibitor

The functional tests described in this chapter are aimed at the identification and characterization of translational inhibitors active *in vitro*. However, before deciding whether further characterization of a given inhibitor is worth pursuing, it might be important to ascertain whether the identified inhibitor has a similar activity *in vivo*. For this purpose, it is common practice to test the *in vivo* effect of the inhibitor on the main macromolecular syntheses that are potentially targeted by the inhibitor. Obviously, these tests can be carried out at any stage of the screening campaign. However, in consideration of the substantial effort required, they are generally performed only on a limited and select number of promising hits. Depending upon the need, the tests described later are carried out with the gram positive *Bacillus subtilis* and/or with the gram negative *E. coli*. However, in principle, these tests can be carried out with any microrganisms that can grow in a defined medium and whose nutritional requirements are known. Furthermore, in some cases and according to the particular experimental needs, it might be useful to carry out the tests on select (antibiotic resistant, permeability) mutants.

3.1.1. Macromolecular syntheses in gram-positive bacteria

B. subtilis ATCC 6633 cells are grown at 37° in 25 ml of Spizizen's medium (1.4% K_2HPO_4, 0.6% KH_2PO_4, 0.2% $(NH_4)_2SO_4$, 0.1% trisodium citrate dehydrate, 0.02% $MgSO_4(H_2O)_7$, 0.5% D-glucose) supplemented with 0.1% casamino acids. Upon reaching $A_{600} = 0.2$ (time $= 0$), the culture is divided into four aliquots of 4 ml; each aliquot receives one of the following precursors: (a) 58.8 μl of [^{3}H] thymidine 19.6 μM (51 Ci/mmol Amersham), 3.2 μl nonradioactive thymidine (1 mg/ml), and 100 μl adenosine (2 mg/ml); (b) 9 μl of [^{3}H] uridine 27.8 μM (36 Ci/mmol Amersham) and 15μl nonradioactive uridine (2 mg/ml); (c) 2.5 μl of Promix containing [^{35}S] Met/Cys 14.3 μM (1000 Ci/mmol Amersham) (d) 10.4 μl of [^{3}H] N-acetylglucosamine 125 μM (8 Ci/mmol Amersham). After 10 min incubation, each culture is divided into two equal aliquots. A given amount of inhibitor (corresponding to 0.5- to 10-fold its MIC) is added to one aliquot, while the control receives the solvent (e.g., DMSO) in which the inhibitor is dissolved. Bacterial growth is allowed to continue and 50-μl samples of each culture are withdrawn at 10-min intervals and mixed with 50-μl of 2% SDS. After vigorous vortex mixing, 3 ml of 10% cold TCA is added to the sample. After standing for at least 30 min in ice, the samples are filtered through glass fiber discs (or 3MM paper discs) and the hot acid (TCA)-insoluble radioactivity (see later) present in each sample determined by liquid scintillation counting. An example of results obtained with a translational inhibitor in a test of this kind is presented in Fig. 12.2.

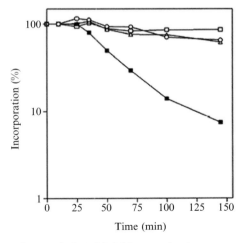

Figure 12.2 Effect of a translational inhibitor on *in vivo* macromolecular syntheses. Levels of (■) Protein, (○) RNA, (□) DNA, and (△) cell wall synthesized by *B. subtilis in vivo* after addition of the antibiotic (time 0). In each case, the level of synthesis in the presence of the antibiotic is normalized with respect to the level obtained in its absence and plotted as a function of the time elapsed after the addition of the antibiotic. The figure is taken from Brandi *et al.* (2006a).

3.1.2. Macromolecular syntheses in gram-negative bacteria

Tests essentially identical to those described previously can be carried out with *E. coli* grown in M9 Minimal Medium (Sambrook and Russell, 2001).

3.2. Tests to detect translational inhibitors

The tests described in the following sections are suitable for small, medium, or large-scale (HTS) screening of inhibitors of mRNA translation as a whole in cell extracts of bacteria, lower (yeast), and higher (HeLa) eukaryotes. In general, these *in vitro* tests could represent the primary screening of a library of entirely uncharacterized compounds or could be preceded by an *in vivo* test aimed at pre-selecting relevant (e.g., microbiologically active) hits. These two strategies obviously emphasize different characteristics of the hits. For instance, while the microbiological screening may overlook interesting molecules that fail to enter the cells, the *in vitro* screening may identify interesting inhibitors having no or very poor microbiological activity and that would need more or less extensive chemical modifications before becoming effective *in vivo*. Either approach, at least in our experience, can be successfully used to identify novel translational inhibitors in a library of natural products.

Depending upon the particular aim of each screening campaign, one type of cell extract is selected and programmed with a suitable template. The hits identified in the initial screening are then generally tested in a secondary screening that allows a better characterization of the inhibitor. For instance, the inhibitors of a natural-like mRNA translation could be tested for their capacity to inhibit: (i) the formation of aminoacyl-tRNAs or (ii) translational steps other than elongation, in case they fail to inhibit poly(U)-dependent polyphenylalanine synthesis. Furthermore, secondary screening tests in which the same compounds are tested in parallel in two types of extracts (bacterial vs. yeast or yeast vs. HeLa cells) can be very useful for establishing the selectivity of a given inhibitor.

3.2.1. Translation of mRNAs (standard) in *E. coli*

The activity of the bacterial translational apparatus can be studied in cellfree systems programmed, depending upon the particular experimental need and design, with any of the mRNAs shown in Fig. 12.1 and listed in Table 12.1. The amount of synthesized product can be assessed using either a radioactive test or, when translation is directed by 027IFCp(A), an immunological test (see later).

Protocol: The reaction mixture to test standard mRNA translation in bacterial extracts contains, in 50 μl of 10 mM Tris–HCl (pH 7.7): 7 mM Mg acetate, 100 mM NH$_4$Cl, 2 mM DTT, 2 mM ATP, 0.4 mM GTP, 10 mM PEP, 0.025 mg/ml PK, 0.12 mM 10-formyl-tetrahydrofolate, 3 μg/μl tRNA (*E. coli* MRE600), an amino acid mixture containing 0.2 mM of all amino acids (with the exception of phenylalanine); 9 μM [^{14}C] phenylalanine and 36 μM nonradioactive phenylalanine, an optimized amount of the S30 cell extract (generally, 2–6 μl/reaction mixture) and $\cong$1–3 μM mRNA (preheated 5 min at 65°). It is advisable to conduct preliminary tests to assess the optimal amount of each template.

If poly(U)-dependent polyphenylalanine synthesis is to be measured, the amount of template to be used is 5 μg/reaction mixture, the Mg acetate concentration is increased to 12 mM, and the amino acid mixture is omitted.

After 30 to 60 min incubation at 37°, 20- to 40-μl aliquots from each reaction mixture are spotted on 3MM paper discs that are dropped into 10% ice-cold TCA and processed according to the "hot TCA procedure" (see later). If the translational product is to be quantified only immunologically, the radioactive precursor ([^{14}C] phenylalanine) is replaced by nonradioactive phenylalanine. Examples of experiments in which the preceding protocol has been used are shown in Fig. 12.3A,B.

3.2.2. Translation of Renilla luciferase mRNAs (luciferase synthesis)

Renilla luciferase (rLuc) is a natural luminescent protein found in *Renilla reniformis* (Lorenz *et al.*, 1991) that has been used as a reporter protein for *in vivo* and *in vitro* analysis of gene expression (see technical manual of Renilla Luciferase assay system, Promega, for refs). Furthermore, the synthesis of rLuc can be the basis for an *in vitro* translation test aimed at determining not only the level of the synthesis (Fig. 12.4A,B) but also the possible occurrence of misreading, since only the correctly and faithfully translated and folded protein binds its co-factor (Coelenterazine) and O$_2$, thereby catalyzing the light emitting reaction (Lorenz *et al.*, 1991). In fact, as seen from Fig. 12.4C, incorporation of a precursor amino acid and expression of the luciferase activity are inhibited with roughly the same efficiency by an antibiotic (GE81112) that does not induce misreading, while the expression of the enzymatic activity of the luciferase is inhibited much more than amino acid incorporation in the presence of increasing concentrations of an aminoglycoside (streptomycin), known to cause misreading. In addition, an altered timing in the appearance of the product (e.g., the existence of an extended lag) can offer useful indications concerning the mechanism of action of an inhibitor suggesting, for instance, that translation initiation is inhibited (Fig. 12.4B).

Protocol: For our experiments, we use the coding sequence of renilla luciferase mRNA preceded by the translation initiation region (TIR) of *infA** (Calogero *et al.*, 1987) (see Fig. 12.1 and Table 12.1); the tests are routinely carried out using 96-well microtiter plates. A typical reaction mixture for rLuc synthesis contains, in 60 μl of 10 mM Tris-HCl (pH 7.7), 10 mM Mg acetate, 15 mM NH$_4$Cl, 180 mM KCl, 1 mM DTT, 1 mM GTP, 2 mM ATP, 10 mM PEP, 0.025 μg/μl PK, 1 μg/μl total

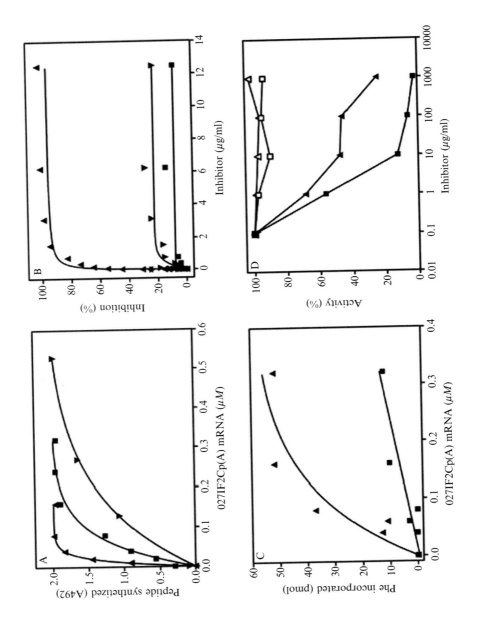

tRNA from *E. coli* MRE600, an amino acid mixture containing 0.2 mM of each amino acid, 0.12 mM 10-formyl-tetrahydrofolate (10-formyl THF, prepared as described later), 0.25 μM rLuc mRNA (preheated 5 min at 65°), and 12 μl of S30 cell extract (corresponding to ~15 pmol of 70S ribosomes).

As an alternative, it is possible to use a mixture of purified 30S and 50S ribosomal subunits or 70S monomers (0.25 μM final concentration) and 4 to 8 μl/reaction tube of S100 post-ribosomal supernatant as well as initiation factors IF1, IF2, and IF3 in a 1.5-to-1 ratio with the ribosomes. After 30 min incubation at 20°, the activity of the synthesized luciferase is determined as described later.

To prepare 10-formyl THF (formyl donor for the formylation of fMet-tRNA$_{fMet}$), 10 mg of 5-formyl-tetrahydrofolate (5-formyl THF) are dissolved in 1 ml of 0.1 M HCl that has been flushed extensively with N_2 and kept overnight at 4° in a closed vial. The resulting yellow suspension of 5,10-methenyl THF is agitated and an aliquot withdrawn for spectrophotometric determination of the concentration (6 mM solution = 85 A_{355}). The expected 5,10-methenyl THF concentration is approximately 20 mM. The cyclic methenyl THF can be stored for several weeks at −20°. The 10-formyl THF is stable for only a few weeks and is freshly prepared hydrolyzing the cyclic methenyl THF. For this purpose, the stock solution is diluted in 100 mM Tris–HCl (pH 8.0) extensively flushed with N_2 and supplemented with 100 mM 2-mercaptoethanol. After 15 min incubation at 20°, the solution is divided in aliquots, which are stored at −20°.

Figure 12.3 Use of "universal" mRNAs for screening translational inhibitors active on the translational apparatus of prokaryotic, lower, and higher eukaryotic cells. (A) Comparison of translational activity obtained with (■) *E. coli*, (▲) *S. cerevisiae*, and (▼) HeLa cell extracts as a function of increasing amounts of 027IF2Cp(A) mRNA offered. The protocols for the preparation of the cell extracts and the optimization of the amounts of extracts used for *in vitro* translation and for the immunological detection (ELISA) of the product are described in the text. (B) Translational inhibitor selectively targeting bacterial translation initiation identified through the comparison of its effects on the 027IF2Cp(A) mRNA-dependent synthesis of IF2C (detected and quantified by immunoblotting) by (▲) *E. coli* and (▼) *S. cerevisiae* cellfree extracts and on (■) poly (U)-dependent polyphenylalanine synthesis (radioactively detected and quantified) by an *E. coli* extract. In all cases, the effect of increasing amounts of the antibiotic is expressed as percentage of inhibition with respect to triplicate controls (in each translational system) that had received no antibiotic. (C) Comparison of (▲) CAP-dependent and (■) CAP-independent translation by *S. cerevisiae* cellfree extracts programmed with 027p(A) mRNA. The translational activity is quantified from the incorporation of radioactive Phe into the acid-insoluble 027 peptide. (D) Differential effect of increasing concentrations of the aminoacylation inhibitors purpuromycin (■,□) and pseudomonic acid (▲,△) on the synthesis of the 027 peptide (open symbols) and IF2C domain (closed symbols) in an *E. coli* cellfree system programmed with 027IF2Cp(A) mRNA. Further details are given in the text.

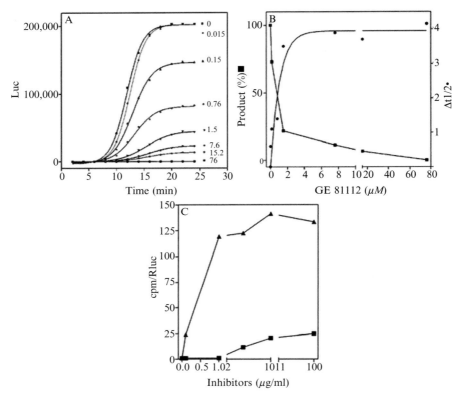

Figure 12.4 Use of rLuc mRNA translation for screening translational inhibitors. (A) Inhibition of luciferase synthesis by increasing amounts of the indicated amounts of antibiotic GE81112 offered; (B) delay in the expression of activity due to the interference of GE81112 with the formation of a 30S initiation complex. Both panels are taken from Brandi *et al.* (2006b); (C) detection of misreading during translation of rLuc mRNA by an *E. coli* cellfree system. In this experiment, the incorporation of a radioactive amino acid ([^{14}C]Phe) in the luciferase product and the luminescence emitted by the synthesized luciferase are quantified and their ratio plotted as a function of increasing concentrations of (▲) streptomycin and (■) GE81112.

3.2.3. Translation of mRNAs (standard) in yeast

As mentioned previously, the translational apparatus of pathogenic lower eukaryotes represents a potentially ideal, yet still underexploited, target for inhibitors capable of bringing under control infections caused by this class of organisms. In fact, unlike bacteria, whose translational apparatus is targeted by approximately half of all known natural antibiotics, there are almost no antibiotics that selectively inhibit protein synthesis in lower eukaryotes such as fungi and protozoa, the main exceptions being the sordarins, which specifically inhibit fungal elongation factor EF2 (Domínguez *et al.*, 1999; Shastry *et al.*, 2001; Søe *et al.*, 2007). Also, the overall number of available antifungal antibiotics is very limited and most are directed against targets other than protein biosynthesis. In fact, the main antimycotic drugs known bind to (*polyene antimycotics*) or inhibit the synthesis of (*allylamines*) ergosterol, a typical constituent of the fungal cell membrane or inhibit the biosynthesis (*echinocandins*) of a cell wall component such as glucan. Furthermore, these antifungals are prone to inducing resistance and are also generally rather toxic. Thus, in consideration of these premises and of the fact that fungal infections are escalating significantly and are expected to continue to increase in the future, the simple discovery of effective inhibitors specific for the translational machinery of lower eukaryotes could be of paramount importance; for this reason, the protocol for measuring the translational activity of yeast extracts presented later should be considered not only a useful tool to ascertain that a bacterial inhibitor is not active in an eukaryotic system, but is also an important instrument for the discovery of translational inhibitors selectively active on lower eukaryotes.

Furthermore, the existence of substantial differences between the translation apparatus of lower and higher eukarya justifies some optimism concerning the perspectives of identifying molecules capable of selectively targeting the protein synthetic machinery of the pathogen without affecting that of the higher eukaryotic host cell. Indeed, fungi are unique not only in their requirement for a third elongation factor named EF-3 (Belfield *et al.*, 1995), but also for other properties of their translational machinery, such as the mRNA "capping" mechanism (see later).

Thus, after the general protocol to test the cap-independent translational activity of yeast extracts, we shall also present an example of how cap-dependent and cap-independent translational activity can be determined *in vitro*, a test that might be very useful for a number of reasons. In fact, while the cap structure plays an important role in determining both stability and translational efficiency of the transcripts in all eukaryotes, the "cap" is introduced at the 5′ end of the mRNAs of lower eukarya through a mechanism that is different from that of higher eukarya. Three enzymatic activities (RNA 5′-triphosphatase, RNA guanylyltransferase, and RNA (G-7-)-methyltransferase) are involved in generating the mRNA cap and the physical organization of the genes encoding these enzymes is different in higher eukaryotic cells on one side and lower eukaryotes and viruses on the other. Furthermore, the catalytic mechanism of mammalian RNA triphosphatases is different from that of the enzymes of the lower eukaryotes such as yeasts, viruses, protozoa, and algae (some of which are also pathogenic) (Hausmann and Shuman, 2005; Shuman, 2001).

Therefore, mRNA capping represents a potentially useful target for inhibitors with antifungal or antiviral activity; indeed, whole-yeast cell-based assays developed to identify and characterize inhibitors of fungal mRNA capping allowed the identification and characterization of sinefungin, an S-adenosylmethionine analog that inhibits mRNA cap methyltransferases with approximately 5- to 10-fold specificity for the yeast ABD1 and fungal CCM1 enzymes over the human Hcm1 enzyme (Chrebet *et al.*, 2005).

Furthermore, even though all caps contain a $m^7G(5')ppp(5')N$ structure, the adjacent nucleotides are generally more extensively methylated in higher eukaryotes (Furuichi and Shatkin, 1989); this property might confer some degree of species-specificity to potential inhibitors of either cap formation or cap binding by the eIF4F complex, notwithstanding the substantial evolutionary conservation of the individual components of this complex and, more particularly, of the cap-binding factor eIF4E (Joshi *et al.*, 2004).

Another possible advantage of detecting inhibitors of cap-dependent but not of cap-independent mRNA translation could be that of detecting compounds such as 4EGI-1. This is a small molecule that binds to the cap-binding protein eIF4E and interferes with its interaction with eIF4-G to form eIF-4F. Since eIF4E appears to be the limiting initiation factor in the cells and its overexpression causes malignant transformation (Lazaris-Karatzas *et al.*, 1990), it is not surprising that inhibiting its function selectively reduces cap-dependent translation of oncogenic proteins and inhibits growth of multiple cancer cell lines (Moerke *et al.*, 2007).

Cap-independent translation   Like bacterial translation, yeast translational activity too can be assessed by measuring the radioactivity incorporated into an acid–insoluble product or by immunological quantification of the product (see later). However, unlike with bacteria, not all templates can be efficiently translated by a yeast extract. In particular, 002mRNA, 022mRNA, and any template with an initiation codon other than AUG (Fig. 12.1 and Table 12.1) are not suitable for translation by yeast (or HeLa) cell extracts. On the other hand, unlike poly(U) and rLuc mRNA, both 027p(A) and 027IF2Cp(A) mRNAs, by virtue of having the "consensus" TIR of bacterial mRNAs, as well as a long poly(A) tail and the eukaryotic consensus signals near the first AUG codon, have the properties of "universal" mRNAs, which can be translated with comparable efficiency by bacterial, lower, and higher eukaryotic extracts (Fig. 12.3A,B). Furthermore, with 027IF2Cp(A) mRNA, the three systems can be assayed and compared in parallel, using the same immunological detection/quantification method (Fig. 12.3A,B).

Protocol: The reaction mixture to test standard, cap-independent mRNA translation in yeast extracts contains, in 50 μl of 33 mM HEPES-KOH (pH 7.4): 160 mM K acetate, 3.3 mM Mg acetate, 3.3 mM DTT, 0.5 mM ATP, 0.1 mM GTP, 30 mM CP, 20 μg/ml CPK, 200 U/ml RNase inhibitor, an amino acid mixture containing 0.2 mM each of all amino acids (minus phenylalanine), 9 μM [^{14}C] phenylalanine, and 36 μM nonradioactive phenylalanine; in addition, the mixture contains optimized amounts of the S30 yeast extract (generally, $0.5–1A_{260}$ / tube) and 027p(A) or 027IF2Cp(A) mRNA (generally, $\cong 0.06\ \mu M$), which are preheated 5 min at $65°$ before use. After 90 to 120 min incubation at $25°$, 20-μl aliquots from each reaction mixture are spotted on 3MM filters and processed by the hot TCA procedure (see later). If the translation product is to be detected immunologically (see later), radioactive phenylalanine is replaced by nonradioactive phenylalanine (45 μM) in the reaction mix.

Examples of experiments in which the preceding protocol has been used are shown in Fig. 12.3A,B (immunological detection) and Fig. 12.3C (radioactive detection). The preceding assay has been validated by showing that it is sensitive to the translational capacity of the cell extract. In fact, it has been shown that compared to the extracts prepared from wt cells, extracts of ribosomal mutants of yeast display a reduced translational activity in a test based on the preceding protocol (data not shown). Furthermore, the preceding test, in addition to being routinely used in a secondary screening assay to assess the prokaryotic selectivity of some newly discovered antibiotics (Fig. 12.3B; see also Brandi et al., 2006a,b) has also been successfully employed in a primary screen that has identified at least one yeast-specific translational inhibitor (not shown).

Cap-dependent vs. cap-independent translation in yeast   Inhibitors of the lower eukaryotic capping enzymes and/or of the cap-eIF4F interaction could be theoretically identified and distinguished from translational inhibitors targeting other steps of protein synthesis by comparison of the level of inhibition of capped vs. noncapped mRNA translation. In fact, only cap-dependent translation is expected to be reduced by an inhibitor of the "capping-cap binding system," while a general translation inhibitor is expected to reduce by the same proportion both cap-dependent and cap-independent translation. Thus, the universal mRNA could be subjected to a preliminary mRNA capping reaction in the presence of potential inhibitors and its translation efficiency compared to that of the same noncapped mRNA. While this system would not distinguish a capping from a cap-binding inhibition, the use of chemically capped mRNA could identify a specific inhibitor of cap-binding activity.

Protocol: The reaction mixtures to test standard, cap-dependent translation or to compare cap-dependent and cap-independent translation are identical to those described previously, except the universal 027p(A) or 027IF2Cp(A) mRNAs used to program translation are subjected to transcriptional precapping (see previous).

That this translational system can indeed detect a different (increased) translational activity when the extracts are programmed with capped mRNA can be seen from the results presented in Fig. 12.3C. On the other hand, a general reduction of the translational efficiency caused by ribosomal mutations affecting the elongation rate was found to reduce to the same extent cap-dependent and cap-independent translation (not shown).

3.2.4. Translation of mRNAs (standard) in HeLa cells

This is a secondary test, the purpose of which is to ascertain that translational inhibitors active on the yeast and/or bacterial translational apparatus are harmless for the human protein synthetic machinery. All the considerations made for the yeast translation apply also to this system.

Protocol: Translation is performed in 15- to 25-μl reaction mixtures containing: 16 mM HEPES-KOH (pH 7.6), 75 mM K acetate, 2.5 mM Mg acetate, 0.1 mM spermidine, 2 mM DTT, 0.8 mM ATP, 0.1 mM GTP, 20 mM CP, 0.1 μg/μl CPK, 0.1 μg/μl bovine liver tRNA, an amino acid mixture containing 0.2 mM each of all amino acids including phenylalanine, an optimized amount of HeLa cell extract (generally, 10–15 μl/25 μl reaction mixture) and $\cong$0.3 μM of preheated (5 min at 65°) 027IF2Cp(A) mRNA. The reaction is allowed to proceed for 60 min at 30° and 10 to 20 μl are used for the immunological quantification of the IF2C domain (see later). An example of an experiment in which the preceding protocol has been used is shown in Fig. 12.3A.

3.2.5. Translational test for the identification of inhibitors of tRNA aminoacylation

In spite of the discovery of a number of natural compounds capable of inhibiting the aminoacylation of the tRNAs, this particular step of translation represents—as mentioned in the introduction—a leading target for novel anti-infectives. In consideration of the potential offered by the large number of evolutionary unrelated, structurally and functionally different enzymes that perform a vital function, aminoacyl-tRNA synthetases can be regarded as a largely underexploited antibiotic target (Kim et al., 2003; Pohlmann and Brotz-Oesterhelt, 2004). For these reasons, the early detection of a tRNA aminoacylation inhibitor within a large number of positive hits in a translational inhibition HTS test might be of great importance and relevance.

If a translation reaction directed by the universal 027IF2Cp(A) mRNA is carried out in the presence of four precharged aminoacyl-tRNAs (fMet-tRNA, Phe-tRNA, Thr-tRNA, and Ile-tRNA) in amounts sufficient to ensure the synthesis of the 027 peptide (which contains only these amino acids) even in the presence of an aminoacyl-tRNA inhibitor, the system will be able to detect an aminoacylation inhibitor in a library of natural or synthetic products through the selective inhibition of IF2C domain synthesis. Thus, if the synthesis of the 027 and IF2C peptides is measured in parallel, a general inhibitor of translation would be expected to inhibit the synthesis of both products, while an aminoacylation inhibitor would inhibit only the synthesis of the C-domain of IF2. A validation of this method is presented in Fig. 12.3D. As seen from the figure, in the presence of two known synthetase inhibitors such as pseudomonic acid, a specific Ile-tRNA synthetase inhibitor (Hughes and Mellows, 1980; Kim et al., 2003) and purpuromycin, a general inhibitor of all synthetases (Kirillov et al., 1997), only the 027 peptide is synthesized while the synthesis of the IF2C domain is progressively inhibited with increasing concentrations of the two aminoacylation inhibitors.

Protocol: The test for the identification of aminoacyl-tRNA synthetase inhibitors requires the availability of precharged fMet-tRNA, Phe-tRNA, Thr-tRNA, and Ile-tRNA, which correspond to the amino acids present in the 027 peptide. The preparation of fMet-tRNA is described in the accompanying chapter by Milon et al., (2007), while the preparation of the other aminoacyl-tRNAs has been described previously.

The translational test is carried out in a 50-μl mix containing 10 mM Tris-HCl (pH 7.5), 12 mM Mg acetate, 100 mM NH$_4$Cl, 2 mM DTT, 2 mM ATP, 0.4 mM GTP, 10 mM PEP, 0.025 mg/ml PK, a mix containing 0.2 mM of each amino acid (except for Phe), 40 pmoles of preheated (5 min at 65°) 027IF2Cp(A) mRNA, an optimized amount (approximately 6.5 μl) of E. coli MRE600 S30 extract, and 2.6 μM fMet-tRNA and 3 μg/μl tRNA mixture in which [^{14}C]Phe-tRNA represents at least ≅2% of the total. After 30 min incubation at 37°, two 20-μl aliquots of the reaction mixtures are withdrawn. One is spotted onto 3MM paper discs and processed by the hot TCA procedure, described later, while the other is used for the immunological quantification of the IF2C produced.

3.3. Product detection

3.3.1. Determination of the level of aminoacylation by the cold TCA procedure

At the end of the aminoacylation reaction, a 5-μl aliquot of the reaction mixture is spotted onto a 3MM paper disc (Schleicher & Schuell) that is immediately placed in 10% ice-cold TCA for 30 min. After three washes in 5% TCA at room temperature for 5 to 10 min, the filter is placed in an ethyl ether:ethanol (1:1) mixture for 10 min and then in ethyl ether for 10 min before being dried under an infrared lamp. The amount of radioactivity precipitated on each filter is finally determined by liquid scintillation counting.

3.3.2. Methods to quantify the translation level

Depending upon the experimental design, the purpose of the experiment, and the nature of the mRNA template used to program the translational systems, different types of product quantification should (or could) be used. Aside from the quantification of the luciferase synthesized, which is carried out via a determination of the luminescence emitted following an enzymatic reaction, two main approaches are available to test the translational activity, one based on the detection of a radioactive product, the other on its immunological detection. The two systems are not mutually incompatible and actually, the test to detect aminoacyl-tRNA synthetase inhibitors makes use of both (see later). Nevertheless, the two detection systems have different characteristics that make them more or less suitable in different circumstances. The radioactive test is faster and very sensitive because the "detection window" to quantify the product can range from a few to several hundred thousand cpm. However, accurate quantification of a radioactive product may be difficult in eukaryotic systems that have quantitatively variable pools of nonradioactive amino acids when extensive dialysis cannot be performed without substantial loss of activity. Furthermore, the nonradioactive translation assay, which is obviously possible only when the template used encodes an antigenic product like IF2C, is suitable for systematic HTS assays in which a massive use of radioactive materials might be problematic.

3.3.3. Determination of the level of translation by the hot TCA procedure

This procedure to test the levels of translation using a radioactive precursor was originally described by Mans and Novelli (1960). At the end of the incubation, aliquots of the reaction mixtures are spotted onto 3MM paper discs, which are immediately dropped into 10% ice-cold TCA and processed as described for the cold TCA procedure, except that the three washes at room temperature in 5% TCA are preceded by a 10-min wash at 90°.

3.3.4. Immunological determination of the level of translation

PBS: 140 mM NaCl; 2.7 mM KCl; 10 mM Na$_2$HPO$_4$; 2 mM KH$_2$PO$_4$ adjusted to pH 7.4 with HCl

AEC stock solution: 0.4 g of 3–amino 9–ethyl carbazole dissolved in 100 ml of N,N dimethyl formamide

Staining solution: 0.67 ml AEC stock solution diluted in 10 ml Na acetate (pH 5.2).

3.3.5. Detection by immunoblotting

Aliquots (20 μl) of the translation reaction are filtered through a 0.45-μM nitrocellulose membrane using a slot-blot (e.g., BioRad) apparatus. After washing in PBS for 45 to 60 min, the membrane is blocked by incubation in PBS containing 3% BSA for 1.5 to 2.0 h and then in PBS containing 0.3% BSA and a suitable dilution of the first anti–IF2C antibody (9F11) for 3 h. The membrane is then subjected to three 10-min washes in PBS containing 0.05% Tween 20 and finally incubated in PBS containing 0.3% BSA and a suitable dilution of the second antibody (HRP-conjugated anti-mouse antibody) for 1 h. After three 10-min washes in PBS containing 0.05% Tween 20, the membrane is incubated in 10 ml of staining solution. The peroxidase reaction is started by addition of 10 μl of 30% H_2O_2. After stopping the antibody-conjugated enzyme reaction in H_2O, the intensity of the stained bands is quantified densitometrically.

3.3.6. Detection by ELISA

Bacterial and eukaryotic translation are stopped by addition of 20 μg/ml kirromycin and 100 μg/ml anisomycin, respectively, at the end of the incubation period. For a large number of samples, as in HTS, after adjustment of the amounts so as to avoid exceeding the detection limit of the ELISA assay, the translation reactions are carried out directly in microtiter plates (Nunc Immuno^{TM} MaxiSorp Plates); in the other cases, aliquots of the reaction mixtures are placed in the microtiter wells. The plates are incubated overnight at room temperature to allow adsorption of the peptide product to the plastic well. After removal by aspiration of the unabsorbed material, the wells are washed with PBS and 150 μl PBS containing 3% BSA are added to each well. After 90 min incubation, 50 μl of the first antibody (9F11) appropriately diluted with PBS containing 0.3% BSA and 0.05% Tween 20 (Sigma) are added. After 90 min, the wells are washed 5 times with 300 μl PBS containing 0.05% Tween and the second antibody (50 μl of HRP conjugated anti-mouse appropriately diluted with PBS containing 0.3% BSA and 0.05% Tween) is added. After 90 min incubation, the wells are washed 5 times with 300 μl PBS containing 0.05% Tween and then incubated for 15 min with 150 μl of 0.1 M Na citrate (pH 5) containing 1 mg/ml o-phenylenediamine (Sigma) and 1:3000 of 30% H_2O_2. The peroxidase reaction is stopped with 20 μl of 4.5 M of H_2SO_4 and the A_{492} is determined with a multiscan MC reader (e.g., Titertek, Flow Laboratories).

3.3.7. Enzymatic (luminescent) determination of the level of translation

These experiments are performed in 96-well microtiter plates (Black and White Isoplate 96-well Wallac) and the amount of rLuc synthesized (see previously) is determined from the amount of light emitted in the visible range, the emission$_{max}$ being at 480$_{nm}$ (Lorenz et al., 1991) using a luminometer (e.g., Microbeta Wallac, Gaithersburg, MD) upon addition of an appropriate luciferase assay reagent (Renilla Luciferase assay system, Promega). The luminescent signal should be recorded immediately after the addition of the reagent because it may slowly decay with time, depending upon experimental conditions. The recording parameters may vary, depending upon the instrument used. However, the operating software provided with most instruments contains preset settings to record luminescence derived from enzymatic reactions.

In addition to the total amount of rLuc synthesized, it is possible to monitor the kinetics of rLuc synthesis following the increasing light emitted by the neo-synthesized rLuc. In this case, coelenterazine (6 μM) is added to the translation reaction mixture and the instrument is set up in a homemade "discontinuous signal registration mode" so as to record the luminescence variation of each well at fixed time intervals (2–4 min). This type of kinetic measurement can yield important information as to the mechanism of action of an inhibitor. For instance, a strong inhibitor of 30S initiation complex formation has been found to lengthen the lag that precedes the expression of the luminescence (Fig. 12.4B). Finally, to determine and quantify antibiotic-induced mRNA misreading with consequent misincorporation of amino acids into rLuc, it is possible to use a combination of the enzymatic and radioactive detection methods. In this experimental setup, rLuc misfolding (Fedorov and Baldwin, 1995; Kolb *et al.*, 2000) caused by misreading is expected to cause a marked decrease of rLuc luminescence and a comparatively less pronounced decrease of the radioactivity incorporated (Fig. 12.4C).

3.4. Partial reactions

3.4.1. 30S and 70S IC formation

fMet-tRNA binding to 30S or 70S ribosomes Each reaction mixture contained, in 40 μl of 20 mM Tris-HCl (pH 7.7), 7 mM Mg acetate, 100 mM NH$_4$Cl, 0.1 mM DTT; 0.5 mM GTP, 30 pmol of *E. coli* 30S ribosomal subunits (or a 1:1 stoichiometric mixture of 30S and 50S subunits), and the desired amounts of the antibiotic under scrutiny. After a brief incubation and unless the activity of one of the initiation factors is to be tested, 45 pmol each of IF1, IF2, and IF3 are added. The binding reaction is started by the addition of 45 pmol each of 022 mRNA and f[^{35}S]Met-tRNA. After 10 min incubation at 37°, the amount of ribosome-bound f[^{35}S]Met-tRNA is determined by filtering 30 μl of each reaction mixture through nitrocellulose discs. Examples of the results obtainable with this method applied to a number of known P-site inhibitors can be found in Brandi *et al.* (2006b). The formation of the 30S and 70S initiation complexes can become the basis for a secondary screening assay. In this case, the binding reactions described previously are adapted to a microtiter format. This can be obtained using a 96-well microtiter filter unit (Multiscreen HTS Millipore MSHVN4B10) connected to a Millipore XX5522050 vacuum pump.

An example of the results that can be obtained using the microtiter format is shown in Fig. 12.5A, which illustrates the inhibition of fMet-tRNA binding to 022 mRNA-programmed 30S ribosomal subunits caused by increasing concentrations of GE81112, the inhibitor of 30SIC formation characterized in Brandi *et al.* (2006b).

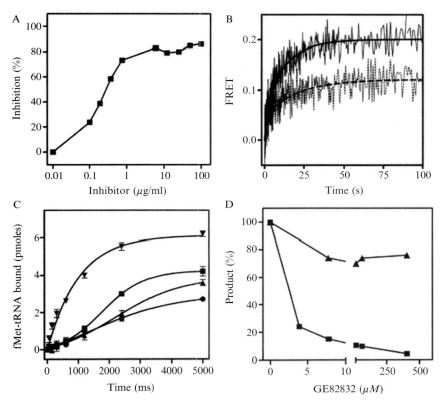

Figure 12.5 Secondary tests for the analysis of specific steps of the translational pathway. Panels (A), (B), and (C) present three different tests to detect the inhibition of fMet-tRNA binding to 30S ribosomal subunits. (A) Binding of radioactive initiator tRNA to 022 mRNA-programmed 30S ribosomal subunits at equilibrium carried out in a 96-well microtiter plate in the presence of increasing concentrations of the antibiotic GE81112; (B) effect of a fixed concentration of GE81112 on fMet-tRNA binding to 002 mRNA-programmed 30S subunits analyzed by transient kinetics using the fluorescence stopped flow technique and a FRET signal between fMet-tRNA (Fluf-Thio-U8) and IF3 (Cys 166 Alexa 555); (C) kinetics of fMet-tRNA binding to 022-mRNA-programmed 30S ribosomal subunits in the absence (▼) or presence of 0.25 μg/ml (■) 0.5 μg/ml (▲) and 1 μg/ml (●) of GE81112 studied by rapid filtration through nitrocellulose membranes. This panel is taken from Brandi *et al.* (2006b). (D) Effect of the translocation inhibitor GE82832 in promoting (▲) initiation dipeptide (fMet-Phe) and (▲) tripeptide (fMet-Phe-Ile) synthesis by 012 mRNA-programmed ribosomes (modified from Brandi *et al.*, 2006a).

Kinetics of fMet-tRNA binding to 30S ribosomal subunit

Inhibition of ribosomal binding of fMet-tRNA by an antibiotic may reduce the level of initiation complex formed at equilibrium. However, if the effect of the inhibitor consists mainly of slowing down the binding reaction, its effect may appear less dramatic after a relatively long incubation time. For this reason, the kinetics of fMet–tRNA binding should also be studied. Since the binding reactions are relatively slow, manual sampling within the seconds range may serve the purpose. However, to obtain better insight into the mechanism of inhibition, manual sampling may be inadequate. The better alternatives are the use of a fast filtration apparatus or study of the transient kinetics using the FRET approach by stopped-flow fluorescence, as described in the chapter by Milon *et al.*, (2007). An example of this approach is shown in Fig. 12.5B. In this experiment, the inhibition of 30SIC formation by the previously mentioned antibiotic GE81112 (Brandi *et al.*, 2006b,c) is monitored by stopped flow-FRET generated upon binding of fMet–tRNA(Fluf-Thio-U8) to 002 mRNA-programmed 30S. The FRET occurs between fluorescein (donor) of fMet–tRNA and Alexa555-derivatized Cys 166 of 30S-bound IF3 acting as acceptor. These binding experiments yield complex curves that can be fit by more than one exponential, which likely reflect the primary 30S-fMet–tRNA interaction, an mRNA-dependent codon–anticodon interaction and a subsequent adjustment of the initiator tRNA that depends upon the nature of the initiation triplet (Milon *et al.* 2007). It seems clear, however, that when the binding reaction is carried out in the presence of a fixed concentration (5 μM) of GE81112, there is a reduction of the level of binding and that at least one of the previously mentioned events is characterized by a slower apparent rate (Fig. 5B).

The rapid filtration experiments are performed on a Bio-Logic SFM-400 apparatus (Bio-Logic Science Instruments, Grenoble, France) in quench flow configuration.

Protocol:

Buffer C: 10 mM Tris-HCl, pH 7.7, 7 mM Mg acetate, 50 mM NH$_4$Cl
Buffer D: 20 mM Hepes, pH 7.1, 7 mM Mg acetate, 80 mM NH$_4$Cl, 0.1 mM DTT

To perform fast nitrocellulose filtration, the apparatus was modified by the addition of two external flow lines (line 3 and line 4) connected to the corresponding reservoir syringes. These lines were positioned inside an umbilical link connected to a circulating bath for temperature regulation. The analyses are routinely carried out at 20°. The lines lead to a mixer followed by an ejection delay line where the reaction mixture is stored for the programmed incubation period (between 80 and 5000 ms). The ejection delay line is placed directly over a filtering apparatus containing a nitrocellulose filter covered by 3 ml of ice-cold Buffer C. At the end of each incubation time, the reaction mixture is immediately pushed into the Buffer C with 1 ml of the same buffer by the programmed action of a third syringe independently connected to the mixer; the line is then completely emptied by the airflow generated by an external pump connected to the mixer. The binding reaction is virtually stopped by the overall 50-fold dilution followed by rapid filtration. The filters are washed twice with 3 ml of Buffer C and the f[^{35}S]Met–tRNA radioactivity associated with the 30SIC is determined by liquid scintillation counting. Duplicate or triplicate time points (preferably in a logarithmic scale) are taken for a total of more than 30 points for each curve. The apparent rates are calculated by numerical integration software (GraphPad Prism), fitting the resulting curves to (one to three) exponential equations.

In a typical experimental setup, Syringe A contains, in 2 ml of Buffer D, 0.5 mM GTP, 1 μM 30S subunits, 1.5 μM each of IF1, IF2, and IF3, and the desired amounts of the antibiotic under scrutiny while Syringe B contains, in 2 ml of Buffer D, 0.5 mM GTP, 2 μM 022mRNA, and 2 μM f[^{35}S]Met–tRNA. Equal volumes (50 μl) of the two solutions are rapidly pushed into the mixing chamber and allowed to age for times ranging between 30 and 5000 ms before being rapidly filtered through a nitrocellulose disk, as described previously.

An example of the results that can be obtained by rapid filtration is shown in Fig. 12.5C, which illustrates the inhibition of fMet–tRNA binding to 022 mRNA-programmed 30S ribosomal subunit by increasing concentrations of GE811112 (Brandi *et al.*, 2006b).

fMet-puromycin formation   In prokaryotes, a 30S initiation complex (30SIC) is formed when, in response to an initiation codon, a molecule of fMet–tRNA is bound to the 30S ribosomal subunit with the help of the three initiation factors (IF1, IF2, and IF3). Under normal conditions, the bound initiator tRNA occupies a position on the small subunit that corresponds to the P-site so that, upon association of the 30SIC with the 50S subunit and upon dissociation of the IF2-fMet–tRNA interaction that frees the acceptor end of the tRNA, fMet–tRNA can function as a donor in the formation of the first peptide bond (Gualerzi *et al.*, 2001; Milon *et al.*, 2007. However, it may be possible, under some conditions, that the position of the bound fMet–tRNA is distorted so that it forms a nonproductive complex (Giuliodori *et al.*, 2007). Presumably, situations of this type could also result from the activity of an antibiotic; there are several known antibiotics capable of inhibiting the activity of the ribosomal peptidyl transferase center (PTC). In either case, fMet–puromycin formation would be inhibited since this reaction requires a functioning PTC and an fMet–tRNA properly placed in the ribosomal P-site. Thus, testing for fMet–puromycin formation could provide important clues concerning the action mechanism of an antibiotic.

Protocol: The primary mixture (100 μl) contains 20 mM Tris-HCl (pH 7.7), 7 mM Mg acetate, 100 mM NH$_4$Cl, 0.1 mM, DTT, 0.5 mM GTP, 30 pmol of 30S ribosomal subunits, and 45 pmol each of 022 mRNA, IF1, IF2, IF3, and f[^{35}S]Met–tRNA. The 30S initiation complex is formed by incubating this mixture for 10 min at 37° and a 10-μl aliquot is withdrawn to

258 Letizia Brandi *et al.*

determine the amount of 30S initiation complex formed. Two 40-μl aliquots are withdrawn from the rest of the mixture and each is mixed with a solution containing puromycin (final concentration 1 mM) and 25 pmol of 50S ribosomal subunits. One sample is incubated at 37° for 45 s and the other for 10 min before stopping the reaction by addition of 500 μl of (NH$_4$) HCO$_3$ (pH 9.0). The f[^{35}S]Met-puromycin formed is extracted by vigorous vortex mixing for 1 min with 1 ml of ethyl acetate; the amount of f[^{35}S]Met-puromycin present in 0.5 ml of the ethyl acetate phase is then determined by liquid scintillation counting. An example of an experiment in which the effect of an antibiotic (thiostrepton) on fMet-puromycin formation has been analyzed can be found in Brandi *et al.* (2004).

3.4.2. Initiation dipeptide and tripeptide formation

The formation of fMet-puromycin described previously is catalyzed by the PTC and occurs between two substrates having different characteristics; one is a ribosome-bound macromolecule endowed with restricted mobility while the other is a small molecule that can diffuse more readily without having the structural constraints imposed upon a bona fide A-site bound aminoacyl-tRNA. Thus, formation of the initiation dipeptide represents a more stringent test for antibiotics targeting translation initiation. In fact, in consideration that this reaction can be made more IF2-dependent than fMet-puromycin formation and notwithstanding the fact that failure to form an initiation dipeptide could stem from the inhibition of the EF-Tu-GTP-aa-tRNA binding to the ribosome (an occurrence that can be easily checked), formation of the initiation dipeptide can be a rather stringent test for the IF2 activity. Indeed, it is not uncommon that antibiotics (e.g., thiostrepton), having little or no effect on fMet-puromycin formation, may severely inhibit initiation dipeptide formation (Brandi *et al.*, 2004).

After formation of the initiation dipeptide, the first EF-G-dependent translocation allows binding of the third aminoacyl-tRNA in the A-site so that a tripeptide is formed. The apparent rate of this event may depend upon the nature of the initiation complex initially formed, being slower, for instance, with those containing mRNAs with an extended SD sequence than with those having either very short or no SD complementarity (C. O. G. and M. Rodnina, unpublished results). Furthermore, very powerful translocation inhibitors may block tripeptide formation to such an extent that they mimic translation initiation inhibitors.

Protocol:

Buffer E: 50 mM Tris-HCl (pH 7.5), 100 mM NH$_4$Cl, 30 mM KCl, 7 mM MgCl$_2$

To facilitate the HPLC analysis of the tripeptide product, these experiments make use of 012 mRNA, a modified form of 022 mRNA in which the third triplet ACG coding for Thr is changed into TTG, coding for Leu

(Fig. 12.1 and Table 12.1). Thirty S initiation complexes are prepared by incubating 10 min at 37° mixtures containing, in 50 μl Buffer E, 1 mM GTP, 0.3 μM 30S subunits, 0.45 μM each of IF1, IF2, IF3, f[^{35}S]Met-tRNA, and 0.9 μM 012 mRNA. A ternary complex containing EF-Tu-GTP-Phe-tRNA is prepared by incubating 10 min at 37° a mixture containing 1 mM GTP, EF-Tu (0.3 μM final concentration), Phe-tRNA (0.3 μM final concentration), 3 mM PEP, and 0.25 mg/ml PK in Buffer E.

The EF-Tu-GTP-Leu-tRNA ternary complex is prepared in the same way except that Phe-tRNA is substituted by 0.3 μM Leu-tRNA. To form the initiation dipeptide, the 30S initiation complex is mixed with an equal volume (40 μl) of a mixture containing the EF-Tu-GTP-Phe-tRNA ternary complex and 50S subunits (0.3 μM final concentration). After 5 min at 37°, tripeptide (fMet-Phe-Leu) formation is triggered by the addition of a mixture of EF-Tu-GTP-Leu-tRNA complex and elongation factor EF-G. After 5 min at 37°, the reaction is quenched with an equal volume of 0.5 M KOH. After 15 min incubation at 37°, the reaction mixture is neutralized with acetic acid and centrifuged at 12,000 rpm for 5 min. Dipeptide and tripeptide formed are analyzed by HPLC on a reverse phase (LiChrosorb RP-8, 5 mM-Merck) column with a linear (0–65%) acetonitrile gradient in 0.1% TFA. The radioactivity present in the individual chromatographic fractions is determined by liquid scintillation counting. The effect of the translocation inhibitor GE82832 (Brandi et $al.$, 2006a) is shown in Fig. 12.5D. It can be seen that this antibiotic is capable of inhibiting the first translocation leading to tripeptide formation without having a substantial effect on initiation dipeptide formation.

3.5. Screening for IF2 inhibitors

Translation initiation factor IF2 is highly conserved among prokaryotes and its activity is strictly bacteria specific. Nevertheless, in spite of the finding that there are molecules like thiostrepton (Brandi et $al.$, 2004), some structural analogues of fMet-adenosine/NacPhe-adenosine (Delle Fratte et $al.$, 2002), and ppGpp (Milon et $al.$, 2006) capable of inhibiting IF2 activity, so far no antibiotic selectively targeting its functions has been found. Thus, IF2 can be regarded as an ideal target for new bacteria-specific drugs for which no resistance mechanisms have yet been developed.

In the following section, we describe protocols for tests aimed at screening for compounds capable of interfering with some of the main activities of this factor, such as (a) recognition and binding of initiator tRNA; (b) codon-dependent ribosomal binding of fMet-tRNA leading to the formation of a 30S or 70S initiation complex; (c) ribosome-dependent hydrolysis of GTP; and (d) accommodation of fMet-tRNA in the ribosomal P-site and formation of the first peptide bond (initiation dipeptide formation).

3.5.1. Inhibition of IF2-dependent translation

The test described here has been developed to determine whether a translational inhibitor that might be affecting initiation as well as other steps (e.g., elongation) might have its preferential target in the initiation functions of IF2. The test is based on the finding that translation of an mRNA having a noncanonical start codon (AUU) is more IF2-dependent than that of a mRNA beginning with the canonical AUG triplet (Giuliodori *et al.*, 2004; La Teana *et al.*, 1993). The test exploits the different IF2-dependence of two mRNAs having identical sequences in both coding and noncoding regions except for the initiation codon (see 027AUG mRNA vs. 027AUU mRNA in Fig. 12.1 and Table 12.1). Translation is carried out in the presence of increasing concentrations of the inhibitor and the ratio between the level of inhibition obtained for each concentration of the inhibitor is plotted. For inhibitors preferentially targeting the IF2 function, the ratio AUU/AUG is expected to increase as a function of increasing the concentration of the antibiotic while, for the other inhibitors, the ratio is expected to remain constant. An example of this type of test is shown in Fig. 12.6A and indicates the preferential inhibition of IF2 activity (compared to those of EF-Tu and EF-G) by ppGpp (Milon *et al.*, 2006).

Protocol: The reactions are carried out in a mixture (30 μl) consisting of 20 mM Tris HCl (pH 7.7), 7 mM Mg acetate, 80 mM NH$_4$Cl, 60 mM KCl, 1 mM DTT, 2 mM ATP, 0.1 mM of each amino acid except for Phe or Met, 40 μM nonradioactive Phe and 0.5 μM [^{3}H]-Phe, 2 μl S100 extract, 0.5 μM of each 70S ribosomes, mRNA (027AUG or 027AUU), IF1, IF2, IF3, and [^{35}S]-fMet-tRNA^{Met}. After 30 min incubation at 37°, aliquots of the reaction mixtures are spotted on 3MM filters which are processed by the hot TCA procedure for determination of the acid–insoluble radioactivity incorporated into the translation product.

3.5.2. IF2-dependent GTPase

GTP is present in millimolar concentrations in the cell and is therefore an unchallenged *in vivo* ligand of IF2, at least under optimal growth conditions (Milon *et al.*, 2006). Thus, IF2 binds to the 30S subunit in its GTP conformation in which the functionally active site of the factor has the highest affinity for the small ribosomal subunit (Caserta *et al.*, 2006). Upon 30SIC joining with the 50S subunit, the GTPase center of 30SIC-bound IF2 is very rapidly activated to hydrolyze GTP. This process is followed by the slower (rate-limiting) dissociation of the γ-phosphate from the complex which, in turn, is necessary to allow IF2 conformational changes, induced by tightening of the intersubunit bridges; this change is required to dissociate the IF2 from the acceptor end of fMet-tRNA, thereby permitting initiation dipeptide formation to occur (Gualerzi *et al.*, 2001; Milon *et al.*, 2007). The very small amount of GTP hydrolyzed by IF2 during this

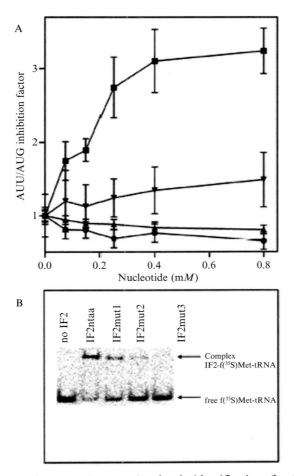

Figure 12.6 Secondary screening tests aimed at the identification of a selective inhibitor of bacterial translation initiation factor IF2. Panel (A), taken from Milon *et al.* (2006), illustrates the use of model mRNAs bearing different initiation codons to detect a selective inhibitor of translation initiation factor IF2. Translation of 027 mRNA bearing either AUG or AUU initiation triplet (see scheme in Fig. 12.1) is carried out in an *E. coli* cellfree extract in the presence of increasing concentrations of (▲) GTP, (●) GDP, (▼) GDPNP, and (■) ppGpp. Since the translation of an mRNA bearing the noncanonical initiation triplet AUU is more IF2-dependent than the translation of an mRNA beginning with the canonical AUG (Giuliodori *et al.*, 2004; LaTeana *et al.*, 1993), the preferential inhibition of translation of the former mRNA (deducible from the increase of the AUU/AGG inhibition ratio) by ppGpp indicates that, unlike the other, this guanine nucleotide preferentially inhibits the initiation function of IF2. Panel (B) illustrates the use of an fMet-tRNA electrophoretic band shift assay to detect and quantify the formation of the binary complex between initiator tRNA and IF2. In the absence of a suitable inhibitor having the capacity of inhibiting this interaction, the test is validated by use of different IF2 molecules having decreasing affinities for the initiator tRNA as a result of different amino acid substitutions within the C-2 domain of the protein.

process is difficult to detect in standard GTPase activity tests that are suitable to measure multiple turnover hydrolysis. Furthermore, the evolutionary conservation of the guanine nucleotide binding domain of IF2 in a large number of proteins belonging to all kingdoms of life makes this domain an improbable target for an IF2-specific inhibitor. On the other hand, multiple turnover GTPase can arise from multiple cycles of IF2 dissociation and reassociation with the 70S ribosomes, each rebinding causing the hydrolysis of additional GTP molecules. Thus, while this activity cannot be easily related to the IF2 function in promoting the formation of a productive 70SIC, it can nevertheless help to identify molecules capable of either tightening or loosening the IF2–ribosome interaction, since these are expected to decrease or increase the multiple turnover GTPase of the factor, respectively. Indeed, both thiostrepton (Brandi *et al.*, 2004) and 23S rRNA mutations (unpublished results from our laboratory) that weaken or distort the IF2–ribosome interaction were found to increase substantially the GTPase activity of IF2. Thus, the GTPase test described later is suitable to detect substances interfering with the normal IF2–ribosome interaction, unlike the fast kinetics method reported in the accompanying article (Milon *et al.*, 2007), which is suitable to test the GTPase activity directly associated with the IF2 function during formation of the 70SIC.

Protocol: The reaction mixture (50 μl) contains 10 mM Tris-HCl (pH 7.7), 10 mM Mg acetate, 60 mM NH$_4$Cl, 6 mM β-mercaptoethanol, 30 pmol each of IF2, 30S and 50S subunits and 50 mM [α-^{32}P]GTP. After 5 to 10 min incubation at 37°, the reaction is stopped by addition of 4 μl 25% (v/v) HCOOH and 5% (v/v) TCA. The precipitated proteins and nucleic acids are removed by centrifugation at 8000 rpm for 5 min and 10 μl aliquots of the resulting supernatants are loaded onto PEI–cellulose sheets. The TLC is developed with 1.5 M KH$_2$PO$_4$ (pH 3.4) to separate [α-^{32}P] GTP from [α-^{32}P]GDP and the dried plates are subjected to autoradiography or molecular imaging (e.g., BioRad) to quantify the radioactivity present in each spot.

3.5.3. IF2-fMet-tRNA interaction

This essential property of IF2 can be tested in at least three different ways, all of which require the availability of f[^{3}H]Met-tRNA and IF2, which are prepared according to the protocol detailed in Milon *et al.* (2007). However, all the tests described in this section can make use of the sturdier and smaller C domain of *Bacillus stearothermophilus* IF2, since this domain contains all molecular determinants for the IF2-fMet-tRNA interaction (Guenneugues *et al.*, 2000; Spurio *et al.*, 2000). The method for the preparation and purification of *B. stearothermophilus* IF2C is essentially that described by Spurio *et al.* (1993). The concentration of the protein can be determined from the extinction coefficient at (1 A_{276} = 0.65 mg·ml^{-1}).

Protection of fMet-tRNA from spontaneous hydrolysis   Buffer F: 100 mM Tris-HCl (pH 8.0), 160 mM NH$_4$Cl, 6 mM Mg acetate, 6 mM 2-mercaptoethanol.

The activity of IF2 in binding fMet-tRNA was measured quantifying the protection conferred by these proteins on the initiator tRNA with respect to spontaneous hydrolysis occurring at alkaline pH (Gualerzi *et al.*, 1991; Petersen *et al.*, 1979). Reaction mixtures (50 μl) in Buffer F contained 22 μM f[^{35}S]Met-tRNA, an appropriate amount of protein that is capable of protecting approximately 80% of the initiator tRNA after 60 min incubation as well as increasing concentrations of the antibiotic to be tested. Samples (20 μl), withdrawn after 0 and 60 min of incubation at 37°, are spotted on Whatman 3MM paper discs for determination of the acid–insoluble radioactivity by the cold TCA procedure, described previously.

Scintillation proximity assay

Buffer G: 50 mM NaHCO$_3$, (pH 8.5) containing 1 mM DTT
Buffer H: 50 mM Tris HCl (pH 7.5), 200 mM NH$_4$Cl, 0.5 mM EDTA, 1 mM DTT, 0.5% glycerol

This test, developed by Delle Fratte *et al.* (2002), allowed the identification, within a chemical library, of a few compounds capable of interfering with the binary interaction between IF2 and fMet-tRNA$_{metf}$. The purified protein (native IF2 or, better, IF2C from *B. stearothermophilus*) is extensively dialyzed against Buffer G. After incubation for 2 h at 20° with a 10-fold molar excess of NHS-LC Biotin (Pierce), the protein is extensively dialyzed against Buffer H. The optimal conditions for this assay have been experimentally determined to be 150 mM NH$_4$Cl, 5 mM Mg^{++}, and a slightly acidic pH (Delle Fratte *et al.*, 2002), namely, in 50 μl HEPES HCl (pH 6.8) containing 150 mM NH$_4$Cl, 5 mM Mg acetate, 0.5 mM DTT, 10% glycerol, 0.7μM f[^{3}H]Met-tRNA, and 0.5 μM bIF2C. After incubation for 1 h at 37°, 100 μg of streptavidin–coated SPA beads (for a total 130 nM biotin-binding capacity) are added to each well, bringing the total volume to 100 μl. To determine the extent of the binary interaction, the microplates are subjected to counting in a top counter (e.g., Canberra Packard). The counts obtained in the absence of the biotinylated factor are subtracted as background.

Electrophoretic band-shift   The reaction mixtures contain 20 μl of 50 mM imidazole-HCl (pH 7.5) buffer, 50 mM NH$_4$Cl, 10 mM Mg acetate, and 1 μM each IF2 and f[^{35}S]Met-tRNA$_{metf}$. After 10 min incubation at 37°,

5 μl of 20 mM MOPS–NaOH (pH 7.5) containing 40% glycerol are added and the samples subjected to electrophoresis on a gel slab constituted by an upper (approximately 2.5 cm) and a lower (approximately 3.5 cm) portion containing 6 and 12% (w/w) polyacrylamide, respectively. After the electrophoretic run ($\cong$2 h at 100 V at 20°) carried out in 20 mM MOPS–NaOH (pH 7.5), the gels are dried and subjected to autoradiography or molecular imaging to determine the amount of radioactivity associated with each electrophoretically resolved band. An example of this type of gel-shift analysis is presented in Fig. 12.6B. As seen from the figure, this method is able to identify IF2 (mutant) molecules having an reduced affinity for fMet-tRNA compared to wtIF2. In the absence of radioactive tRNA proteins, tRNA and complex can be detected by silver staining.

ACKNOWLEDGMENTS

This work was made possible by the financial support of an EU grant, the "Vigoni" Italian–German exchange program, and the Italian MIUR (PRIN 2005 to COG and CLP).

REFERENCES

Barker, J. J. (2006). Antibacterial drug discovery and structure-based design. *Drug Discov. Today* **11**, 391–404.

Belfield, G. P., Ross-Smith, N. J., and Tuite, M. F. (1995). Translation elongation factor-3 (EF-3): An evolving eukaryotic ribosomal protein? *J. Mol. Evol.* **41**, 376–387.

Beyer, D., Kroll, H. P., Endermann, R., Schiffer, G., Siegel, S., Bauser, M., Pohlmann, J., Brands, M., Ziegelbauer, K., Haebich, D., Eymann, C., and Brotz-Oesterhelt, H. (2004). New class of bacterial phenylalanyl-tRNA synthetase inhibitors with high potency and broad-spectrum activity. *Antimicrob. Agents Chemother.* **48**, 525–532.

Brandi, L., Marzi, S., Fabbretti, A., Fleischer, C., Hill, W. E., Gualerzi, C. O., and Lodmell, J. S. (2004). The translation initiation functions of IF2: Targets for thiostrepton inhibition. *J. Mol. Biol.* **335**, 881–894.

Brandi, L., Fabbretti, A., Di Stefano, M., Lazzarini, A., Abbondi, M., and Gualerzi, C. O. (2006a). Characterization of GE82832, a peptide translocation inhibitor interacting with bacterial 30S ribosomal subunits. *RNA* **12**, 1262–1270.

Brandi, L., Fabbretti, A., La Teana, A., Abbondi, M., Losi, D., Donadio, S., and Gualerzi, C. O. (2006b). Specific, efficient, and selective inhibition of prokaryotic translation initiation by a novel peptide antibiotic. *Proc. Nat. Acad. Sci. USA* **103**, 39–44.

Brandi, L., Lazzaroni, A., Cavalletti, L., Abbondi, M., Corti, E., Ciciliato, I., Gastaldo, L., Marazzi, A., Feroggio, M., Maio, A., Colombo, L., Donadio, S., Marinelli, F., Losi, D., Gualerzi, C. O., and Selva, E. (2006c). Novel tetrapeptide inhibitors of bacterial protein synthesis produced by a streptomyces. *Biochemistry* **43**, 3700–3710.

Brandi, L., Dresios, J., and Gualerzi, C. O. (2007). Assays for the identification of inhibitors targeting specific translational steps. *Meth. Mol. Med.* **142**, in press.

Bush, K. (2004). Antibacterial drug discovery in the 21st century. *Clin. Microbiol. Infect.* **10** (Suppl. 4), 10–17.

Calogero, R. A., Pon, C. L., and Gualerzi, C. O. (1987). Chemical synthesis and *in vivo* hyperexpression of a modular gene coding for *E. coli* translational initiation factor IF1. *Mol. Gen. Genet.* **208**, 63–69.

Calogero, R. A., Pon, C. L., Canonaco, M. A., and Gualerzi, C. O. (1988). Selection of the mRNA translation initiation region by *Escherichia coli* ribosomes. *Proc. Natl. Acad. Sci. USA.* **85**, 6427–6431.

Caserta, E., Tomsic, J., Spurio, R., La Teana, A., Pon, C. L., and Gualerzi, C. O. (2006). Translation initiation factor IF2 interacts with the 30S ribosomal subunit via two separate binding sites. *J. Mol. Biol.* **362**, 787–799.

Castañedo, N., Goizueta, R., González, O., Pérez, A., González, J., Silveira, E., Cuesta, M., Martínez, A., Lugo, E., Estrada, E., Carta, A., Navia, O., and Delgado, M. (1996). Universidad Central de Las Villas, assignee. Procedure for obtaining 1-(5-bromophur-2-yl)-bromo-2-nitroethene and its microcide action Cuban patent No. 22 446, European patent No. 0678516, Japan patent No. 3043003, Japan patent No. 2875969, Canada patent No. 2,147,594.

Castañedo, N., Sifontes, S., Monzote, L., López, Y., Montalvo, A., Infante, J., and Olazába, E., inventors; Universidad Central de Las Villas, assignee (2005). Pharmaceutical composition, that included nitrovinylfuran derivatives for the treatment of leishmaniasis and trypanosomiasis Cuban patent No. 2005–0175, PCT Patent No. PCT/CU2006/000009.

Chrebet, G. L., Wisniewski, D., Perkins, A. L., Deng, Q., Kurtz, M. B., Marcy, A., and Parent, S. A. (2005). Cell-based assays to detect inhibitors of fungal mRNA capping enzymes and characterization of Sinefungin as a cap methyltransferase inhibitor. *J. Biomol. Screen.* **10**, 355–364.

Dasso, M. C., and Jackson, R. J. (1989). On the fidelity of mRNA translation in the nuclease-treated rabbit reticulocyte lysate system. *Nucleic Acids Res.* **17**, 3129–3144.

Delle Fratte, S., Piubelli, C., and Domenici, E. (2002). Development of a high-throughput scintillation proximity assay for the identification of C-domain translational initiation factor 2 inhibitors. *J. Biomol. Screen.* **7**, 541–546.

Domínguez, J. M., Gómez-Lorenzo, M. G., and Martín, J. J. (1999). Sordarin inhibits fungal protein synthesis by blocking translocation differently to fusidic acid. *J. Biol. Chem.* **274**, 22423–22427.

Donadio, S., Monciardini, P., Brandi, L., Sosio, M., and Gualerzi, C. O. (2007). Novel assays and novel strains—Promising routes to new antibiotics? *Expert Opinion on Drug Discovery.* Accepted for publication.

Fedorov, A. N., and Baldwin, T. O. (1995). Contribution of co-translational folding to the rate of formation of native protein structure. *Proc. Natl. Acad. Sci. USA* **92**, 1227–1231.

Furuichi, Y., and Shatkin, A. J. (1989). Characterization of cap structures. *Methods Enzymol.* **180**, 164–176.

Gillam, I., Blew, D., Warrington, R. C., von Tigerstrom, M., and Tener, G. M. (1968). A general procedure for the isolation of specific transfer ribonucleic acids. *Biochemistry* **7**, 3459–3468.

Giuliodori, A. M., Brandi, A., Gualerzi, C. O., and Pon, C. L. (2004). Preferential translation of cold-shock mRNAs during cold adaptation. *RNA* **10**, 265–276.

Giuliodori, A. M., Brandi, A., Giangrossi, M., Gualerzi, C. O., and Pon, C. L. (2007). Cold stress induced *de novo* expression of *infC* and role of IF3 in cold-shock translational bias *RNA.* **13**, 1355–1365.

Gualerzi, C. O., Severini, M., Spurio, R., La Teana, A., and Pon, C. L. (1991). Molecular dissection of translation initiation factor IF2. Evidence for two structural and functional domains. *J. Biol. Chem.* **266**, 16356–16362.

Gualerzi, C. O., Brandi, L., Caserta, E., Garofalo, C., Lammi, M., La Teana, A., Petrelli, D., Spurio, R., Tomsic, J., and Pon, C. L. (2001). Initiation factors in the early events of mRNA translation in bacteria. *Cold Spring Harb. Symp. Quant. Biol.* **66**, 363–376.

Guenneugues, M., Meunier, S., Boelens, R., Caserta, E., Brandi, L., Spurio, R., Pon, C. L., and Gualerzi, C. O. (2000). Mapping the fMet-tRNA binding site of initiation factor IF2. *EMBO J.* **19,** 5233–5249.

Hackbarth, C. J., Chen, D. Z., Lewis, J. G., Clark, K., Mangold, J. B., Cramer, J. A., Margolis, P. S., Wang, W., Koehn, J., Wu, C., Lopez, S., and Withers, G., 3rd, *et al.* (2002). N-alkyl urea hydroxamic acids as a new class of peptide deformylase inhibitors with antibacterial activity. *Antimicrob. Agents Chemother.* **46,** 2752–2764.

Hausmann, S., and Shuman, S. (2005). *Giardia lamblia* RNA cap guanine-N2 methyltransferase (Tgs2). *J. Biol. Chem.* **280,** 32101–32106.

Hughes, J., and Mellows, G. (1980). Interaction of pseudomonic acid A with *Escherichia coli* B isoleucyl-tRNA synthetase. *Biochem. J.* **191,** 209–219.

Joshi, B., Cameron, A., and Jagus, R. (2004). Characterization of mammalian eIF4E-family members. *Eur. J. Biochem.* **271,** 2189–2203.

Kaji, A. (1968). Techniques for measuring specific sRNA binding to *Escherichia coli* ribosomes. *Methods Enzymol.* **12B,** 692–699.

Kim, S., Lee, S. W., Choi, E. C., and Choi, S. Y. (2003). Aminoacyl-tRNA synthetases and their inhibitors as a novel family of antibiotics. *Appl. Microbiol. Biotechnol.* **61,** 278–288.

Kirillov, S., Vitali, L. A., Goldstein, B. P., Monti, F., Semenkov, Y., Makhno, V., Ripa, S., Pon, C. L., and Gualerzi, C. O. (1997). Purpuromycin: A new inhibitor of tRNA aminoacylation. *RNA* **3,** 905–913.

Kolb, V. A., Makeyev, E. V., and Spirin, A. S. (2000). Co-translational folding of a eukaryotic multidomain protein in a prokaryotic translation system. *J. Biol. Chem.* **275,** 16597–16601.

La Teana, A., Pon, C. L., and Gualerzi, C. O. (1993). Translation of mRNAs with degenerate initiation triplet AUU displays high initiation factor 2 dependence and is subject to initiation factor 3 repression. *Proc. Natl. Acad. Sci. U. S. A.* **90,** 4161–4165.

Lazaris-Karatzas, A., Montine, K. S., and Sonenberg, N. (1990). Malignant transformation by a eukaryotic initiation factor subunit that binds to mRNA 5′ cap. *Nature* **345,** 544–547.

Lorenz, W. W., McCann, R. O., Longiaru, M., and Cormier, M. J. (1991). Isolation and expression of a cDNA encoding *Renilla reniformis* luciferase. *Proc. Natl. Acad. Sci. USA* **88,** 4438–4442.

Mans, R. J., and Novelli, G. D. (1960). A convenient, rapid, and sensitive method for measuring the incorporation of radioactive amino acids into protein. *Biochem. Biophys. Res. Commun.* **3,** 540–543.

Milon, P., Tischenko, E., Tomšic, J., Caserta, E., Folkers, G., La Teana, A., Rodnina, M. V., Pon, C. L., Boelens, R., and Gualerzi, C. O. (2006). The nucleotide binding site of bacterial translation initiation factor IF2 as a metabolic sensor. *Proc. Nat. Acad. Sci. USA* **103,** 13962–13967.

Milon, P., Konevega, A., Peske, F., Fabbretti, A., Gualerzi, C. O., and Rodnina, M. V. (2007). Transient kinetics, fluorescence, and FRET in studies of bacterial initiation. *Meth. Enzymol.* **430,** 1–30.

Moerke, N. J., Aktas, H., Chen, H., Cantel, S., Reibarkh, M. Y., Fahmy, A., Gross, J. D., Degterev, A., Yuan, J., Chorev, M., Halperin, J. A., and Wagner, G. (2007). Small-molecule inhibition of the interaction between the translation initiation factors eIF4E and eIF4G. *Cell* **128,** 257–267.

Petersen, H. U., Roll, T., Grunberg-Manago, M., and Clark, B. F. (1979). Specific interaction of initiation factor IF2 of *E. coli* with formylmethionyl-tRNA^{fMet}. *Biochem. Biophys. Res. Commun.* **91,** 1068–1074.

Poehlsgaard, J., and Douthwaite, S. (2005). The bacterial ribosome as a target for antibiotics. *Nat. Rev. Microbiol.* **3,** 871–881.

Pohlmann, J., and Brotz-Oesterhelt, H. (2004). New aminoacyl-tRNA synthetase inhibitors as antibacterial agents. *Curr. Drug Targets Infect. Disord.* **4,** 261–272.

Projan, S. J. (2003). Why is big Pharma getting out of antibacterial drug discovery? *Curr. Opin. Microbiol.* **6,** 427–430.

Sambrook, J., and Russell, D. W. (2001). "MolecularCloning: A Laboratory Manual." Cold Spring Harbor Laboratory Press, Cold Spring Harbor, NY.

Shastry, M., Nielsen, J., Ku, T., Hsu, M. J., Liberator, P., Anderson, J., Schmatz, D., and Justice, M. C. (2001). Species-specific inhibition of fungal protein synthesis by sordarin: Identification of a sordarin-specificity region in eukaryotic elongation factor 2. *Microbiology* **147,** 383–390.

Shuman, S. (2001). The mRNA capping apparatus as drug target and guide to eukaryotic phylogeny. *Cold Spring Harb. Symp. Quant. Biol.* **66,** 301–312.

Søe, R., Mosley, R. T., Justice, M., Nielsen-Kahn, J., Shastry, M., and Merrill, A. R. Andersen, G. R. (2007). Sordarin derivatives induce a novel conformation of the yeast ribosome translocation factor eEF2. *J. Biol. Chem.* **282,** 657–666.

Spurio, R., Severini, M., La Teana, A., Canonaco, M. A., Pawlik, R. T., Gualerzi, C. O., and Pon, C. L. (1993). Novel structural and functional aspects of translational initiation factor IF2. *In* "The Translational Apparatus" (K. H. Nierhaus, A. R. Subramanian, V. A. Erdmann, F. Franceschi, and B. Wittmann-Liebold, eds.), pp. 241–252. Plenum Publishing Corp., New York, NY.

Spurio, R., Brandi, L., Caserta, E., Pon, C. L., Gualerzi, C. O., Misselwitz, R., Krafft, C., Welfle, K., and Welfle, H. (2000). The C-terminal sub-domain (IF2 C-2) contains the entire fMet-tRNA binding site of initiation factor IF2. *J. Biol. Chem.* **275,** 2447–2454.

Walsh, C. (2003a). Where will new antibiotics come from? *Nat. Rev. Microbiol.* **1,** 65–70.

Walsh, C. (2003b). "Antibiotics. Actions, origins, resistance." ASM Press, Washington DC.

Wang, J., Soisson, S. M., Young, K., Shoop, W., Kodali, S., Galgoci, A., Painter, R., Parthasarathy, G., Tang, Y. S., Cummings, R., Ha, S., and Dorso, K., *et al.* (2006). Platensimycin is a selective FabF inhibitor with potent antibiotic properties. *Nature* **441,** 358–361.

Young, K., Jayasuriya, H., Ondeyka, J. G., Herath, K., Zhang, C., Kodali, S., Galgoci, A., Painter, R., Brown-Driver, V., Yamamoto, R., Silver, L. L., and Zheng, Y., *et al.* (2006). Discovery of FabH/FabF inhibitors from natural products. *Antimicrob. Agents Chemother.* **50,** 519–526.

CHAPTER THIRTEEN

IDENTIFYING SMALL MOLECULE INHIBITORS OF EUKARYOTIC TRANSLATION INITIATION

Regina Cencic,[*,1] Francis Robert,[*,1] *and* Jerry Pelletier[*,†]

Contents

[*] Department of Biochemistry, McGill University, Montreal, Quebec, Canada
[†] McGill Cancer Center, McGill University, Montreal, Quebec, Canada
[1] Equal contribution by both authors

Methods in Enzymology, Volume 431
ISSN 0076-6879, DOI: 10.1016/S0076-6879(07)31013-6

Abstract

In eukaryotes, translation initiation is rate-limiting with much regulation exerted at the ribosome recruitment and ternary complex (eIF2•GTP•Met-tRNA$_i^{Met}$) formation steps. Although small molecule inhibitors have been extremely useful for chemically dissecting translation, there is a dearth of compounds available to study the initiation phase *in vitro* and *in vivo*. In this chapter, we describe reverse and forward chemical genetic screens developed to identify new inhibitors of translation. The ability to manipulate cell extracts biochemically, and to compare the activity of small molecules on translation of mRNA templates that differ in their factor requirements for ribosome recruitment, facilitates identification of the relevant target.

1. INTRODUCTION

1.1. Translation initiation

A significant portion of our current understanding of the translation process is a consequence of utilizing small molecules to chemically dissect this complex process (Pelletier and Peltz, 2007; Pestka, 1977; Vazquez, 1979). Translation also offers therapeutic opportunities that cannot be achieved using the conventional drug discovery paradigm of enzyme inhibitor screening. Mechanistic and structural differences between eukaryotic and prokaryotic ribosomes are the basis for the selectivity of some of the most potent anti-microbial compounds in clinical use (Pelletier and Peltz, 2007). Sophisticated high throughput screens (HTS) have identified compounds that promote readthrough of nonsense mutations, one of which is in clinical trials for the treatment of genetic disorders, such as cystic fibrosis and Duchenne muscular dystrophy (Welch *et al.*, 2007). Chemical suppressors of ribosomal frameshifting that exhibit antiviral activity have also been identified using a forward chemical genetic screen (Hung *et al.*, 1998).

The ribosome recruitment step of translation initiation is generally rate-limiting and under control of several regulatory mechanisms (Holcik and Sonenberg, 2005). Most eukaryotic cellular mRNA translation is thought to occur by a cap-dependent process, catalyzed by the eukaryotic initiation factor (eIF) 4F complex (Fig. 13.1). eIF4F is composed of three subunits: (i) eIF4E, the cap-binding protein responsible for binding of the complex to the cap structure, (ii) eIF4A, a DEAD-box RNA helicase thought to unwind local RNA structure and facilitate access of the 43S ribosomal

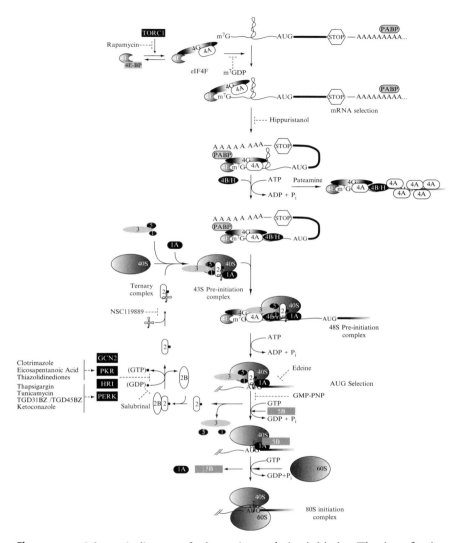

Figure 13.1 Schematic diagram of eukaryotic translation initiation. The sites of action of small molecule inhibitors are shown with dashed lines. Kinases that affect the phosphorylation of 4E-BP and eIF2α, and exert effects on ribosome recruitment and ternary complex formation, respectively, are shown in a black box. See text for details.

complex to the mRNA template and, (iii) eIF4G, a modular scaffold that binds eIF4E and eIF4A and recruits the ribosome to the mRNA via its interactions with eIF3. The RNA binding proteins, eIF4B and eIF4H, modulate eIF4A activity by increasing its affinity for RNA (Rogers *et al.*, 2002) (Fig. 13.1). Once bound to the mRNA, the 40S ribosome and associated initiation factors are thought to scan the 5′ untranslated region (UTR) until the appropriate initiation codon is reached. After binding of the 60S ribosomal subunit to form an 80S complex, polypeptide chain elongation can commence (Fig. 13.1) (Kapp and Lorsch, 2004).

Formation of eIF2•GTP•Met-tRNA$_i^{Met}$ ternary complex is also highly regulated (Holcik and Sonenberg, 2005). The Met-tRNA$_i^{Met}$ is recruited to the 40S ribosome in association with eIF2 and GTP (Fig. 13.1). In response to different types of stress (e.g., hypoxia, amino acid starvation, heat shock, viral infection, ultraviolet light), the α-subunit of eIF2 becomes phosphorylated at residue Ser51. This event prevents the exchange of GDP for GTP during recycling of the eIF2 complex by eIF2B and inhibits the translation of most mRNAs (Hinnebusch, 2000). The phosphorylation of eIF2 α is mediated by four kinases—heme-regulated inhibitor kinase (HRI), protein kinase RNA (PKR), PKR-like endoplasmic reticulum (ER) kinase (PERK), and general control non-derepressible-2 (GCN2) (Fig. 13.1) (Kapp and Lorsch, 2004).

A small number of viral and cellular mRNAs recruit ribosomes via an alternative, cap-independent mechanism (Fig. 13.2) (Holcik and Korneluk, 2000). These mRNAs contain a *cis*-acting structural element known as an internal ribosome entry site (IRES). IRESes directly recruit ribosomes, bypassing the requirement for the mRNA 5′ cap structure and eIF4E, and allowing translation to proceed under physiological conditions under which cap-dependent translation is impaired. They provide valuable tools to characterize small molecule inhibitors of initiation, since mechanistic differences among IRESes can be attributable to IRES trans-acting factors (ITAFs) and eIF requirements (Fig. 13.2) (Pestova *et al.*, 2001; Pisarev

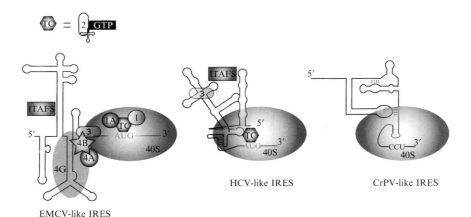

Figure 13.2 Initiation factor requirement for EMCV-, HCV-, and CrPV-like IRESes. See text for details.

et al., 2005). The encephalomyocarditis virus (EMCV) and poliovirus IRESes require the same set of translation initiation factors as capped mRNAs for ribosome recruitment, with the exception of eIF4E and the amino-terminal domain of eIF4G to which eIF4E binds (Fig. 13.2) (Pestova *et al.*, 1996a,b). On the other hand, the hepatitis C virus (HCV) IRES does not require eIF4F, eIF4B, eIF5, eIF1, or eIF1A for formation of the 48S pre-initiation complex (Fig. 13.2). Although 40S ribosomes can directly and specifically bind to the HCV IRES (Pestova *et al.*, 1998), binding of the eIF2-ternary complex leads to correct positioning of the 40S ribosome at the AUG of the IRES. The exact role of eIF3 is still unclear, but its binding to the IRES appears essential for 80S formation (Fig. 13.2) (Kieft *et al.*, 2001; Odreman-Macchioli *et al.*, 2000; Pestova *et al.*, 1998; Sizova *et al.*, 1998). Last, the intergenic region IRES of cricket paralysis virus (CrPV) does not require any of the canonical translation factors for ribosome recruitment (Fig. 13.2) (Jan and Sarnow, 2002; Pestova and Hellen, 2003; Pestova *et al.*, 2004; Schuler *et al.*, 2006; Wilson *et al.*, 2000).

1.2. Translation initiation and cancer

Deregulation of global translation and selective mRNA expression have emerged as important components of cancer etiology and genetic modifiers of the chemosensitivity response. Of all the translation factors, eIF4E is the one that is most implicated in cancer. Overexpression of eIF4E in NIH-3T3 cells is transforming and likely a consequence of increased translational activity that leads to deregulated cell growth (Lazaris-Karatzas *et al.*, 1990). Increased expression of eIF4E in preclinical mouse cancer models accelerates tumorigenesis (Ruggero *et al.*, 2004; Wendel *et al.*, 2004) and modulates chemosensitivity (Wendel *et al.*, 2004). Consistent with this, eIF4E can suppress apoptosis (Li *et al.*, 2003, 2004; Polunovsky *et al.*, 1996) and elevated levels are found in many human cancers (Rosenwald, 2004). As well, eIF4E has been proposed as an independent prognostic tumor marker in breast cancer (Li *et al.*, 1997, 2002). Overexpression of 4E-BP1 (a negative regulator of eIF4E) in eIF4E transformed cells can partially reverse tumorigenicity (Rousseau *et al.*, 1996) and transformed rat fibroblasts expressing an antisense eIF4E mRNA are less tumorigenic when injected into mice (De Benedetti *et al.*, 1991).

1.3. Translation initiation inhibitors

The discovery and characterization of novel small molecule inhibitors that target the ribosome recruitment step of translation initiation is extremely important in order to validate translation initiation as a chemotherapeutic target. Rapamycin, an inhibitor of TOR (target of rapamycin) complex I

(TORCI), affects initiation by altering the availability of eIF4E to assemble in the eIF4F complex (Fig. 13.1). Clotrimazole (Aktas *et al.*, 1998), eicosapentaenoic acid (Palakurthi *et al.*, 2000), and thiazolidinediones (Palakurthi *et al.*, 2001) all appear to affect intracellular Ca^{2+} stores, resulting in phosphorylation of eIF2α by PKR (Fig. 13.1). These compounds show promising results in mouse cancer models (Aktas *et al.*, 1998; Benzaquen *et al.*, 1995; Wendel *et al.*, 2004, 2006) and clinical trials (Faivre *et al.*, 2006). Tunicamycin, an inhibitor of protein glycosylation, and thapsigargin, a compound that discharges intracellular Ca^{2+} stores, both cause phosphorylation of eIF2α in a PERK-dependent fashion (Fig. 13.1) (Harding *et al.*, 2000).

Until recently, there were only a few direct inhibitors of translation initiation known. Cap analogues (such as m^{7}GDP) prevent eIF4E from recognizing the mRNA 7-methyl guanosine cap structure and are useful for studying cap-dependent ribosome recruitment (Fig. 13.1). Their inability to cross the cell membrane, however, limits their application to *in vitro* studies. The inhibitor edeine acts downstream of the ribosome recruitment phase by preventing AUG recognition by 40S ribosomes (Fig. 13.1) (Kozak and Shatkin, 1978). Pactamycin was previously reported to inhibit initiation (Tai *et al.*, 1973; Vazquez, 1979), but other data has questioned this mode of action and suggests that pactamycin inhibits translocation instead (Dinos *et al.*, 2004).

Chemical genetic screens have yielded novel inhibitors of translation initiation. A search for chemical suppressors of ER stress-induced apoptosis identified salubrinal, a small molecule that prevents eIF2α dephosphorylation (Fig. 13.1) (Boyce *et al.*, 2005). TGD31BZ and TGD45BZ were isolated from a small molecule screen in a search for activators of the integrated stress response (ISR) and demonstrated a link between sterol flux and regulation of translation initiation via eIF2α phosphorylation (Fig. 13.1) (Harding *et al.*, 2005). Ketonazole, a known inhibitor of cholesterol biosynthesis and acting at the same step as TGD31BZ and TGD45BZ, also activates the ISR (Harding *et al.*, 2005). A forward chemical genetic screen identified hippuristanol and pateamine, two natural product inhibitors of eIF4A that have very different mechanisms of action (Bordeleau *et al.*, 2005, 2006). Hippuristanol inhibits the RNA binding activity of eIF4A and has been useful for determining the eIF4A requirement of IRESes for initiation, whereas pateamine acts as a chemical inducer of dimerization by stimulating binding of eIF4A to RNA (Bordeleau *et al.*, 2005). In 2006, NSC119889, an inhibitor of ternary complex formation, was identified and used to demonstrate that the HCV IRES is refractory to reduced ternary complex availability (Fig. 13.1) (Robert *et al.*, 2006b). There is an unmet need for additional inhibitors of translation initiation to better understand this complex process. Herein, we describe chemical genetic screens to identify such compounds.

2. A REVERSE CHEMICAL GENETIC ASSAY PROBING EIF4E:EIF4G AND EIF4A:EIF4G INTERACTIONS

2.1. Principle of the approach

The complexity of translation initiation indicates that a large number of reverse chemical genetic assays can be established to identify inhibitors of specific steps. These screens can range from (i) targeting protein–protein interactions implicated in translation initiation (Fig. 13.1); (ii) inhibition of RNA recognition by several of the eIFs (cap recognition by eIF4E, RNA binding by eIF4B or eIF4H, tRNA binding by eIF2, rRNA binding by eIF3, etc.); and (iii) enzyme-based assays (such as scoring for eIF4A helicase or eIF5B GTPase activity). One of the current challenges in chemical biology is to find small molecules that can be used to block protein–protein interactions because most proteins exert their function as members of protein complexes or show altered activities when present in different complexes (Berg, 2003).

eIF4F contains one of two related but modular scaffolding proteins— eIF4GI and eIF4GII. These share 46% identity at the amino acid level and all structural features described for eIF4GI are present in eIF4GII (Gradi *et al.*, 1998; Imataka *et al.*, 1998; Pyronnet *et al.*, 1999). The amino terminal fragment of eIF4G has a small ~10 amino acid region that interacts with eIF4E (Fig. 13.3A) (Lamphear *et al.*, 1995; Mader *et al.*, 1995). A domain at the amino terminus of eIF4G also interacts with the poly A binding protein

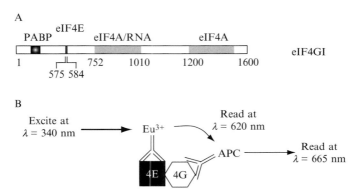

Figure 13.3 TR-FRET assay to monitor eIF4E:eIF4G interaction. (A) Schematic representation of eIF4GI with various domains highlighted, including PABP (radial shaded box), eIF4E (dark grey box), and eIF4A/RNA (light grey box) interacting regions. (B) Schematic diagram of the TR-FRET based assay between eIF4E and eIF4G for monitoring protein–protein interaction.

(PABP) and this interaction is thought to mediate circularization of the mRNA during translation (Fig. 13.1) (Imataka *et al.*, 1998). The middle domain of eIF4G contains binding sites for eIF3 and eIF4A (Imataka and Sonenberg, 1997; Lamphear *et al.*, 1995) and possesses RNA binding activity (Pestova *et al.*, 1996a). The carboxy-terminus of eIF4G contains a second, independent binding site for eIF4A (Fig. 13.3A) (Imataka and Sonenberg, 1997; Lamphear *et al.*, 1995).

The association of eIF4E and eIF4G is regulated via competitive inhibition by 4E-BPs (there are three highly related proteins in mammals) that form inhibitory heterodimers with eIF4E (Fig. 13.1). The $Y-X_4-L-\Phi$ (Φ denotes a hydrophobic amino acid) eIF4E recognition motif, present in both 4E-BP and eIF4G, adopts an L-shaped, extended chain/α-helical conformation and forms tight complexes ($K_D = 10^{-8}M$) with eIF4E (Fletcher and Wagner, 1998; Marcotrigiano *et al.*, 1999). The interaction between eIF4G and eIF4E extends beyond the eIF4E recognition motif with eIF4G forming a right-handed helical ring wrapped around the N-terminus of eIF4E (Gross *et al.*, 2003; Hershey *et al.*, 1999; von der Haar *et al.*, 2006). Although residues from eIF4GI (517–606) contact eIF4E over a large surface area spanning 4400 $\mathring{A}^2$ (Gross *et al.*, 2003; Hershey *et al.*, 1999; von der Haar *et al.*, 2006), oligopeptides containing the $Y-X_4-L-\Phi$ motif are capable of inhibiting cap-dependent translation (Fletcher *et al.*, 1998). In addition, single amino acid mutations within, and adjacent to, the $Y-X_4-L-\Phi$ motif can dramatically impact on binding to eIF4E (Cencic *et al.*, 2007; Marcotrigiano *et al.*, 1999; Poulin *et al.*, 1998), suggesting the presence of "hot spots" within the protein–protein interaction interface. In principle, if such hot spots could be targeted by small organic molecules, the interaction between eIF4E and 4E-BP or eIF4G should be destabilized and translation initiation inhibited.

The binding of eIF4E to 4E-BPs is regulated by phosphorylation (Fig. 13.1). Exposure of cells to hormones and mitogens increases the phosphorylation of 4E-BPs, decreases their affinity for eIF4E, and allows eIF4E to assemble into the eIF4F complex, with concomitant stimulation of translation (for a review, see Gingras *et al.*, 2001). Conversely, deprivation of nutrients or growth factors leads to dephosphorylation of 4E-BP1, increased sequestration of eIF4E from eIF4G, and a decrease in translation initiation (Fig. 13.1). Phosphorylation of 4E-BP1 is mediated by TORCI and regulated translation initiation by altering the availability of eIF4E for assembly into the eIF4F complex (Gingras *et al.*, 2001).

eIF4A can exist as a free form (eIF4A$_f$) and as a subunit of eIF4F (eIF4A$_c$), and is thought to recycle through the eIF4F complex during translation initiation (Rogers *et al.*, 2002; Yoder-Hill *et al.*, 1993). eIF4A$_f$ is incorporated into the eIF4F complex via its interaction with eIF4G (Lamphear *et al.*, 1995; Pause *et al.*, 1994). eIF4G is hypothesized to act as a clamp that stabilizes the closed "active" conformation of eIF4A (Lomakin *et al.*, 2000). This is consistent with the observation that the helicase activity of $eIF4A_c$ is approximately 20-fold higher than that of $eIF4A_f$ and that the middle domain of eIF4G is capable of stimulating eIF4A ATPase activity (Imataka and Sonenberg, 1997; Korneeva et al., 2001).

Here, we discuss the development of time-resolved fluorescence resonance energy transfer (TR-FRET)-based HTS assays for the discovery of small molecules that block the interaction between $eIF4E:eIF4GI_{517-606}$, $eIF4E:eIF4GII_{555-658}$, eIF4E:4E-BP1, $eIF4A:eIF4GI_{688-1023}$, and $eIF4A:eIF4GI_{1203-1600}$ (Fig. 13.1) (Cencic et al., 2007). TR-FRET utilizes time-gated fluorescence intensity measurements to quantitate molecular association or dissociation events. FRET involves energy transfer between a fluorophore and a chromophore. The event is distance and orientation dependent and requires an overlap of the donor emissions and acceptor absorption spectra (Fig. 13.3B). TR-FRET assays are ideally suited for monitoring protein–protein interactions because they can be of homogenous format, are rapid, and minimize background fluorescence interference. The most commonly used labels are the long-lived lanthanide europium (Eu^{3+}) (donor) and the short-lived acceptor protein allophycocyanin (APC). For convenience, these labels are introduced into an HTS assay by using anti-epitope tag antibodies to which they are conjugated. This minimizes the need to chemically modify the protein of interest, but nonetheless, it requires the two protein pairs to contain epitope tags (Fig. 13.3B). Because the emission from the europium complex is long-lived, a time delay is implemented before reading the fluorescence to reduced fluorescence interference from small organic molecules in library collections (Pope, 1999). A general flowchart that we have used to develop TR-FRET assays is presented in Fig. 13.4.

2.2. Material and equipment

Expression vectors: For expression of recombinant His_6-eIF4E and His_6-$eIF4E_{W73A}$ [a mutant that no longer interacts with eIF4G or 4E-BP (Ptushkina et al., 1998; Pyronnet et al., 1999)], the plasmids pProExHta-eIF4Ewt and pProExHta-$eIF4E_{W73A}$ were utilized (kind gift of N. Sonenberg, McGill University, Montreal). Construction of pGEX5X1-$4GI_{517-606}$, pGEX5X1-$4GII_{555-658}$, and pGEX6P-1/h4E-BP1 has been described (Cencic et al., 2007). Plasmids pMA311 and pMA312, which express GST-$eIF4GI_{688-1023}$ and GST-$eIF4GI_{1230-1600}$, respectively, were a kind gift of Dr. S. Tahara (USC-Keck School of Medicine, Los Angeles, CA). The eIF4AI expression vector, pET15b/4AI, has been previously described (Bordeleau et al., 2005). The plasmid pGEX6P1-HMK-$Paip1_{145-415}His$, which encodes GST-$Paip1_{145-415}$-His_6, was a kind gift of N. Sonenberg (McGill University, Montreal).

Affinity resins: m^7GDP-agarose matrix for purification of His_6-eIF4E was prepared as previously reported (Edery et al., 1988). For purification of

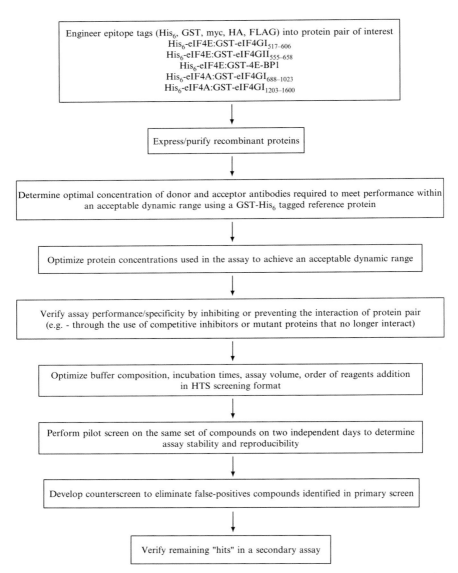

Figure 13.4   Flow chart for the development of TR-FRET assays for HTS. See text for details.

GST- or His$_6$-tagged proteins, Glutathione Sepharose™ 4B beads (GE Healthcare) or Ni^{++}-NTA Agarose (Qiagen) were utilized, respectively. Recombinant His$_6$-eIF4E was stored in 20 mM Hepes-KOH (pH 7.5), 0.2 mM EDTA (pH 8.0), 100 mM KCl. Recombinant eIF4A and eIF4G proteins were stored in 50 mM Tris-HCl (pH 8.0), 50 mM NaCl, 1 mM EDTA (pH 8.0), 5 mM dithiothreitol (DTT), 5% glycerol.

TR-FRET buffer: 20 mM Hepes-KOH (pH 7.4), 100 mM KCl, 1 mM DTT, 0.015% Tween 20, 1 μg/ml IgG (Sigma)

Peptides: Peptides harboring either the eIF4E recognition motif from 4E-BP1 (wt: NH_2- PGGTRIIYDRKFLMECRNSP-COOH) or a mutant with reduced binding affinity (mut: NH_2- PGGTRII**A**DRKF**S**-**S**ECRNSP-COOH; amino acid changes are indicated in bold) were purchased from BioSynthesis Inc. (Lewisville, TX).

Antibodies: The LANCE™ Eu-W1024 labeled anti-6xHis antibody and anti-GST IgG antibody conjugated to SureLight®-APC are from Perkin Elmer (Woodbridge, Ontario). We found the Eu-W1024 labeled anti-6xHis antibody to be very stable when stored in small aliquots at −20° in a nondefrosting freezer. The anti-GST APC-conjugated IgG antibody is not as stable in our hands. It is delivered lyophilized and should be resuspended to a final concentration of 1 mg/ml. Under these conditions, it remains stable at 4° for a maximum of 3 months. We have found that the energy transfer is more efficient when the APC-conjugated antibody is not used immediately after resuspension, but rather following storage at 4° for a few days.

Screening Plates: HE Microplates 96 Black PS (Molecular Devices).

Hardware and Software: The FRET signal was measured using an Analyst HT reader (LJL Biosystems). Data collection using the "Criterion Host" software (LJL Biosystems) involved setting the Z height at 1 mm and utilizing 1 excitation filter (330/80) and 2 emission filters (620/7.5 and 665/7.5). A dichroic filter with a wavelength of 400 nm was used. For the measurement at 620 nm, we employed 100 readings per well, with 10 ms between reading, integration time of 1000 μs, a delay time of 200 μs, and 1000 μs integration time for the fluorescence emission recording. The parameters for the measurement at 665 nm were the same as for 620 nm, except for an integration time of 150 μs and a delay time of 50 μs.

Calculations: To determine the signal (S) to background (B) ratio, $(S/B) = M_{signal}/M_{background}$, where M_{signal} and $M_{background}$ are the mean of the signal and background readings, respectively. The Z' factor is a useful way of assessing the statistical performance of an assay (Zhang *et al.*, 1999) and is defined as $Z' = 1 - 3 \times (SD_{positive\ ctrl} + SD_{background})/|M_{positive\ ctrl} - M_{background}|$, where $SD_{positive\ ctrl}$ and $SD_{background}$ are the standard deviation of the readings obtained with the positive and negative controls, respectively. $M_{positive\ ctrl}$ and $M_{background}$ are the mean of the readings obtained with the positive and negative controls, respectively. The Z factor is a second useful statistical parameter that is used to assess the performance of an assay during small-scale testing and is defined as $Z = 1 - 3 \times (SD_{population} + SD_{background})/|M_{population} - M_{background}|$, where $SD_{population}$ and $M_{population}$

are the standard deviation and mean of the data obtained upon exposure of the assay to a set of small molecules. For both, the Z' and Z factor a value >0.5 defines the assay as being suitable for HTS conditions. The coefficient of variation is given by $CV = 100 \times SD/M$ (%) and provides an excellent assessment of assay stability, precision of liquid handling, and instrument reading, and should be $<10\%$.

2.3. Preparation of recombinant protein

A convenient way of introducing FRET donors and acceptors to determine protein proximity is via labeled antibodies directed against epitope tags present on the protein pair of interest. A combination of fusions should be tested in which the tags are placed at the NH_2- or COOH-termini to determine the protein pair that yields the highest S/B ratio, and hence affords the greatest sensitivity. We have used NH_2-terminal His_6-tagged eIF4E and eIF4A and GST-tagged 4E-BP and eIF4G fragments in our TR-FRET assays.

Recombinant proteins were expressed in *Escherichia coli* strain BL21 (DE3)pLysS, grown in 1 liter of LB containing ampicillin and chloramphenicol at $37°$ to an OD_{600} of 0.6. Expression was induced for 3 h at $30°$ by the addition of 1 mM IPTG. The cells were harvested by centrifugation and frozen once. After thawing, the cell pellets were resuspended in 10 ml storage buffer (for eIF4E: 20 mM Hepes-KOH (pH 7.5), 0.2 mM EDTA (pH 8.0), 100 mM KCl; for eIF4A and eIF4G: 50 mM Tris-HCl (pH 8.0), 50 mM NaCl, 1-mM EDTA (pH 8.0), 5 mM DTT, 5% glycerol). Cells were sonicated 4 times at 50% intensity on ice (samples continuously kept on ice, 40 s each time with 1 min waiting intervals) and centrifuged at 17,000 g for 30 mins at $4°$. Recombinant His_6-eIF4A was incubated end-over-end overnight at $4°$ with Ni^{++}-NTA agarose (Qiagen), washed with 10 resin volumes of storage buffer, and eluted with 50 mM Tris (pH 8.0) containing 100 mM imidazole. Recombinant GST-eIF4G proteins were incubated end-over-end overnight at $4°$ with glutathione Sepharose 4B (GE Healthcare), washed with 10 resin volumes of storage buffer, and eluted with 10 mM reduced glutathione in 50 mM Tris-HCl (pH 8.0). The eluted proteins were concentrated, and imidazole and reduced glutathione were removed from the buffer by exchanging the buffer using Amicon Ultra-4 Centrifugal Filter Devices [cut off 10,000 Da from Fisher (Nepean, Ontario)]. His_6-eIF4E was expressed in the same manner, but purified by m^7GDP chromatography under native conditions to selectively enrich for recombinant protein that is functional for cap recognition (Edery *et al.*, 1988). These procedures typically yielded approximately 1 to 2 mg recombinant protein per liter of culture. All protein preparations were stored as aliquots at $-70°$ and can be freeze–thawed multiple times with no noticeable precipitation.

2.4. Assay development

2.4.1. Optimizing the concentration of Eu^{3+} and APC-conjugated antibodies

The double-tagged GST-Paip1$_{145-415}$-His$_6$ fusion protein provides an excellent positive control for establishing a new TR-FRET assay involving Eu-W1024 labeled anti-6xHis and anti-GST APC-conjugated IgG antibodies. To determine an initial starting working concentration of antibody to be used in assay development, a two-way matrix in a 96-well plate is set up in which the concentrations of Eu^{3+} labeled anti-6xHis and anti-GST APC-conjugated IgG antibodies are varied from 1 to 5 nM and 10 to 100 nM, respectively. Reactions are set up in 40 μl with 75 nM GST-Paip1$_{145-415}$-His, performed in triplicate, and incubated for 3 h at RT. Once a minimal working concentration of antibody is determined, GST-Paip1$_{145-415}$-His$_6$ is replaced by different concentrations of each specific protein partner under consideration for assay development (see later). After determining a concentration and ratio of the protein pairs that provides a good S/B ratio, a second series of finer titrations with the antibodies can be performed in an attempt to lower the amount of these expensive reagents. A final concentration of 1 nM Eu^{3+} labeled anti-6xHis and 50 nM anti-GST APC-conjugated IgG gave a robust S/B ratio within dynamic range when present in reactions containing His$_6$-eIF4E:GST-eIF4GI$_{517-606}$, His$_6$-eIF4E:GST-eIF4GII$_{555-658}$, and His$_6$-eIF4E:GST-4E-BP1. For reactions with His$_6$-eIF4AI:GST-eIF4GI$_{688-1023}$ or His$_6$-eIF4AI:GST-eFI4-GII$_{1203-1600}$, a concentration of 1 nM Eu^{3+} labeled anti-6xHis and 100 nM anti-GST APC-conjugated IgG yielded an optimal S/B.

2.4.2. Optimizing protein concentrations

The concentration of each protein partner that will yield the highest S/B ratio in the TR-FRET assay needs to be empirically determined. This is performed in 96-well plates containing 40-μl reactions in TR-FRET buffer, in which the concentration of each protein is independently varied from 10 nM to 200 nM. In the case of His$_6$-eIF4E:GST-eIF4GI$_{517-606}$, the best S/B ratio was achieved using 10 nM His$_6$-eIF4E and 40 nM GST-eIF4GI$_{517-606}$. For His$_6$-eIF4E:GST-eIF4GII$_{555-658}$ protein partners, 20 to 80 nM of His$_6$-eIF4E and 10 nM of GST-4GII$_{555-658}$ were optimal, whereas for His$_6$-eIF4E:GST-4E-BP1, 80 nM His$_6$-eIF4E and 10 nM 4E-BP1 gave the best S/B. The TR-FRET signal for His$_6$-eIF4A:GST-eIF4GI$_{688-1023}$ was optimal when 20 nM of His$_6$-eIF4A and 40 nM of GST-eIF4GI$_{688-1023}$ were used, whereas for the interaction between His$_6$-eIF4A and GST-eIF4GI$_{1203-1600}$, 100 nM of His$_6$-eIF4A and 200 nM of GST-eIF4GI$_{1203-1600}$ had to be used to give the best S/B ratio.

2.4.3. Assay optimization

Biochemical assays used in HTS campaigns have a higher requirement for stability than those in other areas of research because they have to consistently maintain low variability and a high S/B. In addition, the economics of running HTS campaigns dictates that minimum amounts of reagents be consumed. As such, a number of parameters need to be empirically tested to optimize a particular assay for HTS. The nature of these parameters will differ from assay to assay and for the particular target under consideration, but in general, they will include testing the performance of plates from different sources and the effects of time, volume (loss of liquid due to evaporation over time), chelating agents, reducing agents, glycerol or PEG, detergents, and biochemical parameters (buffer, pH, ionic strength, temperature, divalent cation requirement). In addition, the need for carrier proteins (BSA, IgG), requirement for protease inhibitors, DMSO tolerability, and pre- and post-incubation times should also be investigated. A thorough discussion of these points has been previously presented (Macarron and Hertzberg, 2002). Also, there may be specific issues that relate to a particular biological target under study. Case in point is the finding that cap binding by eIF4E enhances the affinity of eIF4E for 4E-BP and for peptides harboring the eIF4E recognition motif (Shen *et al.*, 2001; Tomoo *et al.*, 2005). In our hands, addition of m^7GDP to TR-FRET assays involving eIF4E and eIF4G did not improve the S/B ratio.

2.4.4. Use of selective inhibitors as assay development tools

It is essential to demonstrate that the observed TR-FRET signal is a consequence of a specific interaction between the protein partners under study. This can be assessed by the use of protein mutants that do not interact when tested in parallel reactions. Additionally, inhibition of the interaction can be achieved with specific inhibitors. As specificity controls in TR-FRET reactions involving eIF4E, we use His_6-$eIF4E_{W73A}$, which has significantly reduced affinity for eIF4GI and eIF4GII (Ptushkina *et al.*, 1998; Pyronnet *et al.*, 1999). For His_6-$eIF4A$:GST-$eIF4GI_{688-1023}$ and His_6-$eIF4A$:GST-$eIF4GI_{1203-1600}$ TR-FRET assays, we use the GST-$eIF4GI_{517-606}$ fragment, which does not interact with eIF4A, as negative control.

For His_6-$eIF4E$:GST-$eIF4GI_{517-606}$, we have also shown that a peptide harboring the eIF4E recognition motif from 4E-BP1 can inhibit the TR-FRET signal in a dose-dependent fashion, whereas a mutant peptide does not when present in the binding reactions at a final concentration of 20 μM (Cencic *et al.*, 2007). The use of a competitive inhibitor is also a useful tool to determine whether the order of addition of compound affects the ability to detect an inhibitor. The simplest assay design is one where both proteins are present in a master mix, which is subsequently dispensed into assay plates, followed by addition of compound, incubation for a set period of time, and direct reading of the signal, with no need for further manipulations. However, compounds may be more effective if present before both protein partners are mixed and the use of a specific inhibitor can allow one to experimentally test this. For His_6-eIF4E:GST-eIF4GI$_{517-606}$, addition of competitive peptide, either during the setting up of the TR-FRET assay or following a 30-min pre-incubation of His_6-eIF4E and GST-eIF4GI$_{517-606}$, indicated little difference in the ability of the peptide to inhibit the eIF4E: GST-eIF4GI$_{517-606}$ interaction (Cencic et al., 2007). Hence, addition of compounds to plates containing predispensed His_6-eIF4E and GST-eIF4GI$_{517-606}$ should be able to detect competitive inhibitors. The use of a selective inhibitor also facilitates determining the relative separation of the signal and background data points in order to determine the Z' factor and CV value. In the HTS screen that we undertook for inhibitors of His_6-eIF4E: GST-eIF4GI$_{517-606}$, we obtained a Z' value of 0.75 and a CV of 6.7%.

2.4.5. Determining assay performance in pilot screens

The final test before starting an HTS campaign is to assess the performance of the assay with a small set of compounds. We typically perform our pilot screens on a set of 960 compounds that represent a set of known therapeutic agents (Gen-Plus collection from MicroSource Discovery Systems, Inc., Gaylordsville, CT). The assay is performed on two different days and the data used to calculate the Z factor and determine the hit rate and reproducibility of the assay. For our HTS screen for inhibitors of His_6-eIF4E: GST-eIF4GI$_{517-606}$, we obtained a Z value of 0.6 (Cencic et al., 2007).

2.4.6. Counterscreens and secondary confirmatory assays

In order to determine whether compounds identified in the primary HTS screen are specific, a counterscreen is required to identify and eliminate false positives that will arise in the primary screen. For protein–protein interaction screens, it is preferable to test an unrelated protein pair that uses the same mode of detection. For our purposes, we adapted a previously described TR-FRET assay that monitors the interaction between bacterial Staphylococcus aureus DnaI and phage protein 77ORF104 (Liu et al., 2004).

A secondary confirmatory assay usually monitors the same activity as in the primary screen but utilizes a different readout. Its purpose is to provide assurance that the small molecules that have passed the counterscreen are behaving as expected. One convenient assay that monitors eIF4E:eIF4G interaction is an ELISA-based assay developed by Kimball et al. (2004). In this sandwich immunoassay, an anti-eIF4E monoclonal antibody (Santa Cruz Biotechnology, Inc.) is used to coat an ELISA microtiter plate. eIF4E is then added, followed by eIF4G, whose presence is detected with a rabbit anti-eIF4G antibody. Utilization of a horseradish peroxidase-labeled anti-rabbit antibody provides sufficient sensitivity to visualize approximately 5 ng of eIF4G.

Once candidate "hits" have been identified, it is important to confirm the identity of the compounds, either by resupply (if possible) or resynthesis. It is not uncommon for compounds in large library collections to decompose, to be impure, or to be mislabeled (despite the guarantees often provided by commercial suppliers).

2.5. Screening for inhibitors of eIF4E:eIF4G$_{517-606}$ interaction

We have performed small molecule screening for His$_6$-eIF4E:GST-eIF4GI$_{517-606}$ inhibitors in 96-well plates, using 10-μl reaction volumes. Although the following protocol focuses on His$_6$-eIF4E:GST-eIF4G interaction, it can also apply to the other protein–protein interactions described previously (His$_6$-eIF4E:GST-eIF4GII$_{555-658}$, His$_6$-eIF4E:GST-4E-BP1, His$_6$-eIF4A:GST-eIF4GI$_{688-1023}$, and His$_6$-eIF4A:GST-eIF4GI$_{1203-1600}$). Here, we describe detailed procedures of our screen.

1. Recombinant His$_6$-eIF4E and GST-eIF4GI$_{517-606}$ proteins are quickly thawed, then diluted into ice-cold TR-FRET buffer to a final concentration of 10 and 40 nM, respectively, and kept on ice. For twelve 96-well plates, we prepare 10.5 ml of master mix containing 3.5 μg of His$_6$-eIF4E and 17 μg GST-eIF4GI$_{517-606}$, 1 nM Eu^{3+} labeled anti-6xHis, and 50 nM APC-labeled anti-GST IgG. One hundred twenty microliters of this master mix are dispensed into 84 wells (column 1, rows A–D; columns 2–11 rows A–H) of a master plate on ice. A 500-μl negative control reaction containing 10 nM of His$_6$-eIF4E$_{W73A}$ instead of His$_6$-eIF4E, is prepared and 120 μl are dispensed in column 12, rows E to H. Two other 500-μl control reactions containing either 0.8 μM wt-BP1 peptide or 0.8 μM mut-BP1 peptide in the presence of His$_6$-eIF4E and GST-eIF4GI$_{517-606}$ are prepared and 120 μl dispensed in column 1, rows E to H and column 12, rows A to D, respectively. Ten microliters are transferred from the master plate into assay plates using a Biomek FX liquid handling station (Beckman Coulter). Pipetting is performed at RT and the entire process takes ~30 min before addition of compounds.
2. Compound libraries are stored at −20° in 96-well polypropylene plates as 10-mM stocks in DMSO. For high-throughput screening, daughter plates are prepared in which compounds are diluted to 2 mM in DMSO. Compounds (100 nl) are transferred into assay plates using stainless steel pins (V&P Scientific) attached to the Biomek FX (Walling *et al.*, 2001), yielding a final compound and DMSO concentration of 20 μM and 1%, respectively. Between transfers, pins are washed with RNAse-free deionized water, ethanol, and 100% DMSO, and dried on a clean 3MM Whatman paper.
3. Plates are incubated at RT for 3 h, after which time they are read using an Analyst HT reader. Data analysis is performed by calculating the ratio

of 665 nm/620 nm and the S/B for all wells, where the background is taken to be the mean 665 nm/620 nm obtained from the four negative control reactions (column 12, Rows E–H: His_6-eIF4E$_{W73A}$: GST-eIF4GI$_{517-606}$). Subsequently, the standard deviation of the population within one plate is determined and the cutoff for hits set to three times the standard deviation from the mean of the population [M_{pop}-3 × SD]. Compounds identified as hits are further analyzed by examining the value of the donor emission (620 nm), which should increase (since the energy is not being absorbed by an acceptor), compared to the value of the donor emission (620 nm) obtained with the positive control (column 1, rows A–D). In addition, the values of the acceptor emission (665 nm) should decrease, indicating the absence of energy transfer. An autofluorescent compound in the chemical library that interferes with the TR-FRET signal can easily be eliminated because its signals at both donor and acceptor emission wavelengths will not lie between those values observed with the positive and the negative controls. Clearly, the TR-FRET assay will not detect highly fluorescent compounds that are true inhibitors of the interaction under study.

3. A FORWARD CHEMICAL GENETIC SCREEN TO IDENTIFY INHIBITORS OF EUKARYOTIC TRANSLATION

3.1. Principle of the approach

The application of forward chemical genetics to studies of translation provides an opportunity to identify small molecules that inhibit or stimulate this process without any underlying assumptions as to which step is most amenable to targeting by the chemical libraries under consideration. The opportunity exists to identify novel factors involved in translation, unravel new activities of known translation initiation factors, or characterize short-lived intermediates that are "frozen" by the small molecule inhibitor. We have undertaken a forward chemical genetic approach to identify small molecules that inhibit or stimulate translation in extracts prepared from Krebs-2 ascites cells (Novac *et al.*, 2004). These screens have led to the identification of several novel inhibitors of translation initiation and elongation (Bordeleau *et al.*, 2005, 2006; Robert *et al.*, 2006a,b).

3.2. Material and equipment

In Vitro Transcription Vectors: Plasmids pKS/FF/EMCV/Ren•p(A)$_{51}$, pKS/FF/HCV/Ren•p(A)$_{51}$, pcDNA3/Ren/P2/FF, and pGL3/Ren/CrPV/FF have been described previously (Fig. 13.5A) (Bordeleau *et al.*, 2005, 2006;

Novac *et al.*, 2004). Plasmids pKS/FF/EMCV/Ren•p(A)$_{51}$, pKS/FF/ HCV/Ren•p(A)$_{51}$, and pGL3/Ren/CrPV/FF are linearized with *Bam* HI, whereas pcDNA3/Ren/P2/FF is linearized with *Xho* I.

10× Transcription buffer: For T7 RNA polymerase (New England Biolabs): 400 mM Tris–HCl (pH 7.9), 60 mM MgCl$_2$, 20 mM spermidine (Sigma), 100 mM DTT. For T3 RNA polymerase (Invitrogen): 400 mM Tris–HCl (pH 7.5), 100 mM MgCl$_2$, 50 mM DTT, 500 μg/ml BSA (Sigma), 10 mM spermidine (Sigma).

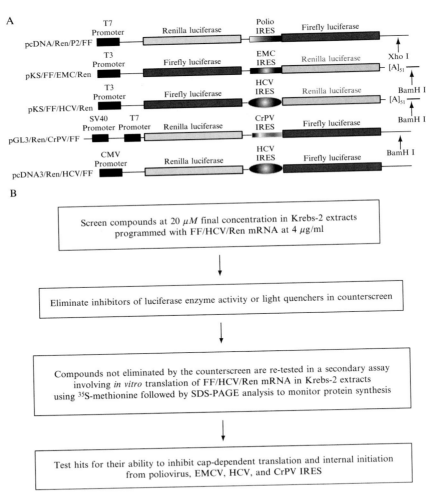

Figure 13.5 (A) Schematic representation of expression plasmids used for *in vitro* transcriptions and *in vivo* expression. (B) Flow chart for the processing of compounds identified in the forward chemical genetic screen used to identify novel translation inhibitors.

Water: We utilize deionized, double distilled RNAse-free water.

dNTPs: 10 mM of ATP, CTP, UTP, and GTP (Amersham Biosciences) with pH adjusted to 7.0 using KOH.

Cap Analogue: 10 mM m^{7}G(5′)ppp(5′)G (New England Biolabs) in water.

RNase inhibitor: RNAguard (GE Healthcare).

RNA Polymerases: T7 RNA polymerase (New England Biolabs) and T3 RNA polymerase (Invitrogen).

Translation extracts: Rabbit reticulocyte lysates (RRL) (Promega), wheat germ (WG) extract (Promega), bacterial S30 extract (Promega). Extracts from Krebs-2 cells were prepared as described (Svitkin and Sonenberg, 2004).

Radioactivity: ^{3}H-CTP (20.3 Ci/mmol) (Perkin-Elmer), ^{35}S-L-Methionine (1175 Ci/mmol), translation grade: 40 μM solution (Easytag; Perkin Elmer).

10× Krebs Translation Mix: 125 mM Hepes-KOH (pH 7.3), 10 mM ATP, 2 mM GTP, 2 mM CTP, 2 mM UTP, 100 mM creatine phosphate (dipotassium salt, Calbiochem), 200 μM of 19 L–amino acids minus methionine (Sigma).

10× Krebs Salt Mix: 500 mM potassium acetate, 7.5 mM MgCl$_2$, 2.5 mM spermidine. The potassium and magnesium concentrations can be varied in the 10× salt mix to optimize for translation of the particular mRNA under study, since the optimum for these can vary among mRNAs (Svitkin and Sonenberg, 2004).

Methionine (Sigma): 40 μM in RNAse-free deionized water.

Creatine Phosphokinase (CPK) (rabbit skeletal muscle; Calbiochem): Resuspended with deionized RNAse free water to a final concentration of 14 mg/ml and stored at −20°.

Luciferase stabilization buffer: 70 mM Hepes-KOH (pH 7.7), 7 mM MgSO$_4$, 3 mM DTT, 10 μg/ml BSA.

Firefly and Renilla luciferase assay reagents: These are prepared as described (Dyer *et al.*, 2000). Firefly reagent: 25 mM glycylglycine, 15 mM K$_x$PO$_4$ (pH 5.1), 4 mM EGTA, 2 mM ATP, 1 mM DTT, 15 mM MgSO$_4$, 0.1 mM coenzyme A, 75 μM luciferine (Promega). The final pH is adjusted to 8.0.

Renilla reagent: 100 mM NaCl, 2.2 mM Na$_2$EDTA, 220 mM K$_x$PO$_4$ (pH 5.1), 0.44 mg/ml BSA, 1.3 mM NaN$_3$, 1.43 μM coelenterazine (Calbiochem). The final pH is adjusted to 5.0.

Anisomycin (Sigma): Stock solution of 10 mM in DMSO stored at −20°.

Luciferase assay tubes: 5-ml translucent plastic tubes for luciferase assays (75 mm × 12 mm, Sarstedt).

Fluostar Optima (BMG Labtech).

Multidrop (Titertek multidrop 384).

Acetate plate seals (Thermo labsystems #3501).

HTS 96-well conical plates: Translucent 96-well plates with conical bottoms (Costar #3897).

HTS Plates for Luciferase Readout: White 96-well flash plates (Nunc 96-micro flat bottom well plates).

Micrococcal Nuclease (Roche): 1 mg/ml (15,000 U/ml). Stored as 50-μl aliquots at $-20°$.

3.3. Assay development

1. Design of the bicistronic reporter mRNA: For HTS screening purposes, we initially used an SP6 promoter-based vector (Novac *et al.*, 2004), but have switched to using pKS/FF/HCV/Ren (Fig. 13.5). We chose to use the firefly and renilla luciferase cistrons, since their products can be assayed sequentially using a simple, cost-effective method that avoids expensive commercial kits (Dyer *et al.*, 2000). Translation of the first cistron is cap-dependent, since we wanted the opportunity to identify inhibitors of cap-dependent ribosome recruitment. Titration with cap analogues was performed to confirm this (Novac *et al.*, 2004). We also were interested in identifying inhibitors of HCV translation initiation; therefore, the HCV IRES was used as the intercistronic region in our bicistronic mRNA (Novac *et al.*, 2004). We additionally engineered a 3' poly(A) tail into our construct because, in eukaryotes, the poly(A) tail and the cap structure cooperate to synergistically enhance translation (Tarun and Sachs, 1996). Titration of the mRNA in Krebs-2 extracts indicated that 4 μg/ml was sufficient to give a S/B of approximately 20 to 50, while maintaining cap-poly(A) synergy.

2. Preparation of Krebs-2 translation extracts: Krebs-2 extracts are an ideal system to screen for compounds that inhibit translation because they faithfully recapitulate the cap dependency and the cap-poly(A) synergism associated with eukaryotic mRNA translation (Svitkin and Sonenberg, 2004), unlike standard rabbit reticulocyte lysates (RRL) (Borman *et al.*, 2000). Furthermore, the translation of many types of IRESes is supported in Krebs-2 extracts. The use of commercially available translation competent extracts prepared from RRL, wheat germ, and *E. coli* is extremely useful in assessing selectivity of inhibitors identified in primary screens.

3.4. *In vitro* synthesis of reporter mRNA

Following linearization of plasmids (Fig. 13.5), the DNA is phenol:chloroform extracted, passed through a Sephadex G50 spin column, and precipitated with 2.5 volume of ethanol and $\frac{1}{10}$th volume of $3M$ NaOAc (pH 5.2). The pelleted DNA is washed with 70% ethanol, dried, and resuspended in water at a final concentration of 1 μg/μl and stored at $-20°$.

Transcription reactions are set up by pipetting 10 μCi H^3-CTP (supplied in 50% ethanol) into an Eppendorf tube and drying down to a volume of about 2 μl using a Sc100 Speed Vac (Savant Instrument Inc.) to eliminate the ethanol. Transcription reactions of 100 μl contain 10 μl of 10 $\times$ transcription buffer, 0.5 mM of each ATP, CTP, and UTP, 0.1 mM of GTP, 0.5 mM m^{7}G (5$'$)ppp(5$'$)G, 100 U RNAguard, 3 μg of linearized plasmid DNA, and 150 U of the RNA polymerase. Reactions are performed at 37° for 2 h, after which the reaction is phenol:chloroform extracted, passed through a Sephadex G50 spin column, and precipitated with one-fifth volume of 10M NH$_4$OAc and 2.5 volumes of ethanol. The RNA is recovered by centrifugation, washed with 70% ethanol, and resuspended in water. One microliter is taken for quantitation by scintillation counting and the mRNA is quantitated, based on the expected yield of 1.1 $\times$ 10^5 cpm/μg.

3.5. Screening for inhibitors of translation

A general flowchart is presented in Fig. 13.5B that we followed for identifying and sorting inhibitors of translation. Shown below is an *in vitro* translation protocol tailored for ten 96-well assay plates (800 compounds), which can be scaled up or down as required. Negative and positive controls are present in wells A1 to D1 and E1 to H1, respectively. Compounds are added to wells A2 to H11. Column 12 is left blank and could be used for additional controls, if desired.

1. Krebs-2 extracts (0.5-ml aliquots) are quickly thawed and placed on ice. Our Krebs-2 extracts are in 50 mM KCl, 50 mM Hepes(pH 7.3), 4.5 mM MgCl$_2$, 4 mM DTT, 100 mM KOAc. For 10 assay plates, 5.5 ml of extract is required because it is present in the master mix at 50% (v/v). A master mix (11 ml) is prepared with 50% (v/v) Krebs-2 extracts (5.5 ml), 10$\times$ translation mix (1.1 ml), 10$\times$ salt mix (1.1 ml), 0.77 ml of 14 mg/ml CPK, 0.55 ml of 40 μM methionine, and mRNA (FF/HCV/Ren) to a final concentration of 4 μg/ml. A 96-well master plate is set up such that each well contains 125 μl of master mix (wells in column 12 are not filled). The master mix in wells A1 to D1 is supplemented with DMSO to a final concentration of 1% (negative control), whereas wells E1 to H1 are supplemented with anisomycin to a final concentration of 50 μM (positive control).

2. Ten microliters of ice-cold master mix is distributed to a 96-well conical plate using a multidrop. Compounds (0.1 μl) are transferred from 2 mM (100% DMSO) daughter plates into translation reactions using stainless steel pins. Between transfers, pins are washed with RNAse-free deionized water, ethanol, and 100% DMSO, and dried on a clean 3 MM Whatman paper.

3. Plates are sealed with an acetate sheet and incubated in a 30° dry air incubator. They are not stacked, but rather make full contact with the metal shelving in the incubator. After 1 h, plates are removed and 25 μl of luciferase stabilization buffer is added to each well using the multidrop. At this point, the plates are either frozen at $-20°$ or the luciferase activity immediately measured. Firefly and renilla luciferase activity is determined by transferring 25 μl of the translation reaction into flash plates using the Biomek FX. Readings are performed with a plate-reading luminometer equipped with two reagent injectors (Fluostar Optima). Dual activity measurements are made by injecting 50 μl firefly luciferase reagent waiting for 2 s, followed by a 10 s measurement period. Fifty microliters of renilla luciferase reagent is then injected, followed by a 2 s delay and a 10 s measurement period.

4. We have found that many compounds identified in our screen are nonspecific inhibitors of luciferase enzyme activity. To eliminate these, we test the "hits" in a luciferase enzyme-based counterscreen. Firefly and renilla luciferase are produced *in vitro* by programming Krebs-2 extracts with FF/HCV/Ren mRNA and allowing the translations to proceed at 30° for 1 h. Ten microliters are then pipetted into a 96-well plate and compound is added to a final concentration of 20 μM (1% DMSO). Luciferase activity is then determined as described previously in step 3. Since compound is added only after the translation reaction is complete, inhibitors of translation should not score positively in this assay. Typically, a 1-ml *in vitro* translation reaction is sufficient to screen 45 candidate hits in duplicate for nonspecific luciferase inhibitory activity. Compounds that inhibit in this counterscreen are eliminated from future analysis.

5. To confirm the activity of potential hits that are not eliminated by the counterscreen, we tested the ability of the compounds to block incorporation of ^{35}S-methionine in Krebs-2 extracts programmed with FF/HCV/sRen mRNA. For this purpose, Krebs-2 extracts are first treated with micrococcal nuclease to eliminate endogenous mRNA. The treatment is performed by adding 1.5 μl of 100 mM CaCl$_2$ and 2 μl of micrococcal nuclease (15 U/μl) to 150 μl of Krebs-2 extracts and incubating at RT for 20 min. The reaction is terminated by the addition of 1.5 μl of 200 mM EGTA and translations are performed using the same reagent composition as described in step 1 previously, with the exception that cold methionine is replaced by 0.5 μl of ^{35}S-translation grade methionine. Translation products are visualized by loading 8 μl of the translation reaction on a 10% polyacrylamide gel, followed by Coomassie staining and destaining (overnight). The gel is treated with En^{3}Hance (Perkin Elmer), dried, and exposed to Kodak X-ray film. The renilla and firefly luciferase proteins migrate at 36 kDa and 61 kDa, respectively.

4. Characterization of Inhibitors of Translation Identified in Chemical Genetic Screens

4.1. Compound sensitivity toward different IRESes

Compounds that pass the counterscreen and secondary assays are then tested for their effects on viral IRES-mediated translation. We test the inhibitory potential of compounds on Krebs-2 extracts programmed with Ren/P2/FF, FF/EMCV/Ren, FF/HCV/Ren, or Ren/CrPV/FF mRNA (Fig. 13.5A). Because the factor requirement of poliovirus, EMCV, HCV, and CrPV IRESes differs for ribosome recruitment (Fig. 13.2), the sensitivity of these IRESes to a specific compound can provide information on the likely biological target(s). In addition to Krebs-2 extracts, we test our compounds for activity in RRL, wheat germ, and bacterial S30 extracts. The latter provides us with an assessment of whether the target is conserved in eukaryotes and prokaryotes. We follow the manufacturer's recommendation for translation when using these extracts (Promega).

4.2. Testing for nucleic acid binding

We have previously found that nucleic acid intercalators can inhibit *in vitro* translation reactions, with some IRESes being more sensitive than others (Malina *et al.*, 2005). It is important therefore to identify such compounds early in the HTS campaign. One assay that we have utilized for this purpose is to measure the change in viscosity of a DNA solution in the presence of small molecule. In this method, three milliliters of 1 mM salmon sperm DNA is incubated for 10 min in the presence of increasing concentrations of compound. The viscosity of each solution is determined by calculating the time required for the solution to flow through a fixed distance in a capillary using an Ostwald viscometer (Technical Glass Products, Inc., New Jersey) (Zimmerman, 1967).

4.3. Monitoring initiation complex formation by sedimentation velocity centrifugation

Our screen has the potential to identify inhibitors of cap-dependent initiation, IRES-mediated initiation, and translation elongation or termination. One assay to identify initiation inhibitors from hits obtained in the primary screen is to assess the ability of a given compound to prevent 48S and/or 80S initiation complex formation on a capped mRNA and on the HCV IRES. Because initiation complexes are formed more efficiently in RRL than in

Krebs-2 extracts, we use the former for this purpose and it is important to confirm that the hits inhibit translation in RRL.

We analyze pre-initiation complexes and initiation complexes as follows:

1. We use the plasmid pSP/CAT cut with the enzyme *Pvu* II as template for *in vitro* transcriptions. This produces an mRNA that is approximately 150 nucleotides in length (Gorman, 1985). For the HCV IRES, we use the plasmid phRL-null, in which the HCV IRES has been cloned into the *Nhe* I site, upstream of the renilla cistron (Robert *et al.*, 2006b). *In vitro* transcriptions are performed as described previously, except that 5 μCi α^{32}P-GTP replaces ^{3}H-CTP in the reaction to label the mRNA.

2. Prepare 10 to 30% glycerol (containing 20 mM Hepes-KOH (pH 7.6), 500 mM NaCl, 30 mM Mg(CH$_3$COO)$_2$, 2 mM DTT) gradients in polyallomer tubes (14 mm $\times$ 95 mm; Beckman) using a gradient-maker apparatus (Hoeffer Scientific Instruments) connected to a peristaltic pump. Twelve milliliter gradients are prepared in these tubes and are stored at 4° until needed. We utilize gradients on the same day that they are prepared.

3. RRL (Promega) is quickly thawed and placed on ice. Twenty microliter reaction mixtures are prepared containing 70% (v/v) RRL, 0.5 μl amino acid mix (minus methionine), 80 mM KOAc, 1.6 μM methionine, 1 mM GMP-PNP (for 48S pre-initiation complex) or 0.6 mM cyclo-heximide (for 80S complex formation), and 20 μM of the small molecule hit under study.

4. The reactions are pre-incubated at 30° for 5 min, followed by the addition of 200,000 cpm of ^{32}P-labeled mRNA, and the incubation continued for 10 min longer at 30°. The reaction is terminated by addition of 75 μl HSB (20 mM Hepes-KOH (pH 7.6), 500 mM NaCl, 30 mM Mg(CH$_3$COO)$_2$, 2 mM DTT) and carefully loaded on the top of glycerol gradients, which are subsequently centrifuged for 3.5 h at 200,000g (39,000 rpm) in an SW40 rotor. We collect 10 drops/fraction, and the 48S and 80S complexes generally peak at fractions 7 to 8 and 11 to 13, respectively, under these conditions.

5. After centrifugation, fractions from the gradients are collected by piercing the bottom of the tube with a Brandel Tube Piercer and introducing a 60% sucrose solution using a peristaltic pump. The gradient is pushed out the top of the tube, which is directly connected to a UV detector [UA-6 UV Detector (ISCO)] and a fraction collector. The 60% pushing solution contains 0.05% (w/v) bromophenol blue that allows one to monitor the interface of pushing solution and sucrose gradient.

6. The radioactivity in each fraction is measured by Cerenkhov counting in a LS650 Multipurpose scintillation counter (Beckman) to determine the distribution of ^{32}P-labeled mRNA throughout the gradient.

4.4. Monitoring cap-dependent RNA binding of eIF4F, eIF4A, and eIF4B

Inhibitors of eIF4F, eIF4A, or eIF4B are expected to impair the ribosome recruitment phase of translation and should not inhibit initiation directed from the HCV or CrPV IRESes (Figs. 13.1 and 13.2). A direct way to monitor the interaction of these factors with mRNA is the chemical cross-linking assay, originally used to identify eIF4E (Sonenberg et al., 1978). The vicinal alcohols present at the 5′ (and 3′) most ribose groups of a cap-labeled mRNA are oxidized to dialdehydes using sodium periodate (Sonenberg and Shatkin, 1977). Schiff bases formed between amino groups in proteins and the RNA are stabilized by reduction using sodium cyanoborohydride (Sonenberg and Shatkin, 1977). The RNA is degraded with RNAse A and because the radiolabel is positioned in the vicinity of the cap structure, only proteins crosslinked to the cap will be visualized following separation on SDS-PAGE and autoradiography. Parallel incubations are performed in the presence of m^{7}GDP (as specific competitor) or GDP (as nonspecific competitor) to identify proteins selectively interacting with the cap structure. Cytoplasmic and nuclear cap binding proteins have been identified using this approach (Goyer et al., 1989; Rozen and Sonenberg, 1987; Sonenberg, 1981; Sonenberg et al., 1978). In this assay, crosslinking of eIF4E is ATP-independent, whereas that of eIF4A and eIF4B is ATP-dependent (Sonenberg, 1981).

Our crosslinking assays are performed as described here.

1. mRNA capping: ^{3}H-labeled in vitro transcribed mRNA is generated as described previously, with the exception that m^{7}GpppG is omitted and the GTP concentration is increased from 0.1 mM to 0.5 mM. After ethanol precipitation, the RNA is resuspended to 1 μg/μl in sterile water. Fifty microliter capping reactions containing 2 μg ^{3}H-RNA, 41U RNAguard (34 U/μl), 2 μl 5 mM SAM [prepared by mixing 10 μl of a 32-mM solution (10% ethanol and 5 mM H$_2$SO$_4$) with 54 μl of 50 mM Tris-HCl (pH 8.0)], 5 μl α-^{32}P-GTP (10 μCi/μl), 1 μl of recombinant capping enzyme (Luo et al., 1995), and 5 μl of 10× capping buffer (10 × capping buffer: 500 mM Tris-HCl (pH 8.0), 12.5 mM MgCl$_2$, 25 mM DTT). The reaction is incubated at 37° for 1 h and then terminated by the addition of EDTA to a final concentration of 18 mM, 10 μg yeast tRNA, and water up to 100 μl. This is followed by phenol/chloroform extraction, purification through a G50 spin column, and ethanol precipitation with 2M NH$_4$OAc. The RNA pellet is washed with 70% ethanol and resuspended in 400 μl 100 mM NaOAc (pH 5.3), 10 mM EDTA.

2. Oxidation of the cap structure: To 400 μl of ^{32}P-cap-labeled mRNA, 1.6 μl of 100 mM NaIO$_4$ is added and left on ice for 2 h, protected from light. The reaction is terminated by addition of 20 μl of 50% glycerol, passed through a G50 spin column and ethanol precipitated. The pelleted

RNA is resuspended in 20 μl of deionized, RNAse-free water. An aliquot is taken for scintillation counting and we generally obtain 50,000 cpm/μg.

3. Chemical crosslinking: The crosslinking reaction is performed in a final volume of 30 μl containing 1× crosslinking buffer [10× crosslinking buffer is 250 mM Hepes-KOH (pH 7.5), 700 μM GTP, 90 mM creatine phosphate (Calbiochem), 110 μM of each L-amino acids minus methionine, 20 mM DTT, 2 mM spermidine (Sigma), 0.6 mM PMSF, 5 mM Mg (OAc)$_2$], 22 μg of CPK, 0.9 mM ATP, 45,000 cpm of ^{32}P cap-labeled, oxidized mRNA, and 20 μg of ribosomal HSW [prepared as described in Pestova *et al.*, 1996a]. The reaction is incubated at 30° for 10 min followed by the addition of 3 μl of 200 mM NaBH$_3$CN (prepared fresh in water) and the incubation continued at 4° overnight. The following day, 10 μg of RNAse A is added and the reaction is incubated for 30 min at 30°. Parallel reactions contain m^{7}GDP or GDP to a final concentration of 0.6 mM. Test reactions with inhibitors will contain compound to a final concentration of 20 μM. The proteins are separated by SDS-PAGE, the gels dried, and radiolabeled proteins visualized by autoradiography.

4.5. Monitoring compound inhibition *in vivo*

We utilize three independent methods to assess a compound's effect *in vivo* on translation—metabolic ^{35}S-methionine/cysteine labeling, polysome profile analysis, and effects on transfected reporter gene activity.

4.5.1. Monitoring ^{35}S-methionine/cysteine incorporation

1. In one set of experiments a titration of compound is performed to assess its potency *in vivo*. HeLa cells are maintained in DMEM supplemented with 10% fetal bovine serum (FBS) at 37° in 5% CO$_2$. One day prior to labeling, the cells are seeded in 24-well plates at approximately 60,000 cells per well. The next day, cells are washed with warm (37°) PBS and the medium replaced with 250 μl of methionine-free DMEM containing 10% dialyzed serum (Invitrogen). After a 15-min incubation at 37°, different concentrations of compound are added to the cells (which can range from 1 nM to 50 μM) and the incubation continued for another 45 min. Anisomycin is used as a positive control at a final concentration of 50 μM. Fifty-five microcuries of ^{35}S-methionine/cysteine [^{35}S-methionine/cysteine express protein labeling mix (1175 Ci/mmol) (Perkin-Elmer)] is added to each well (220 μCi/ml) and the incubation continued for another 15 min.

2. Alternatively, the time of incubation in the presence of compound can be varied, with labeling performed for 15 min at the end of the incubation period.

3. To quantitate the amount of ^{35}S-methionine/cysteine incorporation, the medium is removed, the cells washed with PBS, and overlaid with 250 μl of warm trypsin. After a 4-min incubation at room temperature (which is sufficient to detach the cells, as verified by microscopy), the trypsinized cells are transferred to a 1.5-ml Eppendorf tube, and centrifuged at 2000g at 4°. The supernatant is carefully removed after centrifugation and the cells are placed on dry ice for 5 min.

4. After thawing, the pellet is resuspended in 12 μl of RIPA buffer [10 mM Tris-HCl (pH 7.2), 150 mM NaCl, 0.1% SDS, 1.0% Triton X-100, 1% sodium deoxycholate, 5 mM EDTA].

5. Square pieces (0.5 cm $\times$ 0.5 cm) of Whatmann 3MM paper are pre-blocked with 50 μl of 50$\times$ MEM amino acid solution (Invitrogen) and dried under an Infra-Radiator lamp (Fisher). Whatmann squares are labeled with a pencil, 5 μl of cell lysate is spotted onto each square, and the filter is dried.

6. Spotted filters are placed in a beaker containing 10% TCA+0.1% methionine for 20 min at 4°. The solution is then replaced with 5% TCA and boiled for 15 min. Following 1-min washes each with 5% TCA and 100% ethanol, the filters are dried under an Infra-Radiator lamp. Once dry, they are placed into scintillation vials containing 5 ml of scintillation liquid (MP Biomedicals) and the radioactivity contained in each vial is measured in an LS650 Multipurpose scintillation counter. The results are normalized to the concentration of total protein present in the extract. For this purpose, we use the remaining 6 μl of cell lysate to quantitate the protein concentration using the DC protein assay reagent (Biorad).

4.5.2. Polysome profile analysis

We monitor the ability of compounds to disrupt polysomes from cells, as follows.

1. One day prior to the isolation of polysomes, approximately 5 million HeLa cells are plated into 10-cm dishes. The following day, the media is replaced with fresh DMEM and compound is added to a concentration previously determined to inhibit translation *in vivo* by ^{35}S-methionine metabolic labeling (see previously).

2. Prior to harvesting, the cells are washed twice with 5 ml of cold PBS containing 100 μg/ml of cycloheximide. Cycloheximide binds to the 60S ribosomal subunit and interferes with E site function when it contains deacylated tRNA (Obrig *et al.*, 1971; Pestova and Hellen, 2003), preventing ribosome runoff and improving polysome recovery. After washing, the cells are scraped with a rubber policeman in 1 ml of PBS/cycloheximide solution, transferred to an Eppendorf tube, and harvested by centrifugation for 10 min at 2000g at 4°. The cell pellet is resuspended in 425 μl of hypotonic lysis buffer [5 mM Tris-HCl

(pH 7.5), 2.5 mM MgCl$_2$, 1.5 mM KCl], followed by the addition of 5 μl of 10 mg/ml cycloheximide and 10 μl of 100 mM DTT. After brief vortexing (3 s), 25 μl of 10% Triton X-100 and 25 μl of 10% sodium deoxycholate are added, the sample vortexed again (3 s), and centrifuged for 2 min at 14,000g at 4°.

3. The supernatant is then loaded onto a prechilled 10 to 50% sucrose gradient [20 mM Hepes-KOH (pH 7.6), 100 mM KCl, 5 mM MgCl$_2$ containing either 10 or 50% (w/v) sucrose] and centrifuged for 2 h at 160,000g (35,000 rpm) at 4° in an SW40 rotor. The gradient is collected using a Brandel tube piercer with pushing of the gradient through an ISCO UV [UA-6 UV Detector (ISCO)] detector with constant monitoring of the OD$_{260}$.

4.5.3. *In vivo* cap-dependent and cap-independent translation

We use two bicistronic plasmids to monitor the effects of translation initiation inhibitors *in vivo*—pcDNA-Ren-HCV-FF (Bordeleau *et al.*, 2006) and pGL3-Ren/CrPV/FF (generously given by Dr Peter Sarnow) (Fig. 13.5A).

1. Hela cells are transfected with either pcDNA-Ren-HCV-FF or pGL3-Ren/CrPV/FF in 10-cm plates using Lipofectamine-plus reagent as described by the manufacturer (Invitrogen). Twenty-four hours posttransfection, the cells are trypsinized and re-plated in 24-well plates at 5×10^5 cells/well.

2. The next day (48 h posttransfection), the medium is replaced with fresh DMEM containing different concentrations of compound (nM–μM concentration range). After 12 h of incubation, the cells are washed with PBS, 40 μl of luciferase lysis buffer [100 mM KxPO$_4$ (pH 7.8), 0.2% Triton X-100] is added to each well, and the plate is incubated for 15 min at room temperature with gentle rocking. The cell extract is transferred into Eppendorf tubes and kept on ice.

3. Ten microliters of cell lysate is read for firefly and renilla luciferase activities, as described previously (Section 3.5). The results are normalized to the protein content of each cell lysate.

5. DISCUSSION AND CONCLUDING REMARKS

Using the approaches described herein, we have identified and characterized a number of translation initiation and elongation inhibitors (Bordeleau *et al.*, 2005, 2006; Robert *et al.*, 2006a,b). Two of these, hippuristanol and pateamine, target the helicase eIF4A, whereas a third, NSC119889, inhibits ternary complex formation (Fig. 13.1). Hippuristanol inhibits the RNA binding activity of eIF4A, whereas pateamine stimulates this property. Both compounds inhibit translation and inhibit translation of IRES-driven reporters that require eIF4A for ribosome recruitment (Bordeleau *et al.*, 2005, 2006), making them useful tools to quickly determine eIF4A dependency *in vitro* and *in vivo*. NCS119889 has been useful in demonstrating a reduced dependency for ternary complex availability during HCV-mediated translation initiation. Even though the exact implication of translation in cancer development is not entirely defined, inhibiting this process shows promise for cancer therapy, and clearly it will be exciting to assess whether small molecules identified by screens, such as the ones described herein, show activity in preclinical cancer models.

ACKNOWLEDGMENTS

We are grateful to Lisa Lindqvist for helpful comments on the manuscript. R. C. is supported by a CIHR Cancer Consortium Post-doctoral Training Grant Award. Work in the laboratory is supported by grants from the Canadian Institutes of Health Research, National Cancer Institute of Canada, and NIH (CA114475).

REFERENCES

Aktas, H., Fluckiger, R., Acosta, J. A., Savage, J. M., Palakurthi, S. S., and Halperin, J. A. (1998). Depletion of intracellular Ca2+ stores, phosphorylation of eIF2alpha, and sustained inhibition of translation initiation mediate the anticancer effects of clotrimazole. *Proc. Natl. Acad. Sci. USA* **95**, 8280–8285.

Benzaquen, L. R., Brugnara, C., Byers, H. R., Gatton-Celli, S., and Halperin, J. A. (1995). Clotrimazole inhibits cell proliferation *in vitro* and *in vivo*. *Nat. Med.* **1**, 534–540.

Berg, T. (2003). Modulation of protein–protein interactions with small organic molecules. *Angew Chem. Int. Ed. Engl.* **42**, 2462–2481.

Bordeleau, M. E., Matthews, J., Wojnar, J. M., Lindqvist, L., Novac, O., Jankowsky, E., Sonenberg, N., Northcote, P., Teesdale-Spittle, P., and Pelletier, J. (2005). Stimulation of mammalian translation initiation factor eIF4A activity by a small molecule inhibitor of eukaryotic translation. *Proc. Natl. Acad. Sci. USA* **102**, 10460–10465.

Bordeleau, M.-E., Mori, A., Oberer, M., Lindqvist, L., Chard, L. S., Higa, T., Belsham, G. J., Wagner, G., Tanaka, J., and Pelletier, J. (2006). Functional characterization of IRESes by an inhibitor of the RNA helicase eIF4A. *Nat. Chem. Biol.* **2**, 213–220.

Borman, A. M., Michel, Y. M., and Kean, K. M. (2000). Biochemical characterization of cap-poly(A) synergy in rabbit reticulocyte lysates: The eIF4G-PABP interaction increases the functional affinity of eIF4E for the capped mRNA 5′-end. *Nucleic Acids Res.* **28**, 4068–4075.

Boyce, M., Bryant, K. F., Jousse, C., Long, K., Harding, H. P., Scheuner, D., Kaufman, R. J., Ma, D., Coen, D. M., Ron, D., and Yuan, J. (2005). A selective inhibitor of eIF2alpha dephosphorylation protects cells from ER stress. *Science* **307**, 935–939.

Cencic, R., Yan, Y., and Pelletier, J. (2007). Homogenous time resolved fluorescence assay to identify modulators of cap-dependent translation initiation. *Comb. Chem. High Throughput Screen* **10**, 181–188.

De Benedetti, A., Joshi-Barve, S., Rinker-Schaeffer, C., and Rhoads, R. E. (1991). Expression of antisense RNA against initiation factor eIF-4E mRNA in HeLa cells results in lengthened cell division times, diminished translation rates, and reduced levels of both eIF-4E and the p220 component of eIF-4F. *Mol. Cell Biol.* **11,** 5435–5445.

Dinos, G., Wilson, D. N., Teraoka, Y., Szaflarski, W., Fucini, P., Kalpaxis, D., and Nierhaus, K. H. (2004). Dissecting the ribosomal inhibition mechanisms of edeine and pactamycin: The universally conserved residues G693 and C795 regulate P-site RNA binding. *Mol. Cell* **13,** 113–124.

Dyer, B. W., Ferrer, F. A., Klinedinst, D. K., and Rodriguez, R. (2000). A noncommercial dual luciferase enzyme assay system for reporter gene analysis. *Anal. Biochem.* **282,** 158–161.

Edery, I., Altmann, M., and Sonenberg, N. (1988). High-level synthesis in *Escherichia coli* of functional cap-binding eukaryotic initiation factor eIF-4E and affinity purification using a simplified cap-analog resin. *Gene* **74,** 517–525.

Faivre, S., Kroemer, G., and Raymond, E. (2006). Current development of mTOR inhibitors as anticancer agents. *Nat. Rev. Drug Discov.* **5,** 671–688.

Fletcher, C. M., McGuire, A. M., Gingras, A. C., Li, H., Matsuo, H., Sonenberg, N., and Wagner, G. (1998). 4E binding proteins inhibit the translation factor eIF4E without folded structure. *Biochemistry* **37,** 9–15.

Fletcher, C. M., and Wagner, G. (1998). The interaction of eIF4E with 4E-BP1 is an induced fit to a completely disordered protein. *Protein Sci.* **7,** 1639–1642.

Gingras, A. C., Raught, B., and Sonenberg, N. (2001). Regulation of translation initiation by FRAP/mTOR. *Genes Dev.* **15,** 807–826.

Gorman, C. (1985). High efficiency gene transfer into mammalian cells. *In* Glover, D. (Ed) "DNA cloning II. A Practical Approach," pp. 143–190. IRL Press Ltd., Oxford, UK.

Goyer, C., Altmann, M., Trachsel, H., and Sonenberg, N. (1989). Identification and characterization of cap-binding proteins from yeast. *J. Biol. Chem.* **264,** 7603–7610.

Gradi, A., Imataka, H., Svitkin, Y. V., Rom, E., Raught, B., Morino, S., and Sonenberg, N. (1998). A novel functional human eukaryotic translation initiation factor 4G. *Mol. Cell Biol.* **18,** 334–342.

Gross, J. D., Moerke, N. J., von der Haar, T., Lugovskoy, A. A., Sachs, A. B., McCarthy, J. E., and Wagner, G. (2003). Ribosome loading onto the mRNA cap is driven by conformational coupling between eIF4G and eIF4E. *Cell* **115,** 739–750.

Harding, H. P., Zhang, Y., Bertolotti, A., Zeng, H., and Ron, D. (2000). Perk is essential for translational regulation and cell survival during the unfolded protein response. *Mol. Cell* **5,** 897–904.

Harding, H. P., Zhang, Y., Khersonsky, S., Marciniak, S., Scheuner, D., Kaufman, R. J., Javitt, N., Chang, Y. T., and Ron, D. (2005). Bioactive small molecules reveal antagonism between the integrated stress response and sterol-regulated gene expression. *Cell Metab.* **2,** 361–371.

Hershey, P. E., McWhirter, S. M., Gross, J. D., Wagner, G., Alber, T., and Sachs, A. B. (1999). The cap-binding protein eIF4E promotes folding of a functional domain of yeast translation initiation factor eIF4G1. *J. Biol. Chem.* **274,** 21297–21304.

Hinnebusch, A. G. (2000). "Mechanism and Regulation of Initiator Methionyl-tRNA Binding to Ribosomes." Cold Spring Harbor Laboratories, Cold Spring Harbor, NY.

Holcik, M., and Korneluk, R. G. (2000). Functional characterization of the X-linked inhibitor of apoptosis (XIAP) internal ribosome entry site element: Role of La autoantigen in XIAP translation. *Mol. Cell Biol.* **20,** 4648–4657.

Holcik, M., and Sonenberg, N. (2005). Translational control in stress and apoptosis. *Nat. Rev. Mol. Cell Biol.* **6,** 318–327.

Hung, M., Patel, P., Davis, S., and Green, S. R. (1998). Importance of ribosomal frame-shifting for human immunodeficiency virus type 1 particle assembly and replication. *J. Virol.* **72**, 4819–4824.

Imataka, H., Gradi, A., and Sonenberg, N. (1998). A newly identified N-terminal amino acid sequence of human eIF4G binds poly(A)-binding protein and functions in poly(A)-dependent translation. *EMBO J.* **17**, 7480–7489.

Imataka, H., and Sonenberg, N. (1997). Human eukaryotic translation initiation factor 4G (eIF4G) possesses two separate and independent binding sites for eIF4A. *Mol. Cell Biol.* **17**, 6940–6947.

Jan, E., and Sarnow, P. (2002). Factorless ribosome assembly on the internal ribosome entry site of cricket paralysis virus. *J. Mol. Biol.* **324**, 889–902.

Kapp, L. D., and Lorsch, J. R. (2004). The molecular mechanics of eukaryotic translation. *Annu. Rev. Biochem.* **73**, 657–704.

Kieft, J. S., Zhou, K., Jubin, R., and Doudna, J. A. (2001). Mechanism of ribosome recruitment by hepatitis C IRES RNA. *RNA* **7**, 194–206.

Kimball, S. R., Horetsky, R. L., and Jefferson, L. S. (2004). A microtiter plate assay for assessing the interaction of eukaryotic initiation factor eIF4E with eIF4G and eIF4E binding protein-1. *Anal. Biochem.* **325**, 364–368.

Kozak, M., and Shatkin, A. J. (1978). Migration of 40 S ribosomal subunits on messenger RNA in the presence of edeine. *J. Biol. Chem.* **253**, 6568–6577.

Korneeva, N. L., Lamphear, B. J., Hennigan, F. L., Merrick, W. C., and Rhoads, R. E. (2001). Characterization of the two eIF4A-binding sites on human eIF4G-1. *J. Biol. Chem.* **276**, 2872–2879.

Lamphear, B. J., Kirchweger, R., Skern, T., and Rhoads, R. E. (1995). Mapping of functional domains in eukaryotic protein synthesis initiation factor 4G (eIF4G) with picornaviral proteases. Implications for cap-dependent and cap-independent translational initiation. *J. Biol. Chem.* **270**, 21975–21983.

Lazaris-Karatzas, A., Montine, K. S., and Sonenberg, N. (1990). Malignant transformation by a eukaryotic initiation factor subunit that binds to mRNA 5′ cap. *Nature* **345**, 544–547.

Li, B. D., Gruner, J. S., Abreo, F., Johnson, L. W., Yu, H., Nawas, S., McDonald, J. C., and DeBenedetti, A. (2002). Prospective study of eukaryotic initiation factor 4E protein elevation and breast cancer outcome. *Ann. Surg.* **235**, 732–738.

Li, B. D., Liu, L., Dawson, M., and De Benedetti, A. (1997). Overexpression of eukaryotic initiation factor 4E (eIF4E) in breast carcinoma. *Cancer* **79**, 2385–2390.

Li, S., Perlman, D. M., Peterson, M. S., Burrichter, D., Avdulov, S., Polunovsky, V. A., and Bitterman, P. B. (2004). Translation initiation factor 4E blocks endoplasmic reticulum-mediated apoptosis. *J. Biol. Chem.* **279**, 21312–21317.

Li, S., Takasu, T., Perlman, D. M., Peterson, M. S., Burrichter, D., Avdulov, S., Bitterman, P. B., and Polunovsky, V. A. (2003). Translation factor eIF4E rescues cells from Myc-dependent apoptosis by inhibiting cytochrome c release. *J. Biol. Chem.* **278**, 3015–3022.

Liu, J., Dehbi, M., Moeck, G., Arhin, F., Bauda, P., Bergeron, D., Callejo, M., Ferretti, V., Ha, N., Kwan, T., McCarty, J., Srikumar, R., Williams, D., Wu, J. J., Gros, P., Pelletier, J., and DuBow, M. (2004). Antimicrobial drug discovery through bacteriophage genomics. *Nat. Biotechnol.* **22**, 185–191.

Lomakin, I. B., Hellen, C. U., and Pestova, T. V. (2000). Physical association of eukaryotic initiation factor 4G (eIF4G) with eIF4A strongly enhances binding of eIF4G to the internal ribosomal entry site of encephalomyocarditis virus and is required for internal initiation of translation. *Mol. Cell Biol.* **20**, 6019–6029.

Luo, Y., Mao, X., Deng, L., Cong, P., and Shuman, S. (1995). The D1 and D12 subunits are both essential for the transcription termination factor activity of vaccinia virus capping enzyme. *J. Virol.* **69,** 3852–3856.

Macarron, R., and Hertzberg, R. P. (2002). "Design and Implementation of High Throughput Screening Assays." Humana Press, Totowa, NJ.

Mader, S., Lee, H., Pause, A., and Sonenberg, N. (1995). The translation initiation factor eIF-4E binds to a common motif shared by the translation factor eIF-4 gamma and the translational repressors 4E-binding proteins. *Mol. Cell Biol.* **15,** 4990–4997.

Malina, A., Khan, S., Carlson, C. B., Svitkin, Y., Harvey, I., Sonenberg, N., Beal, P. A., and Pelletier, J. (2005). Inhibitory properties of nucleic acid-binding ligands on protein synthesis. *FEBS Lett.* **579,** 79–89.

Marcotrigiano, J., Gingras, A. C., Sonenberg, N., and Burley, S. K. (1999). Cap-dependent translation initiation in eukaryotes is regulated by a molecular mimic of eIF4G. *Mol. Cell* **3,** 707–716.

Novac, O., Guenier, A. S., and Pelletier, J. (2004). Inhibitors of protein synthesis identified by a high throughput multiplexed translation screen. *Nucleic Acids Res.* **32,** 902–915.

Obrig, T. G., Culp, W. J., McKeehan, W. L., and Hardesty, B. (1971). The mechanism by which cycloheximide and related glutarimide antibiotics inhibit peptide synthesis on reticulocyte ribosomes. *J. Biol. Chem.* **246,** 174–181.

Odreman-Macchioli, F. E., Tisminetzky, S. G., Zotti, M., Baralle, F. E., and Buratti, E. (2000). Influence of correct secondary and tertiary RNA folding on the binding of cellular factors to the HCV IRES. *Nucleic Acids Res.* **28,** 875–885.

Palakurthi, S. S., Aktas, H., Grubissich, L. M., Mortensen, R. M., and Halperin, J. A. (2001). Anticancer effects of thiazolidinediones are independent of peroxisome proliferator-activated receptor gamma and mediated by inhibition of translation initiation. *Cancer Res.* **61,** 6213–6218.

Palakurthi, S. S., Fluckiger, R., Aktas, H., Changolkar, A. K., Shahsafaei, A., Harneit, S., Kilic, E., and Halperin, J. A. (2000). Inhibition of translation initiation mediates the anticancer effect of the n-3 polyunsaturated fatty acid eicosapentaenoic acid. *Cancer Res.* **60,** 2919–2925.

Pause, A., Methot, N., Svitkin, Y., Merrick, W. C., and Sonenberg, N. (1994). Dominant negative mutants of mammalian translation initiation factor eIF-4A define a critical role for eIF-4F in cap-dependent and cap-independent initiation of translation. *EMBO J.* **13,** 1205–1215.

Pelletier, J., and Peltz, S. W. (2007). "Therapeutic Opportunities in Translation." Cold Spring Harbor Laboratories, Cold Spring Harbor, NY.

Pestka, S. (1977). "Inhibitors of Protein Synthesis." Academic Press, New York, NY.

Pestova, T. V., and Hellen, C. U. (2003). Translation elongation after assembly of ribosomes on the Cricket paralysis virus internal ribosomal entry site without initiation factors or initiator tRNA. *Genes Dev.* **17,** 181–186.

Pestova, T. V., Hellen, C. U., and Shatsky, I. N. (1996a). Canonical eukaryotic initiation factors determine initiation of translation by internal ribosomal entry. *Mol. Cell Biol.* **16,** 6859–6869.

Pestova, T. V., Kolupaeva, V. G., Lomakin, I. B., Pilipenko, E. V., Shatsky, I. N., Agol, V. I., and Hellen, C. U. (2001). Molecular mechanisms of translation initiation in eukaryotes. *Proc. Natl. Acad. Sci. USA* **98,** 7029–7036.

Pestova, T. V., Lomakin, I. B., and Hellen, C. U. (2004). Position of the CrPV IRES on the 40S subunit and factor dependence of IRES/80S ribosome assembly. *EMBO Rep.* **5,** 906–913.

Pestova, T. V., Shatsky, I. N., Fletcher, S. P., Jackson, R. J., and Hellen, C. U. (1998). A prokaryotic-like mode of cytoplasmic eukaryotic ribosome binding to the initiation codon during internal translation initiation of hepatitis C and classical swine fever virus RNAs. *Genes Dev.* **12,** 67–83.

Pestova, T. V., Shatsky, I. N., and Hellen, C. U. (1996b). Functional dissection of eukaryotic initiation factor 4F: The 4A subunit and the central domain of the 4G subunit are sufficient to mediate internal entry of 43S preinitiation complexes. *Mol. Cell Biol.* **16,** 6870–6878.

Pisarev, A. V., Shirokikh, N. E., and Hellen, C. U. (2005). Translation initiation by factor-independent binding of eukaryotic ribosomes to internal ribosomal entry sites. *C R Biol.* **328,** 589–605.

Polunovsky, V. A., Rosenwald, I. B., Tan, A. T., White, J., Chiang, L., Sonenberg, N., and Bitterman, P. B. (1996). Translational control of programmed cell death: Eukaryotic translation initiation factor 4E blocks apoptosis in growth-factor-restricted fibroblasts with physiologically expressed or deregulated Myc. *Mol. Cell Biol.* **16,** 6573–6581.

Pope, A. J. (1999). Introduction LANCEtrade mark vs. HTRF(R) technologies (or vice versa). *J. Biomol. Screen.* **4,** 301–302.

Poulin, F., Gingras, A. C., Olsen, H., Chevalier, S., and Sonenberg, N. (1998). 4E-BP3, a new member of the eukaryotic initiation factor 4E-binding protein family. *J. Biol. Chem.* **273,** 14002–14007.

Ptushkina, M., von der Haar, T., Vasilescu, S., Birkenhager, R., and McCarthy, J. E. G. (1998). Cooperative modulation by eIF4G of eIF4E-binding to the mRNA 5′ cap in yeast involves a site partially shared by p20. *EMBO J.* **17,** 4798–4808.

Pyronnet, S., Imataka, H., Gingras, A. C., Fukunaga, R., Hunter, T., and Sonenberg, N. (1999). Human eukaryotic translation initiation factor 4G (eIF4G) recruits mnk1 to phosphorylate eIF4E. *EMBO J.* **18,** 270–279.

Robert, F., Gao, H. Q., Donia, M., Merrick, W. C., Hamann, M. T., and Pelletier, J. (2006a). Chlorolissoclimides: New inhibitors of eukaryotic protein synthesis. *RNA* **12,** 717–725.

Robert, F., Kapp, L. D., Khan, S. N., Acker, M. G., Kolitz, S., Kazemi, S., Kaufman, R. J., Merrick, W. C., Koromilas, A. E., Lorsch, J. R., and Pelletier, J. (2006b). Initiation of protein synthesis by hepatitis C virus is refractory to reduced eIF2.GTP.Met-tRNAiMet ternary complex availability. *Mol. Biol. Cell* **17,** 4632–4644.

Rogers, G. W., Jr., Komar, A. A., and Merrick, W. C. (2002). eIF4A: The godfather of the DEAD box helicases. *Prog. Nucleic Acid Res. Mol. Biol.* **72,** 307–331.

Rosenwald, I. B. (2004). The role of translation in neoplastic transformation from a pathologist's point of view. *Oncogene* **23,** 3230–3247.

Rousseau, D., Gingras, A. C., Pause, A., and Sonenberg, N. (1996). The eIF4E-binding proteins 1 and 2 are negative regulators of cell growth. *Oncogene* **13,** 2415–2420.

Rozen, F., and Sonenberg, N. (1987). Identification of nuclear cap specific proteins in HeLa cells. *Nucleic Acids Res.* **15,** 6489–6500.

Ruggero, D., Montanaro, L., Ma, L., Xu, W., Londei, P., Cordon-Cardo, C., and Pandolfi, P. P. (2004). The translation factor eIF-4E promotes tumor formation and cooperates with c-Myc in lymphomagenesis. *Nat. Med.* **10,** 484–486.

Schuler, M., Connell, S. R., Lescoute, A., Giesebrecht, J., Dabrowski, M., Schroeer, B., Mielke, T., Penczek, P. A., Westhof, E., and Spahn, C. M. (2006). Structure of the ribosome-bound cricket paralysis virus IRES RNA. *Nat. Struct. Mol. Biol.* **13,** 1092–1096.

Shen, X., Tomoo, K., Uchiyama, S., Kobayashi, Y., and Ishida, T. (2001). Structural and thermodynamic behavior of eukaryotic initiation factor 4E in supramolecular formation with 4E-binding protein 1 and mRNA cap analogue, studied by spectroscopic methods. *Chem. Pharm. Bull. (Tokyo)* **49,** 1299–1303.

Sizova, D. V., Kolupaeva, V. G., Pestova, T. V., Shatsky, I. N., and Hellen, C. U. (1998). Specific interaction of eukaryotic translation initiation factor 3 with the 5′ nontranslated regions of hepatitis C virus and classical swine fever virus RNAs. *J. Virol.* **72,** 4775–4782.

Sonenberg, N. (1981). ATP/Mg++-dependent cross-linking of cap binding proteins to the 5′ end of eukaryotic mRNA. *Nucleic Acids Res.* **9,** 1643–1656.

Sonenberg, N., Morgan, M. A., Merrick, W. C., and Shatkin, A. J. (1978). A polypeptide in eukaryotic initiation factors that crosslinks specifically to the 5′-terminal cap in mRNA. *Proc. Natl. Acad. Sci. USA* **75,** 4843–4847.

Sonenberg, N., and Shatkin, A. J. (1977). Reovirus mRNA can be covalently crosslinked via the 5′ cap to proteins in initiation complexes. *Proc. Natl. Acad. Sci. U S A* **74,** 4288–4292.

Svitkin, Y. V., and Sonenberg, N. (2004). An efficient system for cap- and poly(A)-dependent translation *in vitro*. *Methods Mol. Biol.* **257,** 155–170.

Tai, P. C., Wallace, B. J., and Davis, B. D. (1973). Actions of aurintricarboxylate, kasugamycin, and pactamycin on *Escherichia coli* polysomes. *Biochemistry* **12,** 616–620.

Tarun, S. J. J., and Sachs, A. B. (1996). Association of the yeast poly(A) tail binding protein with translation initiation factor eIF4G. *EMBO J.* **15,** 7168–7177.

Tomoo, K., Matsushita, Y., Fujisaki, H., Abiko, F., Shen, X., Taniguchi, T., Miyagawa, H., Kitamura, K., Miura, K., and Ishida, T. (2005). Structural basis for mRNA cap-binding regulation of eukaryotic initiation factor 4E by 4E-binding protein, studied by spectroscopic, X-ray crystal structural, and molecular dynamics simulation methods. *Biochim. Biophys. Acta* **1753,** 191–208.

Vazquez, D. (1979). Inhibitors of protein biosynthesis. *Mol. Biol. Biochem. Biophys.* **30,** 1–312.

von der Haar, T., Oku, Y., Ptushkina, M., Moerke, N., Wagner, G., Gross, J. D., and McCarthy, J. E. (2006). Folding transitions during assembly of the eukaryotic mRNA cap-binding complex. *J. Mol. Biol.* **356,** 982–992.

Walling, L. A., Peters, N. R., Horn, E. J., and King, R. W. (2001). New technologies for chemical genetics. *J. Cell Biochem. Suppl.* **37,** 7–12.

Welch, E. M., Barton, E. R., Zhuo, J., Tomizawa, Y., Friesen, W. J., Trifillis, P., Paushkin, S., Patel, M., Trotta, C. R., Hwang, S., Wilde, R. G., Karp, G., *et al.* (2007). PTC124 targets genetic disorders caused by nonsense mutations. *Nature.* **447,** 87–91.

Wendel, H.-G., de Stanchina, E., Fridman, J. S., Malina, A., Ray, S., Kogan, S., Cordon-Cardo, C., Pelletier, J., and Lowe, S. W. (2004). Survival signaling by Akt and eIF4E in oncogenesis and cancer therapy. *Nature* **428,** 332–337.

Wendel, H. G., Malina, A., Zhao, Z., Zender, L., Kogan, S. C., Cordon-Cardo, C., Pelletier, J., and Lowe, S. W. (2006). Determinants of sensitivity and resistance to rapamycin-chemotherapy drug combinations *in vivo*. *Cancer Res.* **66,** 7639–7646.

Wilson, J. E., Pestova, T. V., Hellen, C. U., and Sarnow, P. (2000). Initiation of protein synthesis from the A site of the ribosome. *Cell* **102,** 511–520.

Yoder-Hill, J., Pause, A., Sonenberg, N., and Merrick, W. C. (1993). The p46 subunit of eukaryotic initiation factor (eIF)-4F exchanges with eIF-4A. *J. Biol. Chem.* **268,** 5566–5573.

Zhang, J.-H., Chung, T. D. Y., and Oldenburg, K. R. (1999). A simple statistical parameter for use in evaluation and validation of high throughput screening assays. *J. Biomol. Screen.* **4,** 67–73.

Zimmerman, E. (1967). Automatic recording of viscosity changes with the Ostwald capillary viscometer. *Anal. Biochem.* **21,** 81–85.

CHAPTER FOURTEEN

Isolation and Identification of Eukaryotic Initiation Factor 4A as a Molecular Target for the Marine Natural Product Pateamine A

Woon-Kai Low,* Yongjun Dang,* Tilman Schneider-Poetsch,*
Zonggao Shi,* Nam Song Choi,[†] Robert M. Rzasa,[†]
Helene A. Shea,[†] Shukun Li,[†] Kaapjoo Park,[†] Gil Ma,[†]
Daniel Romo,[†] and Jun O. Liu*,[‡,§]

Contents

Abstract

Natural products continue to demonstrate their utility both as therapeutics and as molecular probes for the discovery and mechanistic deconvolution of various cellular processes. However, this utility is dampened by the inherent difficulties involved in isolating and characterizing new bioactive natural products, in

* Department of Pharmacology, Johns Hopkins School of Medicine, Baltimore, Maryland
† Department of Chemistry, Texas A&M University, College Station, Texas
‡ Solomon H. Snyder Department of Neuroscience, Johns Hopkins School of Medicine, Baltimore, Maryland
§ Department of Oncology, Johns Hopkins School of Medicine, Baltimore, Maryland

Methods in Enzymology, Volume 431
ISSN 0076-6879, DOI: 10.1016/S0076-6879(07)31014-8

obtaining sufficient quantities of purified compound for further biological studies, and in developing bioactive probes. Key to characterizing the biological
activity of natural products is the identification of the molecular target(s) within
the cell. The marine sponge-derived natural product Pateamine A (PatA) has
been found to be an inhibitor of eukaryotic translation initiation. Herein, we
describe the methods utilized for identification of the eukaryotic translation
initiation factor 4A (eIF4A) as one of the primary protein targets of PatA. We
begin by describing the synthesis of an active biotin conjugate of PatA (B-PatA),
made possible by total synthesis, followed by its use for affinity purification of
PatA binding proteins from cellular lysates. We have attempted to present the
methodology as a general technique for the identification of protein targets for
small molecules including natural products.

1. INTRODUCTION

Natural products have not only offered important compounds for the
discovery and development of new therapeutic agents, particularly antibiotic and anticancer drugs, but have also served as increasingly useful probes
to manipulate and elucidate the mechanisms of various biological processes,
from signal transduction to cell cycle progression. For example, the application of the immunosuppressive natural products cyclosporine A, FK506,
and rapamycin led to the discoveries of the protein phosphatase calcineurin
(Liu *et al.*, 1991) and the PI3 kinase homolog mTOR (Brown *et al.*, 1994;
Chiu *et al.*, 1994; Heitman *et al.*, 1991; Sabatini *et al.*, 1994) in mediating
calcium–dependent intracellular signaling process and cytokine receptor-
mediated signaling and nutrient-sensing signaling pathways, respectively.
The use of trapoxin allowed for the first isolation of the founding members
of histone deacetylases (Taunton *et al.*, 1996). Examination of the mode of
action of the fungal metabolites fumagillin and ovalicin has unraveled a
previously unappreciated function of type 2 methionine aminopeptidase in
angiogenesis and T lymphocyte activation (Griffith *et al.*, 1997; Sin *et al.*,
1997; Turk *et al.*, 1998). Due, in part, to their ability to be delivered to cells
or tissues in a temporally and spatially controlled manner compared with
genetic methods such as gene knockout or RNA interference, natural
products with potent biological activities have become a powerful set of
alternative experimental tools in biology.

A rate-limiting step in the use of natural products as probes in biology
is the chemical synthesis of useful bioactive probes, such as affinity
reagents or fluorescent probes, that can be readily applied by biochemists
and cell biologists. The ability to generate useful probes derived from
natural products is completely dependent on a prior knowledge of the

structure-and-activity relationship for a given natural product to find a viable position for tethering or modification of the compound to produce assayable functional groups, such as the aforementioned affinity tags or fluorescent probes. Close collaboration between synthetic chemists and biologists is often necessary to carry out such studies. There are multiple approaches to identifying the molecular targets of natural products. The most common method for target identification is affinity chromatography (Cuatrecasas, 1970). In this method, a small molecule ligand can be covalently immobilized onto a solid matrix, such as agarose, which can then be used to selectively retain the binding protein(s) of the ligand (Lefkowitz et al., 1972). An inherent problem associated with the conventional affinity chromatography method is that once a natural product is immobilized onto a solid phase, it is not possible to characterize the affinity matrix chemically or biologically, that is, in a cellular assay to control for loss of biological activity caused by the tethering process. In addition, if a natural product of interest binds to its protein target covalently, the conventional affinity chromatography approach may not be feasible or practical, because it will be impossible to release bound protein from the matrix, complicating bound protein identification, or a second cleavage step would be required to release the bound protein. Therefore, an alternative is to prepare a conjugate between a natural product of interest and biotin (Manz et al., 1983). The resultant biotin–natural product conjugate can be used in conjunction with the commercially available streptavidin or avidin matrices to immobilize the biotin–ligand conjugate along with the binding proteins. The use of biotin-conjugated natural products, along with streptavidin or avidin beads, offers several advantages over conventional affinity chromatography. First, the biotin–natural product conjugates can be fully characterized for both structure and activity to ensure that attachment of the biotin does not abrogate the biological activity of the natural products. Second, the bound proteins can be released from the beads whether the interaction is covalent or not, because the biotin–streptavidin or biotin–avidin interactions are noncovalent. However, even though biotin–avidin interactions are not covalent, the interaction is the strongest known noncovalent interaction between a protein and ligand ($K_a = 10^{15} \, M^{-1}$) and is robust under a wide variety of conditions, such as buffer composition, pH, and temperature, such that in a typical streptavidin biotin–conjugate affinity capture assay, the experimenters need not concern themselves with complications arising from loss of biotin–avidin (streptavidin) interaction. It is worth pointing out in passing that while the biotin-mediated affinity binding method is a most commonly used technique for isolating binding proteins for small molecules, there are a number of alternative methods for target identification, including genetic complementation of resistant phenotypes in yeast and mammalian cells (Heitman et al., 1991; Peterson et al., 2006), the yeast three-hybrid system (Licitra and Liu, 1996), the use of mRNA expression profiling (Marton *et al.*, 1998), protein chips (Zhu and Snyder, 2003), and the phage-display system (Sche *et al.*, 1999). Occasionally, phenotypic similarity in established genetic mutants and small molecules has also been used to effectively identify the molecular target (Mayer *et al.*, 1999). But this latter approach cannot be generalized for natural product target identification.

Given the essential role of translation in cell survival and proliferation, one might expect that there would be a plethora of natural products that interfere with protein translation. This is indeed the case for natural products that inhibit prokaryotic translation, which constitute important classes of antibiotic drugs (Katz and Ashley, 2005; Pestka, 1977). In contrast, there are far fewer natural products known to block the translation process in eukaryotes. Moreover, most of the known natural product inhibitors of eukaryotic translation affect elongation (Cundliffe *et al.*, 1974; Fresno *et al.*, 1977; Goldberg and Mitsugi, 1966; Obrig *et al.*, 1971; SirDeshpande and Toogood, 1995; Wei *et al.*, 1974). One underlying reason for the relative scarcity of natural products that inhibit eukaryotic translation has been the lack of systematic screening of natural products for translation inhibitors until 2004 (Novac *et al.*, 2004). Undoubtedly, another underlying cause is that the mechanisms of action for the majority of bioactive natural products remain unknown. It would not be surprising that several anti-proliferative natural products, in fact, target the eukaryotic translation machinery, as we unexpectedly encountered during the mechanistic investigation of the marine sponge-derived natural product pateamine A (PatA) (Low *et al.*, 2005), a result also found independently by a deliberate screening effort in search of protein translation initiation inhibitors (Bordeleau *et al.*, 2005).

PatA was initially discovered as a potent antiproliferative agent against the murine P388 tumor cell line (Northcote *et al.*, 1991). Subsequently, it was found to possess potent immunosuppressive activity in animal models and later at the cellular level (Romo *et al.*, 1998). Although PatA undoubtedly inhibits T cell receptor-mediated signaling leading to interleukin (IL)-2 production, the overall immunosuppressive effect of PatA is likely to result from its effect on eukaryotic translation initiation and, hence, T lymphocyte proliferation. Nevertheless, the IL-2 reporter assay proved useful in our initial structure/activity studies, which enabled the synthesis of an active biotin-PatA (B-PatA) conjugate (see later), as the potencies of PatA analogs used in the studies were similar to their potency in blocking cell proliferation and translation. In what follows, we trace the route of our collaborative efforts combining synthetic organic chemistry and protein biochemistry that led to the detection, isolation, and identification of two PatA-interacting proteins, one of which is the eukaryotic translation initiation factor (eIF)-4A.

2. TOTAL SYNTHESIS OF PATEAMINE A

The provocative initial biological activities reported for PatA, primarily the 20,000-fold difference in toxicity toward cancer cells (P388) versus a quiescent cell line (BSC) (Northcote *et al.*, 1991), led us to undertake a total synthesis of this natural product that would ultimately enable detailed mode of action studies. We envisioned the synthesis and subsequent union of three principal fragments, namely, enyne acid **4**, β-lactam **10**, and dienylstannane **14** (Fig. 14.1). A crucial aspect of this plan was a late-stage Stille coupling to append the expected labile trienyl

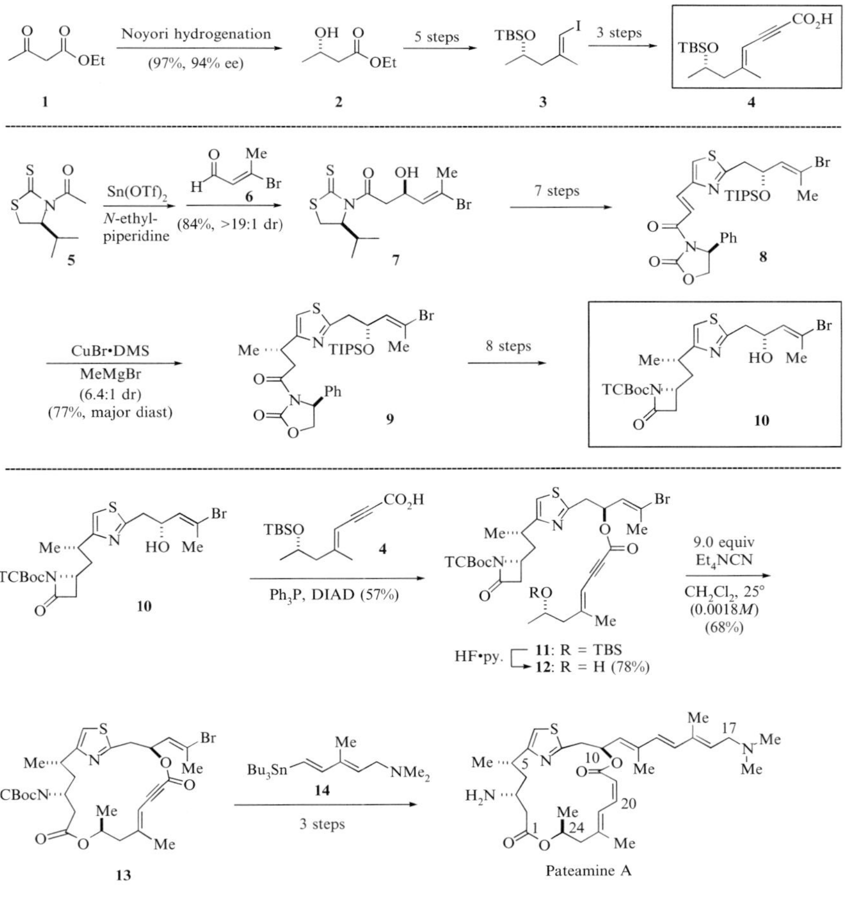

Figure 14.1 Total synthesis of (-)-Pateamine A. TBS, t-butyldimethylsilyl; TIPS, triisopropyl silyl; TCBoc, trichloro t-butoxycarbamate; DIAD, diisopropyl azodicarboxylate.

amine side chain bearing a triallylic acetate moiety, to introduce the polar dimethyl amino group late in the synthesis, and also to allow a variety of side chains to be introduced onto the macrocyclic core structure during subsequent structure/activity relationship studies. An enyne was used to allow a late-stage introduction of the *E, Z*-dienoate to minimize the potential for olefin isomerization. A key strategy to be employed to construct the 19-membered dilactone macrolide of pateamine A was a β-lactam-based macrolactonization, wherein a β-lactam would be used to install the C3-amino group and then serve as an activated acyl group for macrocyclization.

Synthesis of the enyne acid fragment **4** began with (*S*)-2-hydroxybutyrate **2** available on large scale by modified Noyori hydrogenation of ethyl acetoacetate (97% yield, 94% ee) (Noyori *et al.*, 1987; Taber and Silverberg, 1991). Standard manipulations passing through vinyl iodide **3** and subsequent homologation with propargyl alcohol provided enyne acid **4** following a two-step oxidation.

The synthesis of the most complex fragment, β-lactam **10**, employed a Nagao acetate aldol reaction with known aldehyde **6** (Fischer *et al.*, 1990) to give an 84% yield of the aldol adduct **7** in a highly diastereoselective fashion (>19:1 diastereomic ratio, dr) (Nagao *et al.*, 1986). Standard manipulations allowed for a modified Hantzsch thiazole synthesis, leading to thiazole **6** (Aguilar and Meyers, 1994). The second stereocenter at C5 was introduced by an asymmetric conjugate addition with good diastereoselectivity (6.4:1, dr) and the major diastereomer could be isolated in 77% yield (Li *et al.*, 1993; Nicolas *et al.*, 1993). The application of a second Nagao acetate aldol allowed for introduction of the C3 stereocenter and the method of Miller involving Mitsunobu inversion provided β-lactam **10** (Guzzo and Miller, 1994).

A Mitsunobu process simultaneously coupled the enyne acid fragment **4** to β-lactam **10** and inverted the C10 stereochemistry to the required (*S*)-configured ester **11** in 93% yield. A deprotection provided alcohol **12**, the key β-lactam-based macrolactonization substrate, which, under conditions similar to those reported by Palomo for intermolecular alcoholysis of β-lactams (Ojima *et al.*, 1992, 1993; Palomo *et al.*, 1995), provided the desired core macrocycle **13** of PatA **13** (Hesse, 1991; Manhas *et al.*, 1988; Wasserman, 1987). Subsequent Lindlar hydrogenation gave the required *E, Z*-dienoate. A Stille reaction and final deprotection cleanly provided (-)-PatA that was identical in all respects to the natural product (Romo *et al.*, 1998; Rzasa *et al.*, 1998). This first total synthesis confirmed the relative and absolute configuration of the natural product and paved the way for synthesis of derivatives for probing the mode of action of this natural product.

3. SYNTHESIS OF DERIVATIVES OF PATEAMINE A AND STRUCTURE ACTIVITY RELATIONSHIP STUDIES

An initial step to begin mode of action studies was the identification of a viable site of attachment for a biotin probe that would not significantly abrogate biological activity. Derivatives **17** to **22** were prepared to determine the structural tolerance of the terminal functional group on the sidechain of PatA and also to potentially improve the stability of the acid labile triallylic acetate moiety by removal of one unsaturation (Fig. 14.2) (Romo et al., 2004). In initial SAR studies, C3-Boc PatA was found to retain bioactivity (only ~4-fold decrease in activity) and improve stability and, therefore, for ease of handling, this amino protecting group was retained in all derivatives. In addition, derivative **19** bearing the identical side-chain found in PatA with the exception of one unsaturation and the C16 methyl group was prepared. The effects on biological activity of a more rigid macrocycle (enyne vs dienoate) for BocPatA (enyne **25**) in conjunction with modified terminal functional groups on the diene were investigated by synthesis of enynes **20** to **22**. These derivatives were readily prepared by omission of the Lindlar reduction step in the synthetic sequence. Two additional C3-amino derivatives were prepared with the expectation that they should have similar potency to the C3-Boc derivative. In this regard, the C3-phenyl carbamate **26** and the

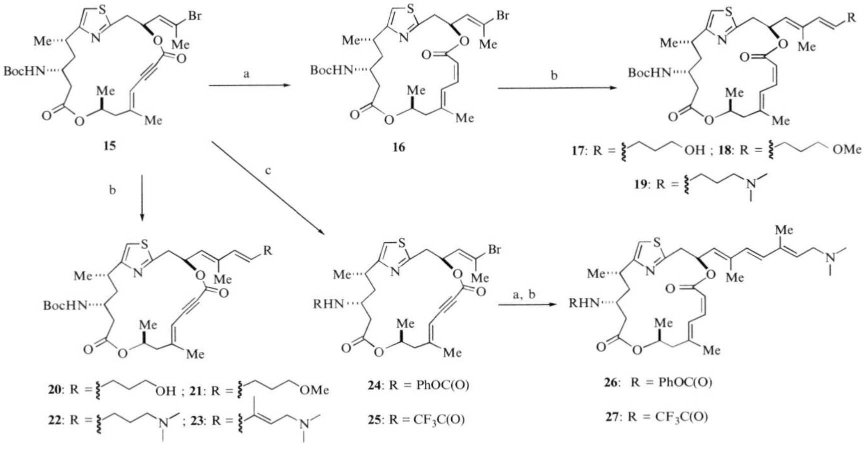

Figure 14.2 Synthesis of PatA derivatives. (a) Pd(CaCO$_3$)/Pb, H$_2$, MeOH, 12h, 99% b. E-Bu$_3$SnCHCHR, 10 mol% [Pd$_2$dba$_3$•CHCl$_3$: AsPPh$_3$ = 1:8], THF, 25°, 10–18 h, 11–76% c.(i) 20% TFA, CH$_2$Cl$_2$, 0°, 15 h, 95% (ii) PhCOCl or (CF$_3$CO)$_2$O, DMAP, py., CH$_2$Cl$_2$, 25°, 5h, 99% [Boc = t-butyl carbamate].

C3-trifluoroacetamide **27** were synthesized by deprotection of Boc-macrocycle **15**, followed by acylation, to give macrocycles **24** and **25**. Subsequent Lindlar reduction and Stille reaction gave the C3-acylated derivatives **26** and **27**.

3.1. Synthesis of PatA derivatives

The IC_{50} values for natural and synthetic PatA were determined to be 0.33 and 4.0 nM, respectively (Romo *et al.*, 1998, 2004), with the ~10-fold variation likely arising from minor differences in the assays that were performed at different times (Table 14.1). Furthermore, the fact that C3-Boc-PatA **25** retains a significant amount of activity suggested this site for attachment of biochemical probes for detection and isolation of the molecular target of PatA. The inactivity (>1000 nM) observed for the Boc-macrocycle **16** and the triene **24** suggests that both the macrocycle and trienyl amine side chain are required for binding to putative target(s).

As seen in Table 14.1, the more modestly altered derivatives (**17–27**) were also, in general, less potent than PatA. As expected, the C3-phenyl carbamate derivative **26** possessed an activity that is approximately 4-fold lower that of PatA (4 nM) in analogy to BocPatA **25**. What is not readily explained is the reduced activity of the trifluoroacetamide **25** (~303 nM). Although we cannot offer an explanation at this time, we note that introduction of fluorine atoms into protein ligands often lead to results that are not readily rationalized (Filler *et al.*, 1993). A possible trend is observed upon comparison of dienoate macrocycles **17** to **19** and enynoate macrocycles **20 to 23**. Enyne derivatives having an expectedly more rigid macrocycle than the natural product and bearing oxygen rather than nitrogen at the terminus of the side chain (e.g., **20** and **21**) were found to have activities in the IL-2 reporter gene assay in the range of 55 and 335 nM, respectively. However, enyne derivatives **22** and **23** with side chains more closely resembling the natural product (e.g., amino end groups) had no activity. Conversely, the dienoate derivatives (**17** and **18**) that should have macrocyclic conformations similar to the natural product but bearing oxygen rather than nitrogen at the terminus of the side chain had very low activity. However, once nitrogen is introduced into the side chain, as in derivative **19**, activity is restored (328 nM) to some degree. Thus, it would appear that an oxygenated side chain, regardless of hydrogen bonding capabilities, compensates to some extent for the change in macrocycle conformation that occurs upon introduction of an enyne. However, oxygen rather than nitrogen on the side chain leads to low activity when coupled to the natural dienoate-containing macrocycle. Further analysis and understanding of these results in regard to their relevance for protein ligand interactions must await structural

Table 14.1 IL-2 reporter gene assay (transfected Jurkat T cells) activity of pateamine A and derivatives

cmpd.	R¹	IC$_{50}$ (nM)	cmpd.	R¹	R²	R³	IC$_{50}$ (nM)
20	OH	340 ± 180	Synthetic (−)-PatA		NH$_2$	Me	4.0 ± 0.94[c] (2004) 0.33 ± 0.03 (1998)
21	OMe	55 ± 16	Natural (−)-PatA (25)	"	NH$_2$	Me	0.45 ± 0.04
22		NA[a]	Boc–PatA	"	NHBoc	Me	2.1 ± 0.5
23		NA[a]	17	OH	NHBoc	Me	>1000[b]

(continued)

Table 14.1 (*continued*)

cmpd.	R^1	R^2	R^3	IC$_{50}$ (nM)
18	(chain terminating in OMe)	NHBoc	Me	>1000[b]
19	(chain terminating in N)	NHBoc	Me	330 ± 120
26	″	NHC(O)OPh	Me	15 ± 6.1
27	″	NHC(O)CF$_3$	Me	300 ± 93

cmpd.	R^1	IC$_{50}$ (nM)
		> 1000
Boc–Macrocycle (**16**)		>1000
Triene (**24**)		

[a] Not active.

[b] Inhibition activity was observed, but it did not reach 50% even with the highest concentration tested.

[c] It should be noted that the IC$_{50}$ value for PatA in this particular assay is ten fold higher than that previously reported. It seems that Jurkat cells appear to vary in their sensitivity to PatA, depending in part on the number of passages they have undergone. All IC$_{50}$ values listed in this table were determined using the same population of Jurkat T cells.

characterization of the interactions of the putative cellular protein receptor(s) with these PatA derivatives.

4. STRUCTURAL ANALYSIS OF PatA LEADING TO DMDA-PatA AND VIABLE POSITIONS FOR DERIVATIZATION

As described previously, several lines of evidence suggested that the C3 amino group was a suitable site for derivatization. Most importantly, the C3-Boc derivative only dropped approximately 4- to 5-fold in activity with an IC_{50} of 2.0 nM and this data, in conjunction with the synthesis of the additional PatA derivatives described previously (Fig. 14.2), and structural analysis of the PatA structure collectively, began to point to the possibility of separate binding and scaffolding domains in the PatA structure (Fig. 14.3) (Romo *et al.*, 2004). Importantly, this analysis led to the development and synthesis of the first bioactive biotin–PatA conjugate (see later) and a highly potent, simplified derivative, des-methyl des-amino pateamine A (DMDAPatA).

5. SYNTHESIS OF A BIOACTIVE BIOTIN-PATEAMINE A (B-PatA) CONJUGATE

Our synthetic studies toward PatA confirmed the expected and significant instability of the trienyl side chain due to the labile nature of the triallylic acetate. While we were able to successfully acylate the C3-amino group of PatA, the instability of the resulting products and complex nature of the reaction mixture ultimately led us to an alternative strategy, made possible by our total synthesis. Namely, we were able to take advantage of the robust nature of the macrocyle **29** prior to attachment of the trienyl amine side chain. Thus, amino macrocycle **29** was alkylated with iodoacetamide reagent **30** (Molecular Probes) to obtain the biotinylated macrocycle **31** (Fig. 14.4). Attachment of the trienyl amine side chain by a final stage Stille coupling provided the desired biotin conjugate **32** (B-PatA) that was utilized in affinity chromatography experiments, described later. This final reaction step, the Stille coupling, continues to be an extremely challenging reaction; only one postdoctoral fellow has been able to isolate this particular biotin–patA conjugate to date! The bioactivity of this conjugate further verified its structure in conjunction with ^{1}H NMR and high-resolution mass spectral data and enabled many subsequent mechanistic studies, described later.

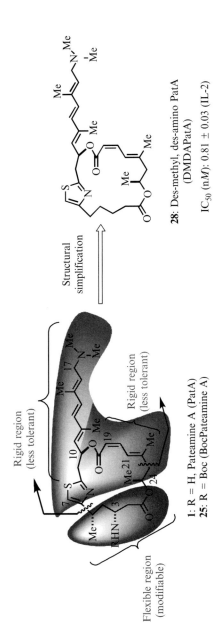

Figure 14.3 Structures of Pateamine A (1), Boc–pateamine A (25), and DMDA-PatA (28). The putative binding (rigid regions, in red) and scaffolding (flexible and modifiable, in blue) domains are indicated and suggest possible sites for modification, namely, the C3-amino group. (See color insert.)

28: Des-methyl, des-amino PatA
(DMDAPatA)
IC$_{50}$ (nM): 0.81 ± 0.03 (IL-2)

Structural
simplification

1: R = H, Pateamine A (PatA)
25: R = Boc (BocPateamine A)

Rigid region
(less tolerant)

Rigid region
(less tolerant)

Flexible region
(modifiable)

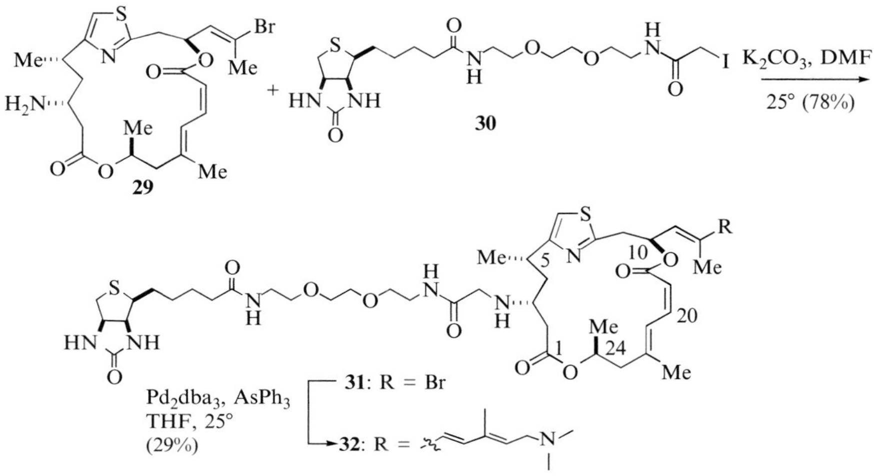

Figure 14.4 Synthesis of a biotinylated Pateamine A. (B–PatA).

6. AFFINITY PULL-DOWN OF PATA-BINDING PROTEINS

For the capture of small-molecule target proteins using biotin con-jugates from cellular lysates, the key factor that must be addressed is the proper preparation of lysates such that the target protein is accessible and available for interaction with the small molecule. However, because a priori knowledge of the target protein is usually lacking, several experiments should be performed prior to the selection of cell line(s) and the lysate preparation method for affinity purification, such that the biotin-conjugate maintains biological activity, at least ensuring that the target protein(s) are present. The acquisition of an active B-PatA conjugate (see IL-2 reporter assays previously described) that retains significant biological activity thus allowed for the identification of PatA-binding proteins.

6.1. Preparation of cell lysate

Although a number of cell lines were shown to be sensitive to inhibition by PatA (Low *et al.*, 2005), we selected RKO cells (IC_{50} of $\sim$0.4 nM in cell proliferation assay) to prepare lysates for the isolation and identification of target protein(s). We often select RKO cells for target identification of small molecules using biotin-conjugates, because they appear to be particularly suitable for target protein isolation.

RKO cells were cultured in DMEM medium (high glucose) supple-mented with fetal bovine serum and antibiotics (Invitrogen) in eight 150 $\times$ 25 mm cell culture dishes to 50 to 60% confluency. Upon removal of growth media, the cells were washed with PBS (10 ml/plate). The washed cells were detached from the plates by incubating with 5 ml trypsin/EDTA at 37° for 3 to 5 min, which can be followed visually under a microscope. The detached cell suspensions were pooled in a pre-weighed 50-mL conical vial. The vial was centrifuged (1000–2000g) in a tabletop centrifuge at room temperature and the supernatant was removed by aspiration. The cell pellet was washed with 20 ml of PBS by resuspension via gentle pipetting, followed by centrifugation and aspiration of PBS. After removal of PBS, the vial was weighed to determine the "wet weight" of cells collected. At this stage, cells can be stored at −80° with minimal detrimental effect, although in our experience it is best to work with cells immediately.

From this point on, all steps should be performed on ice or in a cold room. The wet weight of the cells was determined based on the pre-weighed conical vial weight, and then cells were suspended in a lysis buffer at 5 ml/gram of cells (wet weight). The lysis buffer routinely used in our lab for capture of targets of small molecule compounds contains 20 mM Tris•HCl (pH 7.4), 100 mM KCl, and 0.2% Triton-X-100 supplemented with protease inhibitors that were obtained from Roche Diagnostics in tablet form that are dissolved in a predetermined volume to yield the indicated concentrations of each protease inhibitor (Roche Diagnostics). The cell suspension was incubated on ice for 20 min, followed by 30 strokes using a Teflon dounce homogenizer and another incubation for 20 min on ice. The homogenate was transferred to 1.5-ml microfuge tubes and centrifuged at maximum speed in an Eppendorf microcentrifuge for 10 min. Supernatant was then transferred to 13 × 51 mm polycarbonate centrifuge tubes, and centrifuged for 30 min at 50,000g in a TLA 100.3 rotor in a TL-100 ultracentrifuge (Beckman) at 4°. The cleared supernatant was transferred to a 15-ml Falcon centrifuge tube that was prechilled on ice. The total protein concentration should be in the range of 5 to 10 mg/ml. Protein concentration can be determined using protein assay kits readily available from commercial sources, such as Bio-Rad's D$_C$ Protein Assay (Bio-Rad).

6.2. Affinity capture of PatA-binding proteins using biotin-PatA and streptavidin-agarose

6.2.1. Reagents

Biotin–PatA (100 μM stock solution in DMSO)
PatA (2 mM stock solution in DMSO)
Streptavidin–agarose (Pierce)

6.2.2. Procedures

The overall experimental procedure for an initial detection of PatA binding proteins is schematically shown in Fig. 14.5. Thus, three parallel pull-down experiments were carried out simultaneously. The first is a control in which

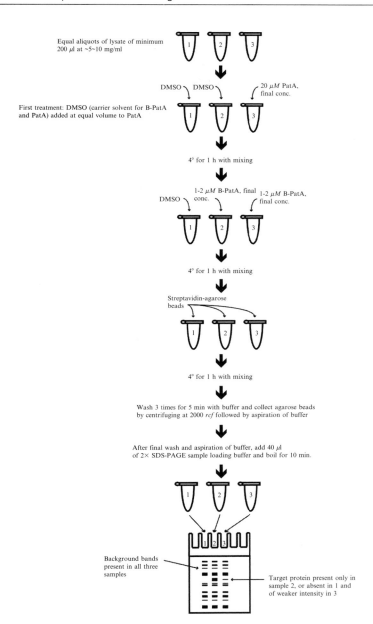

Figure 14.5 Schematic of the procedure for the use of biotin–PatA to detect PatA-binding proteins.

cell lysate was incubated with streptavidin–agarose in the absence of B-PatA to visualize all proteins that can be captured by the streptavidin–agarose (nonspecific binding), most of which are presumably endogenous biotinylated proteins (Griffith *et al.*, 1997) or nonspecific interactions with agarose. The second sample contained B-PatA, which was expected to retain putative PatA-binding proteins on the streptavidin-agarose. The third is a competition experiment in which the lysate was pre-incubated with an excessive amount of free PatA (typically, 5–20 times molar excess to the B-PatA used). After SDS-PAGE, the gel was stained with silver for detection of proteins.

1. Split supernatant into three equal 200-μl aliquots in 1.5-ml microfuge tubes. The protein concentrations should be between 5 and 10 mg/ml.
2. Aliquot 2 μl DMSO into sample 1 (carrier control) and sample 2 (affinity purification), and 2 μM 2 mM PatA into sample 3 (competition), followed by incubation at 4° with mixing for 1 h.
3. Add 2 μl DMSO into sample 1, and 2 μl 100 μM B-PatA stock solution into samples 2 and 3. A second incubation at 4° for 1 h with mixing was performed. B-PatA will efficiently capture its binding proteins at 1 μM final concentrations; however, increasing B-PatA concentrations will increase the amount of captured protein.
4. The streptavidin-agarose should be well suspended prior to use to ensure equal aliquoting of the resin. Pipette equal volume of slurry into 3 microfuge tubes and centrifuge at 2000 to 3000g in a microcentrifuge to collect the resin at the bottom of the tube. Depending on the percentage of resin in the slurry, the volume to aliquot will vary. We routinely use Immunopure immobilized Streptavidin (Pierce), with a settled resin volume of 10 to 20 μl. Once equality of the resin in each tube is ensured, the resin is washed 3 times for 5 min with 1 ml of lysis buffer, as commercial slurries are often stored in an ethanol mixture containing preservatives. After washing and collection of the resin to the bottom of the tubes by centrifugation, visual inspection was proven sufficient to ensure equality of resin volumes in our hands. Although not necessary, results can be enhanced by preblocking the streptavidin resin using a solution of BSA (5 mg/ml) in lysis buffer for 30 min with mixing at 4°, which is sufficient to reduce nonspecific background protein binding, although controls must be performed to determine that no unwanted effects occur, that is, target protein is the same size as BSA or unwanted nonspecific BSA-compound interactions take place. No such problems were observed for PatA.
5. Collect the samples from step 3 to the bottom of each respective microfuge tube by a quick centrifugation step (approximately 15 s) at maximum speed in a microfuge. For the streptavidin-agarose, collect resin to the bottom of each microfuge tube, and remove supernatant by aspiration. If pre-incubation with BSA was used, it is not necessary to perform a washing step. For this final aspiration prior to transfer of lysate to resin-containing tubes, extra care should be taken to aspirate to the surface of the settled resin

without removal of resin. Transfer each sample into the appropriately labeled streptavidin–containing tubes and incubate the mixtures at 4° with mixing for another hour.

6. Collect the resin in each tube by centrifuging as described previously. At this stage, it would be preferable to perform collection of resin by centrifuging in a cooled microfuge (4°) or in a microfuge placed in a cold room. Lysate was removed by aspiration, again taking care not to disturb the settled resin, followed by a wash with 1 ml of lysis buffer with mixing at 4° for 5 min. Wash the beads twice more (total of three washes), and after the final aspiration, resuspend the beads in 2× (20–30 μl) SDS-PAGE sample loading buffer containing β-mercaptoethanol followed by heating at ~100° for 10 min, which denatures the streptavidin tetramer and releases the captured proteins.

7. Spin each tube at maximal speed in a microcentrifuge. Load the boiled supernatant into a standard SDS-PAGE gel. For B-PatA, target proteins are easily detected using a 12% polyacrylamide gel. However, without a priori knowledge of the target protein molecular weight, this may have to be repeated a few times using various gel concentrations to allow for identification of target protein(s). It may not be necessary to load the entire sample into a single gel, and half of the aliquot can be saved by freezing and storing at −20° for later evaluation. However, for situations where the target is unknown, this may limit the range of detection. Furthermore, special care should be taken to maintain the equality of the three samples if only aliquots are used for SDS-PAGE.

8. Stain the gel with silver using standard procedures. Careful examination of the gel should reveal bands that appear in sample 2 and not sample 1, which can be assumed to be captured specifically by the biotin-conjugate. Furthermore, if these same bands are not present, or are lower in intensity in sample 3, they can be assumed to be specifically captured in sample 2, but lost in sample 3 due to occupation of the binding-site by the free compound during the pre-incubation step.

As is apparent from Fig. 14.6, two additional protein bands with apparent molecular weights of 48 and 38 kDa were seen in sample 2 but were absent in sample 1. Moreover, the intensity for both bands was significantly reduced in sample 3, suggesting that they were retained on the streptavidin–agarose due to specific binding to PatA.

7. IDENTIFICATION OF PatA-BINDING PROTEINS

To identify the two newly detected putative PatA-binding proteins, the pull-down experiment was scaled up approximately 10-fold with proportional increases in the amount of cell lysate and B-PatA used with

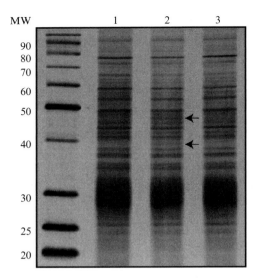

Figure 14.6 Silver-stained SDS-PAGE gel of PatA binding proteins. Lane 1, sample 1: nonspecific proteins captured by the streptavidin–agarose resin; Lane 2, sample 2: proteins affinity captured by the presence of B-PatA; Lane 3, sample 3: affinity capture of target proteins was blocked by prior addition of free PatA before incubation with B-PatA. The two arrows point to two proteins specifically detected in sample 2 versus sample 1, which were also lost due to competition in sample 3, with apparent molecular weights of 38 and 48 kDa.

SDS-PAGE performed using a 16 × 18 cm and 1.5 mm thick gel along with molecular weight markers to help locate the putative PatA-binding proteins. The gel was stained with a modified silver-staining protocol to maintain the integrity of the stained proteins (Shevchenko *et al.*, 1996). The 48- and 38-kDa protein bands were excised from the gel and were identified by using in-gel tryptic digestion and peptide identification by MALDI-TOF using standard procedures. The molecular weights of multiple peptides from each protein led to their identification as human eukaryotic translation initiator (eIF)-4AI or II (eIF4AI and II share 90% identity and thus could not be distinguished by the MS technique utilized) and a multifunctional protein known variably as Serine/Threonine kinase Receptor Associated Protein (STRAP), upstream of N-Ras-interacting protein (UNRIP), and MAP kinase Activator with WD repeats (hMAWD) (Hunt *et al.*, 1999; Matsuda *et al.*, 2000). The identity of both proteins was confirmed by performing the affinity capture again at the mini-gel scale, followed by transfer to nitrocellulose membrane and detection of eIF4AI and STRAP by immunoblotting using specific antibodies.

It should be pointed out for comparison that conventional chromatography was used to immobilize PatA onto preactivated sepharose, presumably via the only primary amino group on PatA (Bordeleau, 2005).

It is extraordinary that eIF4A was found among a number of other proteins bound to the PatA beads without the knowledge of how much PatA is attached to the beads and, more importantly, whether such attachment would lead to inactivation of PatA. In retrospect, it is fortuitous that the amino group used to immobilize PatA onto solid phase can indeed tolerate further chemical modifications without a significant loss of activity.

8. Concluding Remarks

In this chapter, we described experimental protocols for preparing an affinity probe for the natural product PatA and for using the biotin-conjugate, B-PatA to detect, isolate, and eventually identify two PatA-binding proteins. Further studies have confirmed the interaction between eIF4A and PatA and have indicated that binding of PatA to eIF4A is a necessary step for the inhibition of eukaryotic translation initiation and cell proliferation by PatA (Bordeleau et al., 2005; Low et al., 2005). The interaction between the 38-kDs STRAP/Unrip/hMAWD remains to be characterized. We hope to have conveyed the key steps required to identifying target proteins for natural products that inhibit translation and other biological processes. Furthermore, the total synthesis of the natural product has paved the way to accessing synthetic derivatives, a systematic structure/activity relationship study to find a suitable position in the natural product for the attachment of biotin and other probes, the synthesis of a biotin conjugate that retains the biological activity of the natural product, and the use of biotin conjugate to detect, isolate, and identify the putative binding protein(s) of the natural product of interest. In hindsight, it may have been more relevant to perform the initial SAR assays using some eukaryotic translation assays, or a translation initiation assay. However, this exemplifies some of the difficulties and challenges of the initial studies of natural products without a priori knowledge of targeted biological processes or protein(s). Nevertheless, the IL-2 reporter assay did allow for rigorous SAR and confirmation of activity for B-PatA. While the synthetic procedures will vary depending on the structures of the natural products under scrutiny, the general principles involving SAR of modified analogs, followed by attachment of a biotin moiety and affinity purification, are likely to be applicable for protein target identification for a wide variety of natural products.

ACKNOWLEDGMENTS

We are grateful to Drs. Jon Lorsch, Rachael Green, and William Merrick for generous provision of advice. Financial support from the NCI, the Keck Center (J. O. L.), NIGMS (D. R.), and a Fellowship from CIHR (W. K. L.) are gratefully acknowledged.

REFERENCES

Aguilar, E., and Meyers, A. I. (1994). Reinvestigation of a modified Hantzsch thiazole synthesis. *Tetrahedron Lett.* **35,** 2473–2476.

Bordeleau, M. E., Matthews, J., Wojnar, J. M., Lindqvist, L., Novac, O., Jankowsky, E., Sonenberg, N., Northcote, P., Teesdale-Spittle, P., and Pelletier, J. (2005). Stimulation of mammalian translation initiation factor eIF4A activity by a small molecule inhibitor of eukaryotic translation. *Proc. Natl. Acad. Sci. USA* **102,** 10460–10465.

Brown, E. J., Albers, M. W., Shin, T. B., Ichikawa, K., Keith, C. T., Lane, W. S., and Schreiber, S. L. (1994). A mammalian protein targeted by G1-arresting rapamycin-receptor complex. *Nature* **369,** 756–758.

Chiu, M. I., Katz, H., and Berlin, V. (1994). RAPT1, a mammalian homolog of yeast Tor, interacts with the FKBP12/rapamycin complex. *Proc. Natl. Acad. Sci. USA* **91,** 12574–12578.

Cuatrecasas, P. (1970). Protein purification by affinity chromatography. Derivatizations of agarose and polyacrylamide beads. *J. Biol. Chem.* **245,** 3059–3065.

Cundliffe, E., Cannon, M., and Davies, J. (1974). Mechanism of inhibition of eukaryotic protein synthesis b trichothecene fungal toxins. *Proc. Natl. Acad. Sci. USA* **71,** 30–34.

Filler, R., Kobayashi, Y., and Yagupolskii, L. M., Editors (1993). Organofluorine compounds in medicinal chemistry and biomedical applications. *Stud. Org. Chem. (Amst.)* **48.**

Fischer, H., Klippe, M., Lerche, H., Severin, T., and Wanninger, G. (1990). Electrophilic beta-bromination and nucleophilic alpha-methoxylation of alpha,beta-unsaturated carbonyl compounds. *Chem. Ber.* **123,** 399–404.

Fresno, M., Jimenez, A., and Vazquez, D. (1977). Inhibition of translation in eukaryotic systems by harringtonine. *Eur. J. Biochem.* **72,** 323–330.

Goldberg, I. H., and Mitsugi, K. (1966). Sparsomycin, an inhibitor of aminoacyl transfer to polypeptide. *Biochem. Biophys. Res. Commun.* **23,** 453–459.

Griffith, E. C., Su, Z., Turk, B. E., Chen, S., Chang, Y.-W., Wu, Z., Biemann, K., and Liu, J. O. (1997). Methionine aminopeptidase (type 2) is the common target for angiogenesis inhibitors AGM-1470 and ovalicin. *Chem. & Biol.* **4,** 461–471.

Guzzo, P. R., and Miller, M. J. (1994). Catalytic asymmetric synthesis of the carbacephem framework. *J. Org. Chem.* **59,** 4862–4867.

Heitman, J., Movva, N. R., and Hall, M. N. (1991). Targets for cell cycle arrest by the immunosuppressant rapamycin in yeast. *Science* **253,** 905–909.

Hesse, M. (1991). "Ring Enlargements in Organic Chemistry." VCH Publishers, New York.

Hunt, S. L., Hsuan, J. J., Totty, N., and Jackson, R. J. (1999). Unr, a cellular cytoplasmic RNA-binding protein with five cold-shock domains, is required for internal initiation of translation of human rhinovirus RNA. *Genes Dev.* **13,** 437–448.

Katz, L., and Ashley, G. W. (2005). Translation and protein synthesis: Macrolides. *Chem. Rev.* **105,** 499–528.

Lefkowitz, R. J., Haber, E., and O'Hara, D. (1972). Identification of the cardiac beta-adrenergic receptor protein: Solubilization and purification by affinity chromatography. *Proc. Natl. Acad. Sci. USA* **69,** 2828–2832.

Li, G., Patel, D., and Hruby, V. J. (1993). Asymmetric synthesis of (2R, 3S) and (2S, 3R) precursors of b-Methyl-histindine, -Phenylalanine, and -Tyrosine. *Tetrahedron: Asymmetry* **4,** 2315–2318.

Licitra, E. J., and Liu, J. O. (1996). A three-hybrid system for detecting small ligand–protein receptor interactions. *Proc. Natl. Acad. Sci. USA* **93,** 12817–12821.

Liu, J., Farmer, J. D., Lane, W. S., Friedman, J., Weissman, I., and Schreiber, S. L. (1991). Calcineurin is a common target of cyclophilin-cyclosporin A and FKBP-FK506 complexes. *Cell* **66,** 807–815.

Low, W. K., Dang, Y., Schneider-Poetsch, T., Shi, Z., Choi, N. S., Merrick, W. C., Romo, D., and Liu, J. O. (2005). Inhibition of eukaryotic translation initiation by the marine natural product pateamine A. *Mol. Cell.* **20,** 709–722.

Manhas, M. S., Wagle, D. R., Chiang, J., and Bose, A. K. (1988). Conversion of b-lactams to versatile synthons via molecular rearrangement and lactam cleavage. *Heterocycles* **27,** 1755–1802.

Manz, B., Heubner, A., Kohler, I., Grill, H. J., and Pollow, K. (1983). Synthesis of biotin-labeled dexamethasone derivatives. Novel hormone-affinity probes. *Eur. J. Biochem.* **131,** 333–338.

Marton, M. J., DeRisi, J. L., Bennett, H. A., Iyer, V. R., Meyer, M. R., Roberts, C. J., Stoughton, R., Burchard, J., Slade, D., Dai, H., Bassett, D. E., Jr., Hrtwell, L. H., *et al.* (1998). Drug target validatio anidentification of secondary drug target effects using DNA microarrays. *Nat. Med.* **4,** 1293–1301.

Matsuda, S., Katsumata, R., Okuda, T., Yamamoto, T., Miyazaki, K., Senga, T., Machida, K., Thant, A. A., Nakatsugawa, S., and Hamaguchi, M. (2000). Molecular cloning and characterization of human MAWD, a novel protein containing WD-40 repeats frequently overexpressed in breast cancer. *Cancer Res.* **60,** 13–17.

Mayer, T. U., Kapoor, T. M., Haggarty, S. J., King, R. W., Schreiber, S. L., and Mitchison, T. J. (1999). Small molecule inhibitor of mitotic spindle bipolarity identified in a phenotype-based screen. *Science* **286,** 971–974.

Nagao, Y., Hagiwara, Y., Kumagai, T., Ochiai, M., Inoue, T., Hashimoto, K., and Fujita, E. (1986). New C4-chiral 1,2-thiazolidine-2-thiones: Excellent chiral auxiliaries for highly diastereocontrolled aldol-type reactions of acetic acid and a,b-unsaturated aldehydes. *J. Org. Chem.* **51,** 2391–2393.

Nicolas, E., Russell, K. C., and Hruby, V. J. (1993). Asymmetric 1,4-addition of organo-cuprates to chiral a, b-unsaturated N-Acyl-4-phenyl-2-oxazolidinones: A new approach to the synthesis of chiral b-branched carboxylic acids. *J. Org. Chem.* **58,** 766–770.

Northcote, P. T., Blunt, J. W., and Munro, M. H. G. (1991). Pateamine: A potent cytotoxin from the New Zealand marine sponge, Mycale sp. *Tetrahedron Lett.* **32,** 6411–6414.

Novac, O., Guenier, A. S., and Pelletier, J. (2004). Inhibitors of protein synthesis identified by a high throughput multiplexed translation screen. *Nucleic Acids Res.* **32,** 902–915.

Noyori, R., Ohkuma, T., Kitamura, M., Takaya, H., Sayo, N., Kumobayashi, H., and Akutagawa, S. (1987). Asymmetric hydrogenation of b-keto carboxylic esters. A practical, purely chemical access to b-hydroxy esters in high enantiomeric purity. *J. Am. Chem. Soc.* **109,** 5856–5858.

Obrig, T. G., Culp, W. J., McKeehan, W. L., and Hardesty, B. (1971). The mechanism by which cycloheximide and related glutarimide antibiotics inhibit peptide synthesis on reticulocyte ribosomes. *J. Biol. Chem.* **246,** 174–181.

Ojima, I., Habus, I., Zhao, M., Zucco, M., Park, Y. H., Sun, C. M., and Brigaud, T. (1992). New and efficient approaches to the semisynthesis of taxol and its C-13 side chain analogs by means of b-lactam synthon method. *Tetrahedron* **48,** 6985–7012.

Ojima, I., Sun, C. M., Zucco, M., Park, Y. H., Duclos, O., and Kuduk, S. (1993). A highly efficient route to taxotere by the b-lactam synthon method. *Tetrahedron Lett.* **34,** 4149–4152.

Palomo, C., Aizpurua, J. M., Cuevas, C., Mielgo, A., and Galarza, R. (1995). A mild method for the alcholoysis of b-Lactams. *Tetrahedron Lett.* **36,** 9027–9030.

Pestka, S. (1977). "Inhibitors of Protein Synthesis, Academic Press." New York, NY.

Peterson, J. R., Lebensohn, A. M., Pelish, H. E., and Kirschner, M. W. (2006). Biochemical suppression of small-molecule inhibitors: A strategy to identify inhibitor targets and signaling pathway components. *Chem. Biol.* **13,** 443–452.

Romo, D., Choi Nam, S., Li, S., Buchler, I., Shi, Z., and Liu Jun, O. (2004). Evidence for separate binding and scaffolding domains in the immunosuppressive and antitumor marine natural product, pateamine a: Design, synthesis, and activity studies leading to a potent simplified derivative. *J. Am. Chem. Soc.* **126**, 10582–10588.

Romo, D., Rzasa, R. M., Shea, H. A., Park, K., Langenhan, J. M., Sun, L., Akhiezer, A., and Liu, J. O. (1998). Total synthesis and immunosuppressive activity of (-)-Pateamine A and related compounds: Implementation of a b-lactam-based macrocyclization. *J. Am. Chem. Soc.* **120**, 12237–12254.

Rzasa, R. M., Shea, H. A., and Romo, D. (1998). Total synthesis of the novel, immunosuppressive agent (-)-Pateamine A from *Mycale* sp. Employing a b-lactam-based macrocyclization. *J. Am. Chem. Soc.* **120**, 591–592.

Sabatini, D. M., Erdjument-Bromage, H., Lui, M., Tempst, P., and Snyder, S. H. (1994). RAFT1: A mammalian protein that binds to FKBP12 in a rapamycin-dependent fashion and is homologous to yeast TORs. *Cell* **78**, 35–43.

Sche, P. P., McKenzie, K. M., White, J. D., and Austin, D. J. (1999). Display cloning: Functional identification of natural product receptors using cDNA-phage display. *Chem. Biol.* **6**, 707–716.

Shevchenko, A., Wilm, M., Vorm, O., and Mann, M. (1996). Mass spectrometric sequencing of proteins silver-stained polyacrylamide gels. *Anal. Chem.* **68**, 850–858.

Sin, N., Meng, L., Wang, M. Q. W., Wen, J. J., Bornmann, W. G., and Crews, C. M. (1997). The anti-angiogenic agent fumagillin covalently binds and inhibits the methionine aminopeptidase, MetAP-2. *Proc. Natl. Acad. Sci. USA* **94**, 6099–6103.

SirDeshpande, B. V., and Toogood, P. L. (1995). Mechanism of protein synthesis inhibition by didemnin B *in vitro. Biochemistry* **34**, 9177–9184.

Taber, D. F., and Silverberg, L. J. (1991). Enantioselective reduction of b-keto esters. *Tetrahedron Lett.* **32**, 4227–4230.

Taunton, J., Hassig, C. A., and Schreiber, S. L. (1996). A mammalian histone deacetylase related to the yeast transcriptional regulator Rpd3p. *Science* **272**, 408–411.

Turk, B. E., Su, Z., and Liu, J. O. (1998). Synthetic analogues of TNP-470 and ovalicin reveal a common molecular basis for inhibition of angiogenesis and immunosuppression. *Bioorg. Med. Chem.* **6**, 1163–1169.

Wasserman, H. H. (1987). New methods in the formation of macrocyclic lactams and lactones of biological interest. *Aldrichimica Acta* **20**, 63–74.

Wei, C. M., Hansen, B. S., Vaughan, M. H., Jr., and McLaughlin, C. S. (1974). Mechanism of action of the mycotoxin trichodermin, a 12,13-epoxytrichothecene. *Proc. Natl. Acad. Sci. USA* **71**, 713–717.

Zhu, H., and Snyder, M. (2003). Protein chip technology. *Curr. Opin. Chem. Biol.* **7**, 55–63.

Author Index

Subject Index

Protein–protein
interactions analysis

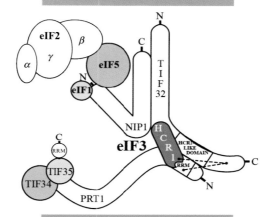

"*in vivo*" deletion analysis

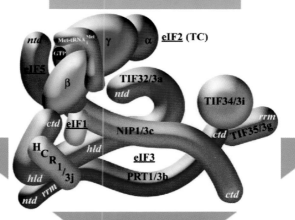

"3D" architecture

Klaus H. Nielsen and Leoš Valášek, Figure 2.2 Schematic illustrating the gain of knowledge after employing the *in vivo* deletion analysis approach. (A) Summary of protein–protein interactions within the yeast eIF3 complex (adopted with permission from Valášek *et al.*, 2001). The eIF3 subunits, as well as eIF5, eIF2, and eIF1 factors, are shown as various shapes with sizes roughly proportional to their molecular weights. Points of overlap between the various shapes indicate sites of known protein–protein interaction. (B) A 3D model of the MFC based on a comprehensive deletion analysis of subunit interactions (adopted with permission from Valášek *et al.*, 2003). The labeled protein subunits are shown roughly in proportion to their molecular weights. The degree of overlap between two different subunits depicts the extent of their interacting surfaces.

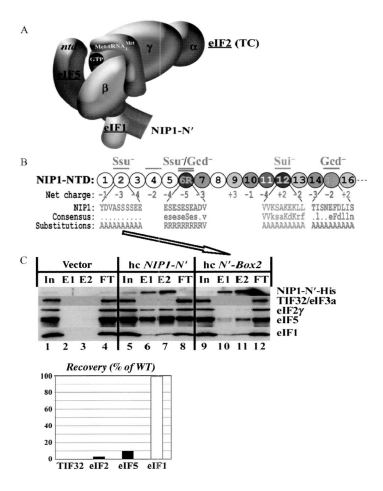

Klaus H. Nielsen and Leoš Valášek, Figure 2.3 An example of the Ni–affinity chromatography showing that the *NIP1-Box2* mutation diminishes binding of eIF5 and eIF2 to the NIP1-NTD *in vivo* (adopted with permission from Valášek *et al.*, 2004). (A) A 3D model of the NIP1-NTD subcomplex with TC, and eIF1 and eIF5. (B) Schematics illustrating the application of clustered 10-Ala mutagenesis (CAM) to the example of the N-terminal domain of NIP1. The sequence of the first 160 amino acids of NIP1 is shown as numbered circles (Boxes 1–16), each of them composed of 10 residues substituted with a stretch of 10 alanines. Different shades of gray indicate the degree of identities between the NIP1-NTD and the N-termini of its various homologues. Color-coded bars above the circles indicate the phenotypes associated with amino acid substitutions in the corresponding boxes: Ssu⁻ (*s*uppressor of *Su*i⁻), Gcd⁻ (*g*eneral *c*ontrol *d*erepressed), and Sui⁻ (*s*uppressor of *i*nitiation codon). Blown-up segments in blue, green, and yellow indicate the amino acid sequences, a consensus sequence derived from sequence alignments, and the substitutions made in the corresponding boxes of the NIP1-NTD. Net charge of individual boxes is indicated below each of them. (C) WCEs prepared from the cells overexpressing either wild-type or mutant form of the NIP1-NTD were incubated with Ni^{2+}-NTA-silica resin, and the bound proteins were eluted and subjected to Western blot analysis using antibodies against the His_8 epitope (to detect the NIP1-NTD polypeptides) or with antibodies against the other factors listed to the right of the blots. Lanes 1, 5, and 9 contained 3% of the input WCEs (In); lanes 2, 6, and 10 contained 15% of the first fractions eluted from the resin (E1); lanes 3, 7, and 11 contained 30% of the same fractions as lanes 2, 6, and 10 (E2); and lanes 4, 8, and 12 contained 3% of the flow-through fractions (FT). The Western signals for eIF2, eIF1, and eIF5 in the E1 and E2 fractions for the Box2 mutant (lanes 10–11) were quantified, combined, normalized for the amounts of the NIP1-NTD-Box2 fragment in these fractions, and the averaged values from 3–5 independent experiments were plotted in the histogram on the right as percentages of the corresponding values calculated for the WT NIP1-NTD (fractions 6–7).

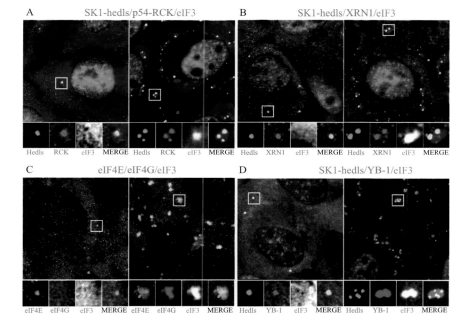

Nancy Kedersha and Paul Anderson, Figure 5.1 SGs and PBs detected using different antibodies. U2OS cells were untreated (left panels) or arsenite treated (0.5 m*M*, 45 min) (right panels), then fixed and stained using the described protocol. Enlarged views of boxed regions are displayed underneath the corresponding panels, showing the separate views of the same field. (A) Untreated (left panels) or arsenite-treated (right panels) cells stained for Hedls/S6K1 in green, p54-RCK in red, and eIF3b in blue. Hedls and p54-RCK stain PBs exclusively, whereas eIF3 is specific for SGs. (B) Untreated (left panels) or arsenite-treated (right panels) cells stained for Hedls/S6K1 in green, XRN1 in red, and eIF3b in blue. Hedls and XRN1 stain PBs exclusively, whereas eIF3 is specific for SGs. (C) Untreated (left panels) or arsenite-treated (right panels) cells stained for eIF4E in green, eIF4G in red, and eIF3b in blue. Note that eIF4E (green) is present in PBs in unstressed cells (left panel), but also is detectable in SGs in arsenite-treated cells (right panels). In contrast, eIF4G and eIF3 remain exclusively associated with SGs. (D) Untreated (left panels) or arsenite-treated (right panels) cells stained for Hedls/S6K1 in green, YB-1 in red, and eIF3b in blue. Note that YB-1 is detectable in PBs in unstressed cells, but is largely relocalized to SGs upon to this particular stress.

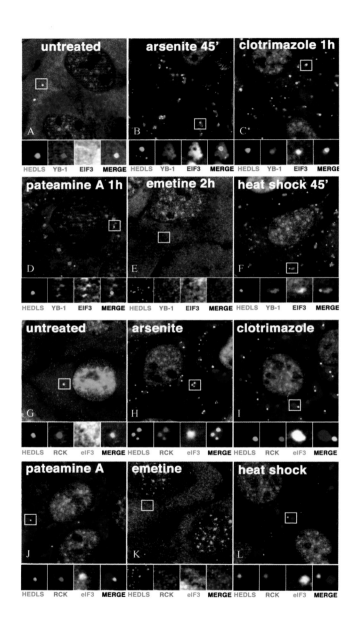

Nancy Kedersha and Paul Anderson, Figure 5.2 SG and PB formation in response to different stresses. U2OS cells were subjected to some of the stresses described in Table 5.1, then stained as indicated. Untreated cells display few PBs and no SGs, whereas arsenite treatment (0.5 mM, 45 min, panels B and H) strongly induces assembly of both PBs and SGs. Clotrimazole (C, I), pateamine A (D, J), and heat shock (F, L) induce more SGs than PBs, whereas emitine treatment (E, K) abolishes PBs and does not induce SGs. Note that YB-1 is present in PBs in unstressed cells (A, red), but relocalizes to SGs upon most stress conditions (C, D, F, whereas p54/RCK (red, G, H, I, J, L) remains predominantly associated with PBs regardless of stress.

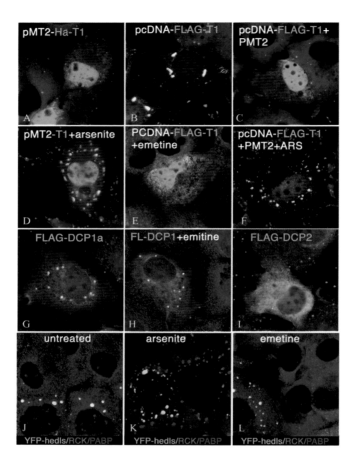

Nancy Kedersha and Paul Anderson, Figure 5.3 Overexpression of SG/PB marker proteins. COS7 cells (panels A–I) were transiently transfected with different vectors containing the same coding region of TIA-1, treated, and stained as indicated in the individual panels. Overexpressed PB markers such as DCP1a (G, H) and hedls/GE-1 (panels J, K, L) induce formation of abnormally large PBs (compare yellow foci to red foci in non-transfected cells) that are resistant to emetine, which dissolves normal PBs (absence of red foci in panel H). In contrast, overexpressed DCP2 (I) is not detected in PBs but appears diffuse. (J–L) U2OS cells stably expressing YFP-Hedls/GE-1 (green, panels J, K, L) were co-cultured with wild-type U2OS cells (lacking green), treated as indicated, and counterstained for the PB marker RCK (red) and the SG marker PABP-1 (blue). YFP-hedls PBs appear yellow, endogenous PBs in untransfected cells appear red. As with overexpressed DCP1, the YFP-hedls cells exhibit huge PB-like structures which are not disassembled upon emitine treatment (panel L, comparing large yellow foci to absent red foci).

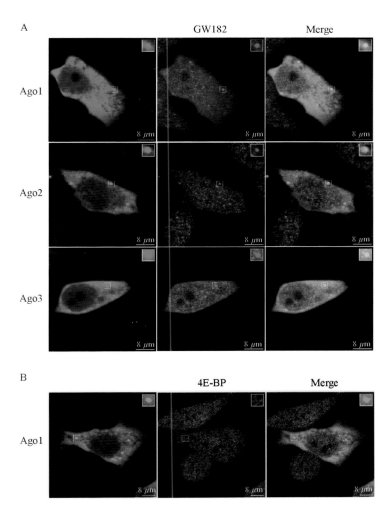

Jennifer L. Clancy *et al.*, **Figure 6.5**   Enrichment of Argonaute proteins in P-bodies. HeLa cells were transfected with plasmids expressing Flag-tagged Argonaute 1, 2, or 3 protein and analyzed by confocal microscopy after 24 h. (A) The Argonaute proteins were detected using an anti-Flag antibody (green), while GW182 was detected with anti-GW182 antibody (red). Co-localization is indicated by the yellow color in the merged panels. Inserts show regions of co-localization at higher magnification. (B) The protein 4E-BP (red) was used as a specificity control and did not co-localize with Ago1 foci (green) (insert shows higher magnification of an Ago1 foci).

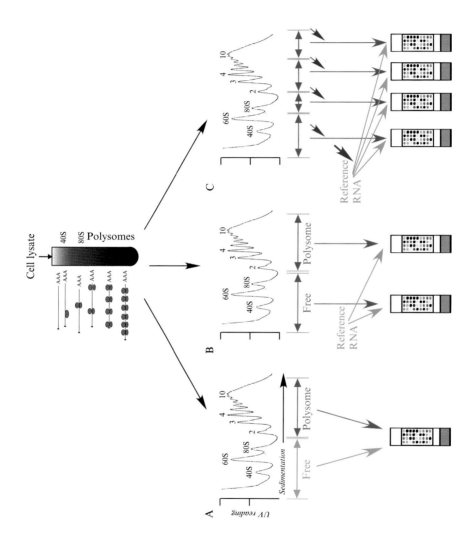

Daniel Melamed and Yoav Arava, Figure 10.1 Experimental schemes for microarray analysis. All experimental schemes start with a separation step of the cell lysate by velocity sedimentation in a sucrose gradient (top scheme). Collection of the desired fractions is assisted by a continuous ultraviolet (UV) reading of the gradient (an example of such UV reading is shown in each section). This allows determination of the sedimentation position of the 40S, 60S, 80S, and polyribosomal complexes (2, 3, and more). Three general ways for fraction collection and analysis are presented (sections A, B, and C): (A) Collection of two fractions (free and polysomes) and direct comparison between them, with the free mRNA fraction labeled with green dye and the polysome fraction labeled with red dye. (B) Collection of two fractions and indirect comparison between them by utilizing an unfractionated reference RNA. (C) Collection of multiple fractions (four in this case), where each fraction is compared to an unfractionated reference sample. The blue arrows indicate the addition of spike-in RNA to each fraction and to the reference RNA.

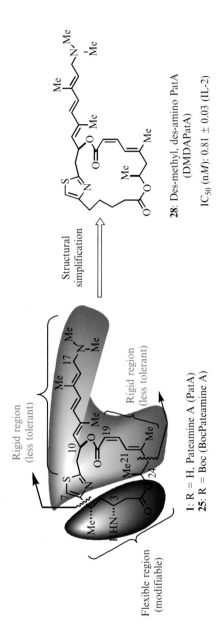

28: Des-methyl, des-amino PatA (DMDAPatA)

IC_{50} (nM): 0.81 ± 0.03 (IL-2)

Structural simplification

1: R = H, Pateamine A (PatA)
25: R = Boc (BocPateamine A)

Rigid region (less tolerant)

Rigid region (less tolerant)

Flexible region (modifiable)

Woon-Kai Low et al., Figure 14.3 Structures of Pateamine A (1), Boc–pateamine A (25), and DMDA-PatA (28). The putative binding (rigid regions, in red) and scaffolding (flexible and modifiable, in blue) domains are indicated and suggest possible sites for modification, namely, the C3–amino group.